130
Advances in Biochemical Engineering/Biotechnology

Series editor

T. Scheper, Hannover, Germany

For further volumes:
http://www.springer.com/series/10

Aims and Scope

This book series reviews current trends in modern biotechnology and biochemical engineering. Its aim is to cover all aspects of these interdisciplinary disciplines, where knowledge, methods and expertise are required from chemistry, biochemistry, microbiology, molecular biology, chemical engineering and computer science.

Volumes are organized topically and provide a comprehensive discussion of developments in the field over the past 3–5 years. The series also discusses new discoveries and applications. Special volumes are dedicated to selected topics which focus on new biotechnological products and new processes for their synthesis and purification.

In general, volumes are edited by well-known guest editors. The series editor and publisher will, however, always be pleased to receive suggestions and supplementary information. Manuscripts are accepted in English.

In references, Advances in Biochemical Engineering/Biotechnology is abbreviated as *Adv. Biochem. Engin./Biotechnol.* and cited as a journal.

Birgit Weyand · Massimo Dominici
Ralf Hass · Roland Jacobs
Cornelia Kasper
Editors

Mesenchymal Stem Cells: Basics and Clinical Application II

With contributions by

Irina Eberle · Mohsen Moslem · Reinhard Henschler · Tobias Cantz
Beate Wagner · Mei Ling Lim · Philipp Jungebluth · Paolo Macchiarini
Birgit Weyand · Peter M. Vogt · Michael Jagodzinski · C. Haasper
K. M. Fawzy El-Sayed · C. Dörfer · F. Fändrich · F. Gieseler
M. H. Moustafa · H. Ungefroren · Khondoker M. Akram · Sohel Samad
Monica Spiteri · Nicholas R. Forsyth · Claudia Papewalis · Daniela Topolar
Barbara Götz · Stefan Schönberger · Dagmar Dilloo · Aline M. Betancourt
Grazia Maria Spaggiari · Lorenzo Moretta · Naomi D'souza · Jorge Sans Burns
Giulia Grisendi · Olivia Candini · Elena Veronesi · Serena Piccinno
Edwin M. Horwitz · Paolo Paolucci · Pierfranco Conte · Massimo Dominici
Michela Pozzobon · Martina Piccoli · Paolo De Coppi · Giovanni Migliaccio
Cristina Pintus · Kevin Cieslak

Springer

Editors
Birgit Weyand
Abteilung für Plastische und Handchirurgie
Medizinische Hochschule Hannover
Hannover
Germany

Roland Jacobs
Klinik für Immunologie und Rheumatologie
Medizinische Hochschule Hannover
Hannover
Germany

Massimo Dominici
Integrated Department of Oncology
Haematology and Respiratory Diseases
University of Modena and Reggio Emilia
Modena
Italy

Cornelia Kasper
Department für Biotechnologie
Universität für Bodenkultur
Wien
Austria

Ralf Hass
AG Biochemie und Tumorbiologie
Klinik für Frauenheilkunde und
Geburtshilfe
Medizinische Hochschule Hannover
Hannover
Germany

ISSN 0724-6145 ISSN 1616-8542 (electronic)
ISBN 978-3-642-37943-7 ISBN 978-3-642-37944-4 (eBook)
DOI 10.1007/978-3-642-37944-4
Springer Heidelberg New York Dordrecht London

Library of Congress Control Number: 2012955477

Printed on acid-free paper

Springer is part of Springer Science+Business Media (www.springer.com)

Preface

Mesenchymal stem cells (MSC) represent one of the most interesting progenitors to date, due to their biodiverse functionalities. Among the fascinating multiple properties of MSC are their supportive roles in wound healing and in the regeneration of damaged tissues and organs. This implies the capacity of MSC to migrate towards injured tissue, to undergo differentiation, to modulate the activation of immune cells and to activate endothelial cells contributing to both angiogenesis and neo-vascularisation. Together with their self-renewal capability, the maintenance of stem cell homeostasis, the release of several bioactive compounds like chemokines, cytokines, micro RNAs and exosomes, MSC can be certainly considered as cellular all-round supporters.

These multi-functional MSC properties are highlighted in the present volume. While some chapters are focused on differentiation capacities of MSC, even beyond the more consolidated mesodermal lineages, others provide novel insights into the stimulatory signals involved in MSC survival and trafficking. Moreover, the MSC role in regulating cancer progression for novel therapeutics is assessed. In-depth molecular analyses of MSC functions are also covered, additionally including initial characterisations of distinct proteomic patterns that are specific for discrete MSC populations. Technical aspects for the isolation and enrichment of selected MSC populations are here additionally addressed in relationship to new cell sources and in the attempt to open new therapeutic platforms for potential clinical applications.

Although MSC research is progressively bridging to more consolidated clinical applications, it still represents a dynamically developing field, where a variety of intriguing aspects remain to be addressed. We feel this volume represents a comprehensive summary gathering a panel of up-to-date articles which combine the diverse MSC biological functionalities and their potential in translational cell therapy, as highlighted from different angles with a broad interdisciplinary perspective.

Birgit Weyand
Massimo Dominici
Ralf Hass
Roland Jacobs
Cornelia Kasper

Contents

Adv Biochem Eng Biotechnol (2013) 130: 1–17
DOI: 10.1007/10_2012_156

Published Online: 23 August 2012

Engineered MSCs from Patient-Specific iPS Cells

Irina Eberle, Mohsen Moslem, Reinhard Henschler and Tobias Cantz

Abstract Mesenchymal stroma/stem cells (MSCs) represent a heterogenic cell population that can be isolated from various tissues of the body or can be generated from pluripotent stem cells by in vitro differentiation. Various promising pre-clinical and clinical studies suggest that MSCs might stimulate endogenous regeneration and/or act as anti-inflammatory agents, which could be of high therapeutic relevance for a number of diseases, including graft-versus-host disease after allogeneic hematopoietic stem cell transplantation, inflammatory bowel diseases, or some forms of liver failure. Notably, conflicting results of various studies illustrated that the source of MSCs, the cultivation condition, and the way of administration have important effects on the desired clinical effect. Some of the involved molecular pathways have recently been elucidated and an artificial modulation of these pathways by engineered MSCs might result in superfunctional MSCs for enhanced endogenous regeneration or anti-inflammatory response. In this review, we summarize important findings of conventional MSCs for applications in gastroenterology and we describe the state-of-the-art for the generation of patient-derived iPS cells that eventually might provide genetically engineered superfunctional iPS cells for advanced cell therapies.

Keywords Cell transplantation · Induced pluripotent stem cells (iPSC) · Inflammatory bowel diseases · Liver diseases · Mesenchymal stromal (stem) cells (MSC)

Irina Eberle and Mohsen Moslem contributed equally

I. Eberle · M. Moslem · T. Cantz (✉)
Junior Research Group Stem Cell Biology, OE 8881, Cluster-of-Excellence REBIRTH, Hannover Medical School, Carl-Neuberg-Str. 1, 30625 Hannover, Germany
e-mail: cantz.tobias@mh-hannover.de

I. Eberle · R. Henschler
DRK-Institute of Transfusion Medicine and Immune Hematology, Frankfurt Germany

Contents

1 Introduction

Mesenchymal stromal cells (MSCs) are a heterogeneous cell population that can be obtained from various tissues, such as the bone marrow or adipose tissue. Recently several studies have shown other sources of mesenchymal cells, obtained from amniotic membrane [15], dental pulp [37], and also nonmesodermal tissue origins such as spleen, liver, kidney, and lung (Anker et al. [39] with similar characteristics to bone marrow-derived MSCs, which show a characteristic surface marker profile consisting of CD-45$^-$, CD-31$^-$, and CD-90$^+$ cells. These findings might suggest a possible common niche for all of these cells, in which extracellular matrix compositions, signaling molecules, cell–cell and cell–matrix interactions, and O_2 tension would be comparable [15]. By far, the most extensively studied source of MSCs is the bone marrow, because earlier studies addressing hematopoietic stem cells within the bone marrow niche resulted in a profound insight into the biology of these mesenchymal cells. Due to the fact that MSCs harbor clonally expandable cells, which could be differentiated towards adipogenic, chondrogenic, and osteogenic tissues, these cells also could be considered as mesenchymal stem cells. It is noteworthy that the abbreviation "MSC" is not uniformly used for either the term "mesenchymal stromal cells" or the term "mesenchymal stem cells" and often both aspects are not fully distinguished in the respective publication. Considering differentiation processes and further cellular fate changes upon extended in vitro culture, a pure population of mesenchymal stem cells might be hard to obtain or even to propagate and, probably, most cultures of MSCs contain stem cells as well as more differentiated stromal cells.

MSCs are meant to be beneficial in the repair of connective tissue injuries such as wound healing, osteogenic deficiencies, and also cartilage repair [4]. However, some reports suggest that MSCs could be an alternative source to repair a variety of other degenerative tissue lesions and might allow new therapeutic strategies for the treatment of neurological disorders [70], myocardial infarction [64], liver injuries [24, 58], and urological diseases [84]. This list of studies is by far not complete and, more important, the reproducibility as well as clinical impact of these studies is controversially discussed [55] as further described below in the context of gastrointestinal disorders.

Encouraging results from animal models and some clinical trials clearly support the research on MSCs, but also raise concerns about the lack of high cell numbers and the lack of a homogeneous cell population. Interestingly, embryonic stem cells or other pluripotent stem cells harbor the capabilities of unlimited self-renewing and differentiation potential into all somatic cell types [86], including mesenchymal stromal/stem cells. Among the various strategies to obtain MSCs from pluripotent stem cells, most protocols were first evaluated with established human embryonic stem cell lines. However, the generation of patient-derived pluripotent stem cells became feasible after pioneering studies of Shinya Yamanaka, who demonstrated that a set of four transcription factors can convert somatic cells into pluripotent stem cells [83]. Such patient-derived induced pluripotent stem cells [11] offer unique opportunities for applications in personalized medicine and allow the generation of high numbers of pluripotent starting cells, when large-scale cultivation systems were applied [62]. Such pluripotent stem cell lines can be differentiated towards functional MSC-like cells and it remains to be analyzed to which extent those MSC preparations harbor therapeutic effects and consist of a more defined homogeneous cell population.

Obviously, such engineered MSCs need to be investigated in further pre-clinical studies but may possess the potential to overcome some of the limitations that raise profound concerns of the clinical MSC applications at present.

2 Sources and Diversity of MSCs

In spite of data from over 100 trials employing MSCs in different clinical settings, correlation of MSC properties to clinical efficacy is limited due to considerable diversity of the applied MSC populations [88]. Also the broad spectrum of functional activity is not adequately reflected by the internationally agreed minimal set of consensus released criteria [22] and a defined subset of surface molecules that exactly characterize the MSCs' phenotype does not exist. All spindle-shaped cells attaching to plastic surfaces and expressing surface markers like CD-29, CD-105, CD-73, CD-44, but not hematopoietic markers (CD-34, CD-45, and CD-14) are considered as MSCs, if their ability to differentiate into mesodermal lineages (adipogenic, chondrogenic, and osteogenic differentiation) is provided. It is speculated that slightly different subtypes exist, which vary in their phenotypes due to extra-cellular or intra-cellular signaling. Nevertheless, they still possess multipotentiality as demonstrated by in vitro differentiation and in xenografts [52].

There are several cultivation protocols available that allow a proper in vitro expansion of MSCs, which makes them a readily accessible cell source for stem cell research. Independent of the harvesting sites, namely bone marrow, fat tissue, and umbilical cord blood, MSCs show some in vitro expansion ability even to clinical scale with minimal shortfalls in stemness [15]. But there are contradicting reports concerning karyotype aberrations of in vitro expanded cells, as early as five to nine passages after MSC harvesting [59, 94].

Transcriptome analysis revealed that genes typically expressed in MSCs are cytoskeletal (vimentin and myosin) and cytolytic or extracellular proteins (Collagene I, III, VI, and different matrix metalloproteinases), cell adhesion molecules (fibronectin and integrins), cytokines (IL-11, HGF, TGF-β), and also receptors (IL-1R and IL-10R). However, among the various abilities of MSCs, their homing properties in different tissues and the sectretion of bioactive compounds such as angiogenic (VEGF), antiapoptotic (HGF), and mitogenic (IGF-I) factors [15] are important variables influencing their potential therapeutic applications. The high homing properties of MSCs are dependent on the chemokine receptor CXCR4 expression on the cell surface and its interaction with SDF-1α stimuli from injured tissue in a gradient-dependent manner that attracts MSCs and promotes further cell interactions [4]. Despite all beneficial effects of MSCs in degenerative diseases there are several issues that could interfere with the MSCs' potential to ameliorate the respective disorder. For instance, aging adversely affects MSCs self-renewal, proliferation, telomerase length, and differentiation capacity [95]. Furthermore, impaired antioxidant activity and the lack of appropriate cytoskeleton properties could lead to malfunction of MSCs in therapeutic settings [43, 45].

However, the effects of MSC therapies are transient and require repeated transplantations and therefore a high number of cells. Due to the fact that MSCs show an impaired growth and increased senescence during in vitro propagation, the proper cell amount for clinical treatment might be a major obstacle. To overcome these obstacles, several groups have reported the derivation of MSC populations from self-renewing human embryonic stem cells by numerous methods [38, 61, 89], which are discussed below.

3 Therapeutic Applications of MSCs in Gastroenterology

3.1 MSCs in Graft-Versus-Host Disease

One of the most critical side effects of allogeneic hematopoietic stem cell transplantation for the treatment of leukemia or other life-threatening hematopoietic diseases is the development of an acute graft-versus-host disease (GvHD) resulting in a high morbidity and mortality [80]. Hereby, graft-derived T cells trigger the induction of GvHD after activation by host-related major histocompatibility class I or II antigens as well as minor antigenic peptides [27]. GvHD mainly targets the skin, intestine, liver, and the hematopoietic system and is routinely treated with immunosuppressive drugs such as cyclosporine or methotrexate [79]. However, various advanced treatment regimes were available using therapeutic antibodies against interleukin-2 [3], tumor necrosis factor α (TNFα; [46], or against CD-147 [20].

Inspired by various animal studies including baboons [8], third-party MSCs have found ready entry into a series of clinical trials for prevention of severe acute GvHD [69] and striking response rates as high as 50–90 % have been reported with noteworthy resolution of refractory intestinal GvHD [49]. It is also noteworthy that in a randomized trial patients suffering from grade II–IV acute GvHD received two transfusions of a commercially produced MSC preparation (Prochymal™, Osiris Therapeutics, Columbia, MD, USA). Of the 32 treated patients 94 % showed an initial response and as many as 77 % remained in a complete response state [44]. Despite these promising results some follow-up studies questioned the dramatic effect of MSCs on prevention and treatment of GvHD, as larger clinical trials failed to show a beneficial effect on the most common skin GvHD. However, results from GvHD phenotypes, which are more difficult to treat and which affect mainly the intestine and the liver, showed an improved response rate over placebo [2].

3.2 MSCs in Inflammatory Bowel Disease

Idiopathic inflammatory bowel disorders (IBD) such as Crohn's disease and ulcerative colitis are highly debilitating diseases of the gut that have remained largely resistant to definitive medical therapy [26, 99]. The pathophysiology of Crohn's disease includes an exaggerated infiltration of macrophages and neutrophilic granulocytes, which is triggered by activated T-helper cells. These cells produce uncontrolled amounts of inflammatory cytokines and chemokines resulting in tissue destruction of the large intestine. For instance, excessive production of IFN-γ and IL-17 by T cells and IL-12 or IL-23 by monocytes is responsible for an acute inflammation and the production of other cytokines such as TNF-α [81]. Based on the finding that an imbalance of effector T cells and suppressive regulatory T cells causes an expansion of self-reactive T cells [10], there is conclusive evidence that Crohn's disease is related to a failure of the mucosal immune system. Consequently, the therapeutic challenge applying MSCs for the treatment of IBD is twofold: curbing the inflammatory attack may be considered as the main action, but, secondly, the regeneration of a large organ such as the intestinal mucosa requires additional tissue-trophic measures to re-establish the protective mucosal barrier.

So far, the published literature on clinical evaluation of MSC-based therapy is comparatively sparse with the main evidence stemming from local application to perianal fistulas and i.v. applications pilotized in small numbers of patients [31, 71]. In a phase I clinical trial it was demonstrated that MSCs derived from the bone marrow of refractory Crohn's disease patients have identical characteristics compared to MSCs from healthy donors and have intact immunomodulatory capacities in vitro. Furthermore, administration of autologous bone marrow-derived MSCs was safe and feasible in the treatment of refractory Crohn's disease [23]. In addition, a more recent study demonstrated the feasibility of ex vivo expanding autologous bone marrow-derived MSCs and the safety of their intra-fistular injections in patients with Crohn's disease. Moreover, the authors described a promoting effect of MSCs on in vivo differentiation of regulatory T cells [17]. Osiris Therapeutics (Columbia, MD, USA) also initiated clinical trials for Crohn's disease using their MSC preparation Proychymal™. However, the placebo group also showed improvements and the treated arm of the study failed to meet the primary endpoint [2]. In conclusion, the lack of knowledge about the direct and indirect effects of different MSC preparations hampers the evaluation of these early clinical trials and more basic research on paracrine effectors and cellular mechanisms contributing to the MSCs' immunomodulatory effect are necessary.

3.3 MSCs in Liver Regeneration

Liver, as the second largest organ in the body serves crucial roles in the human homeostasis and its malfunction could be life-threatening. The high mortality rate because of liver deformities that led to 1.4 million deaths annually has not been avoided by liver transplantation which is the most efficient therapy so far [68]. In addition to stem cell mobilizing strategies [56] and bioartificial liver devices (BAL) [78], several alternative cell-based therapies have been investigated to recover unstable conditions in chronic liver disorders as well as during metabolic or acute liver failure. In general the disorders are treated by transplantation of bone marrow hematopoietic, mesenchymal, and mononuclear cell populations (for review see [93] and [74]). The first evidence indicating MSC infusion in mice models could recover liver failure was suggested by Petersen et al. showing the presence of bone marrow-derived hepatic cells from sex-mismatched donors in the recipient mice livers [65]. These data were substantiated by findings of other groups [47, 85], but later analyses questioned the initial hypothesis of a direct transdifferentiation and rather demonstrated that the transplanted cells fuse with host hepatocytes [12, 75, 96]. Nevertheless, some studies described functional integration of MSCs into injured liver after their in vitro specification towards hepatic cells [5, 6]. On the other hand the inhibitory signals of MSCs over hepatic stellate cells (mostly responsible for extracellular matrix accumulation [29]) inhibited the proliferation and triggered their apoptosis [90]. Also secretion of anti-inflammatory cytokines such as TNF-α, IL-1, IL-10, HGF [90], and matrix metalloproteinases regulation [53] could react as an anti-fibrogenic treatment in chronic liver injuries.

Clearly, the clinical relevance of these findings is still very controversially discussed and most pre-clinical and clinical studies indicate that the MSC therapy is a transient treatment, which may have to be applied repeatedly in order to treat the respective disorders. For instance, in the case of chronic fibrogenesis the cell infusion may be crucial for preventing the turnover of new fibers [73]. Therefore, an application of cells with similar characteristics to those of MSCs with a higher self-renewal capacity and a reduced senescence behavior would be an appropriate approach to support more long-term effects and the possibility of providing transplants from the same batch of initially transplanted cells.

4 MSCs from Patient-Derived Pluripotent Stem Cells

4.1 Generation of Patient-Specific Pluripotent Stem Cells

Pluripotent stem cells such as human embryonic stem cells (ESCs) harbor an unlimited self-renewing capability and have the potential to differentiate into all cells of the three germ layers (ectoderm, mesoderm, and endoderm) as well as into germ cells [86]. Various attempts were undertaken to derive pluripotent cells from adult individuals. The early strategies were strongly influenced by the technique of somatic cell nuclear transfer (SCNT) resulting in a cloned embryo, as first demonstrated for mammals by the birth of the sheep "Dolly" [98]. This technique, however, lacks feasibility with human cells and is ethically heavily disputed because the derivation of SCNT-derived cells implicates the destruction of a human embryo. Nevertheless, other strategies were exploited to "re-program" somatic cells towards pluripotent stem cells, either by cell fusion, by application of ESC extracts, or by using a defined set of transcription factors. In the groundbreaking study of Takahashi and Yamanaka in [83], they successfully reprogrammed mouse fibroblast by introducing ectopic defined transcription factors (Oct-4, Sox-2, Klf-4, and c-Myc known as OSKM) into the cells via retroviral transduction [83]. The oncogenic nature of c-Myc and Klf-4 urged other scientists to reprogram cells with other transcription factors such as Oct-4, Sox-2, Nanog, and Lin-28 (OSNL; [104]).

By this direct reprogramming method cell colonies with similar morphology and genetically similar information to ESCs were generated and termed induced pluripotent stem cells. Later on, numerous reprogramming studies used different cell types of ectodermal (keratinocytes [1] and neural progenitor cells [25]), mesodermal (B cells [36], or cord blood [35, 105]), and endodermal (hepatocytes [77]) origin. In addition to the cell type the composition and stoichiometry of the reprogramming factor cocktail affects successful reprogramming. Considering the stoichiometric variability caused by either high or low transgenic expression of each factor in the reprogramming cocktail, several studies ruled out the importance of a dominating Oct4 expression level [63, 87]. Therefore, a polycistronic reprogramming construct that ensures the expression of all four factors in a defined and preferential stoichiometric ratio is of high relevance for the generation of fully reprogrammed iPSC as described by various reports [13, 97]. So far several combinations of these factors along with the other factors such as Esrrb have been used for reprogramming [33]. Furthermore, small molecules, microRNAs [67], and epigenetic modifiers are used to increase reprogramming efficacy. For instance, PD0325901 and CHIR99021 as inhibitors of MEK and GSK3 pathways increased the ratio of pluripotent cells. It has been demonstrated that members of the microRNA-290 cluster are cell cycle regulators of ESCs and could also increase iPSCs colony numbers [41]. In addition other microRNAs, such as microRNA-130b, -301b, and -721 strongly supported the generation of iPSCs [66]. Moreover, DNA methyltransferase inhibitors such as AZA and RG-180 and also histone deacetylase inhibitors such as VPA could increase reprogramming efficacy when used along with OSKM factors [33].

Direct reprogramming has provided a working methodology to induce somatic cells to go back to their embryonic state by viral integration. However, integrative and nonintegrative methods used for reprogramming is a challenging criterion for safe iPSC production. For viral integration methods classic γ-retroviral and newer lentiviral vectors are used. Because of their well-understood biology and high transduction efficacy γ-retroviral vectors are commonly used for gene transfer systems. Despite high transduction efficacy the smallest result is that γ-retroviral vectors just transduce dividing cells. However, lentiviruses, a subclass of retroviruses, can infect both dividing and nondividing cells with high transduction efficacy. Although by using these integrating vectors iPSCs could be generated very efficiently, the viral vectors' integration in the host cell genome may cause genetic mutagenesis and genomic instability [76, 103]. Nevertheless, several mouse iPSC lines were generated using integrating vectors and further applied to tetraploid embryo aggregation experiments, which resulted in fully iPSC-derived viable mice [9, 100]. To overcome potential issues with integrated reprogramming transgenes several research groups developed nonintegrating reprogramming approaches that could overcome these limitations ([60], [40], [102] and [77]) via transient viral, episomal, modified mRNA, and protein delivery.

However, reprogramming efficiency is extremely low when compared to viral transduction [102]. The same problem exists with the adenoviral delivery system, the efficiency of which is comparable to episomal vector transfection [107]. Most recently viral vectors with floxed transgenes, which could be efficiently removed [92] or piggyback transposone/transposase-based systems [42] were studied to provide clinically applicable iPSC preparations. In this line, delivery of the reprogramming factors with nonintegrating Sendai viruses seems to be a promising alternative, as high reprogramming efficiencies were obtained with this reprogramming setting [30].

4.2 Differentiation of Human iPSCs into MSCs

As outlined above, the therapeutic effect of MSC preparations may depend on the source of MSCs, the in vitro expansion of MSCs, and from batch to batch on preparation variations of MSCs. Therefore, MSCs derived from a self-renewing stem cell source may be a more suitable option. Currently the best investigated source of nontransformed self-renewing stem cells are embryonic stem cells. A number of reports described the in vitro differentiation of human ESCs into mesenchymal cells, which were very similar to primary MSCs. Some reports just applied spontaneous differentiation approaches and basically scraped out differentiating mesenchymal cells from human ESC colonies [61]. Other groups co-cultivated human ESCs with mouse bone marrow stroma cells, namely OP9 cells [7, 89] or isolated migrating cells from embryoid bodies [38]. A more defined MSC-like cell population was obtained after sorting of CD-105^{+} and CD-24^{-} cells [50]. Also a directed differentiation using the TGFβ inhibitor SB-431542 was successfully described recently [72]. By the inhibition of the SMAD-2/3 pathway the study could show an efficient differentiation of hESCs into MSCs. Mostly, MSCs derived from hESC exhibited a normal karyotype and were very similar if not functionally identical to human bone marrow-derived MSCs concerning their immunophenotype and the thus-far investigated functions [19]. Some groups reported a favorable higher proliferation capability of ESC-derived MSCs compared to human bone marrow-derived MSCs [72, 101]. Moreover, the differentiated cells lacked the expression of remaining pluripotency markers and lost the potential of teratoma formation, when those cells were transplanted into immunodeficient mice. However, the transplanted cells produced homogeneous tissues of mesenchymal appearance [34, 48]. In contrast to bone marrow- or adipose tissue-derived MSCs the hESC-derived MSCs did not show any signs of senescence and grew for multiple passages in vitro [38]. This observation might be the determining aspect for using the cells in future cell- and gene-therapy approaches. Another advantage of ESC-derived over the adipose tissue-derived MSCs might be their increased immunosuppressive properties against T lymphocytes [72]. This observation might be important for studying allograft rejection or applying ESC-derived MSCs in inflammatory bowel diseases.

Another source of pluripotent stem cells are the ethically less concerned induced pluripotent stem cells. As discussed above, iPSCs can be derived from a variety of somatic cell types that are easily obtainable from patients. Recent studies investigated human iPSC-derived MSCs (hiPSC-MSCs) in different degenerative diseases. The first study was done by Lian et al. in [51] in which they generated MSCs from hiPSCs with similar characteristics of human bone marrow-derived MSCs in terms of surface marker expression and differentiation potential. The cells could also substitute the therapeutic ability of classical MSCs in the hind limb ischemia model in mice, where significantly attenuated injury was promoted by increased vascular and muscle regeneration [51]. Additionally, hiPSC-MSCs displayed a remarkable immunosuppressive nature by inhibiting NK-cell proliferation and allograft rejection [32].

Some other studies have used additional supplements in order to differentiate iPSCs in functional MSCs. Villa-Diaz et al. introduced a biocompatible synthetic polymer (PMEDSAH) and xenogene-free culture media for differentiation of human iPSCs towards MSCs. Those cells were then applied in a mouse model with osteogenic calvaria defects, where the integrated cells could significantly recover the defect by regenerating new bone tissue compared to the control group [91]. In another study human ESCs and human iPSCs were differentiated into MSCs by using collagen type I coated plates [54]. In an electrophysiological study, patch clamp analysis demonstrated that hiPSCs-MSCs and human bone marrow-derived MSCs exhibited highly similar ion channel properties [106]. Recently the TGFβ inhibitor SB-431542 was successfully used in order to induce the differentiation of iPSCs into MSCs directly. Cells generated after 10 days of treatment have shown MSC characteristics in terms of immunophenotype and differentiation potential [16].

4.3 Large-Scale Cultivation

Controlled scalable expansion culture and a well-ordered differentiation process are challenges for translational clinical therapies, whenever high amounts of cells need to be transplanted. Although human iPSC-derived MSCs (hiPSC-MSCs) have generated great interest in possible clinical applications using iPSCs in regenerative medicine, the actual number of cells that could be cultivated after differentiation with traditional culture methods would be very low (Cormier et al. [18]. Currently cells aimed to be transplanted to patients are produced and cultured in static flasks. But this cultivation system results in a low amount, heterogeneity of cells, increased risk of contamination, and low cell yield due to the lack of real-time controlled parameters within culture media (including O_2 and nutrient concentration, pH, osmolarity, metabolic waste concentration, shear stress, and cell density). One probable resolution to get a sufficient amount of cells is to change the culture conditions towards suspension culture in bioreactors, which might allow the scaling up of the number of these cells in vitro. In this regard all culture parameters must be controlled in a bioreactor in order to get a tangible number of cells [28, 82]. Stirred suspension bioreactors (SSBs) have provided a dynamic condition to produce cell-based products in a safe, robust, and cost-effective manner. SSBs have been developed for many experiments, in which a large amount of cells is required and they were also successfully used for the expansion of undifferentiated pluripotent human stem cells [62, 108]. However, it is unclear if pluripotent cells that were expanded in such a bioreactor system can also be differentiated towards a MSC-like phenotype in a SSB or in another suspension culture system. To investigate these issues, further studies on robust differentiation protocols providing ESC- or iPSC-derived MSCs in suspension cultures or on the amplification of initially differentiated mesenchymal precursor cells in a bioreactor system capable of promoting MSC expansion [21], might be of high impact.

5 Conclusion and Outlook

In spite of data from over 100 trials employing MSCs in different clinical settings correlation of MSC properties to clinical efficacy is limited due to the considerable diversity of the applied MSC-populations [88]. Also the internationally agreed minimal set of consensus criteria for the definition and characterization of MSCs is not able to reflect adequately the broad spectrum of functional activity in the diverse contexts of tissue regeneration and anti-inflammatory therapies. Thus, engineered MSCs from well-characterized iPSC lines may not only solve the problem posed by the principally limited expansion capacity of MSCs, but may also serve as a homogeneous source of MSCs with more defined therapeutic characteristics. In this regard, one could even think of artificial iPSC-derived MSCs that overexpress a distinct therapeutically relevant transgene. One example for such a therapeutic transgene could be indoleamine-2,3-dioxygenase (IDO), which is induced by interferon-gamma (IFN-gamma) and which catalyzes the conversion from tryptophan to kynurenine and has been identified as a T cell inhibitory effector pathway in professional antigen-presenting cells [57]. Engineered MSCs, which were derived from iPSC lines harboring such a constitutively expressed IDO transgene, could serve as artificial MSCs for the treatment of inflammatory disease, where IDO-mediated T cell inhibition could further support the therapeutic effect of MSCs.

As the systemic application of MSCs might be hampered by an impaired pulmonary passage, MSCs could also be genetically modified to overcome such a limitation. For example, Rap1 [14], a member of the GTPase family of proteins with regulatory effects on multiple adhesion molecules, could be knocked-down in engineered MSCs, which then should gain an enhanced bioavailability after intravenous administration due to an improved pulmonary passage.

In conclusion, engineering MSCs from pluripotent stem cells such as iPSCs could generate advanced cellular therapies for the treatment of a variety of diseases, including intestinal GvHD and inflammatory bowel diseases, as well as some forms of liver failure.

Acknowledgments We are very grateful to the members of our lab providing intensive discussions and much critical input, which contributed to this review. M.M. and T.C. receive funding from the cluster of excellence REBIRTH (REgenerative BIology and Reconstructive THerapy), which is funded by the German Research Foundation (DFG; EXC 62/1). I.E. receives funding from the Centre for Cell and Gene Therapy of the LOEWE initiative. Further funding is provided by the Federal Ministry of Education and Research through grants 01GM1110A, 01GP1007C, as well as by the José Carreras leukemia foundation (DJCLS R11/22f).

References

1. Aasen T, Raya A, Barrero MJ, Garreta E, Consiglio A, Gonzalez F, Vassena R, Bilic J, Pekarik V, Tiscornia G, Edel M, Boue S, Izpisua Belmonte JC (2008) Efficient and rapid generation of induced pluripotent stem cells from human keratinocytes. Nat Biotechnol 26(11):1276–1284
2. Allison M (2009) Genzyme backs Osiris, despite Prochymal flop. Nat Biotechnol 27(11):966–967
3. Anasetti C, Hansen JA, Waldmann TA, Appelbaum FR, Davis J, Deeg HJ, Doney K, Martin PJ, Nash R, Storb R et al (1994) Treatment of acute graft-versus-host disease with

humanized anti-Tac: an antibody that binds to the interleukin-2 receptor. Blood 84(4):1320–1327
4. Augello A, Kurth TB, De Bari C (2010) Mesenchymal stem cells: a perspective from in vitro cultures to in vivo migration and niches. Eur Cell Mater 20:121–133
5. Aurich H, Sgodda M, Kaltwasser P, Vetter M, Weise A, Liehr T, Brulport M, Hengstler JG, Dollinger MM, Fleig WE, Christ B (2009) Hepatocyte differentiation of mesenchymal stem cells from human adipose tissue in vitro promotes hepatic integration in vivo. Gut 58(4):570–581
6. Aurich I, Mueller LP, Aurich H, Luetzkendorf J, Tisljar K, Dollinger MM, Schormann W, Walldorf J, Hengstler JG, Fleig WE, Christ B (2007) Functional integration of hepatocytes derived from human mesenchymal stem cells into mouse livers. Gut 56(3):405–415
7. Barberi T, Willis LM, Socci ND, Studer L (2005) Derivation of multipotent mesenchymal precursors from human embryonic stem cells. PLoS Med 2(6):e161
8. Bartholomew A, Sturgeon C, Siatskas M, Ferrer K, McIntosh K, Patil S, Hardy W, Devine S, Ucker D, Deans R, Moseley A, Hoffman R (2002) Mesenchymal stem cells suppress lymphocyte proliferation in vitro and prolong skin graft survival in vivo. Exp Hematol 30(1):42–48
9. Boland MJ, Hazen JL, Nazor KL, Rodriguez AR, Gifford W, Martin G, Kupriyanov S, Baldwin KK (2009) Adult mice generated from induced pluripotent stem cells. Nature 461(7260):91–94
10. Bouma G, Strober W (2003) The immunological and genetic basis of inflammatory bowel disease. Nat Rev Immunol 3(7):521–533
11. Cantz T, Martin U (2010) Induced pluripotent stem cells: characteristics and perspectives. Adv Biochem Eng Biotechnol 123:107–126
12. Cantz T, Sharma AD, Jochheim-Richter A, Arseniev L, Klein C, Manns MP, Ott M (2004) Reevaluation of bone marrow-derived cells as a source for hepatocyte regeneration. Cell Transp 13(6):659–666
13. Carey BW, Markoulaki S, Hanna JH, Faddah DA, Buganim Y, Kim J, Ganz K, Steine EJ, Cassady JP, Creyghton MP, Welstead GG, Gao Q, Jaenisch R (2011) Reprogramming factor stoichiometry influences the epigenetic state and biological properties of induced pluripotent stem cells. Cell Stem Cell 9(6):588–598
14. Carmona G, Gottig S, Orlandi A, Scheele J, Bauerle T, Jugold M, Kiessling F, Henschler R, Zeiher AM, Dimmeler S, Chavakis E (2009) Role of the small GTPase Rap1 for integrin activity regulation in endothelial cells and angiogenesis. Blood 113(2):488–497
15. Chamberlain G, Fox J, Ashton B, Middleton J (2007) Concise review: mesenchymal stem cells: their phenotype, differentiation capacity, immunological features, and potential for homing. Stem Cells 25(11):2739–2749
16. Chen YS, Pelekanos RA, Ellis RL, Horne R, Wolvetang EJ, Fisk NM (2012) Small molecule mesengenic induction of human induced pluripotent stem cells to generate mesenchymal stem/stromal cells. Stem Cells Trans Med 1:83–95
17. Ciccocioppo R, Bernardo ME, Sgarella A, Maccario R, Avanzini MA, Ubezio C, Minelli A, Alvisi C, Vanoli A, Calliada F, Dionigi P, Perotti C, Locatelli F, Corazza GR (2011) Autologous bone marrow-derived mesenchymal stromal cells in the treatment of fistulising Crohn's disease. Gut 60(6):788–798
18. Cormier JT, zur Nieden NI, Rancourt DE, Kallos MS (2006) Expansion of undifferentiated murine embryonic stem cells as aggregates in suspension culture bioreactors. Tissue Eng 12 (11):3233–3245
19. de Peppo GM, Svensson S, Lenneras M, Synnergren J, Stenberg J, Strehl R, Hyllner J, Thomsen P, Karlsson C (2010) Human embryonic mesodermal progenitors highly resemble human mesenchymal stem cells and display high potential for tissue engineering applications. Tissue Eng Part A 16(7):2161–2182
20. Deeg HJ, Blazar BR, Bolwell BJ, Long GD, Schuening F, Cunningham J, Rifkin RM, Abhyankar S, Briggs AD, Burt R, Lipani J, Roskos LK, White JM, Havrilla N, Schwab G,

Heslop HE (2001) Treatment of steroid-refractory acute graft-versus-host disease with anti-CD147 monoclonal antibody ABX-CBL. Blood 98(7):2052–2058
21. Diederichs S, Roker S, Marten D, Peterbauer A, Scheper T, van Griensven M, Kasper C (2009) Dynamic cultivation of human mesenchymal stem cells in a rotating bed bioreactor system based on the Z RP platform. Biotechnol Prog 25(6):1762–1771
22. Dominici M, Le Blanc K, Mueller I, Slaper-Cortenbach I, Marini F, Krause D, Deans R, Keating A, Prockop D, Horwitz E (2006) Minimal criteria for defining multipotent mesenchymal stromal cells. The International Society for Cellular Therapy position statement. Cytotherapy 8(4):315–317
23. Duijvestein M, Vos AC, Roelofs H, Wildenberg ME, Wendrich BB, Verspaget HW, Kooy-Winkelaar EM, Koning F, Zwaginga JJ, Fidder HH, Verhaar AP, Fibbe WE, van den Brink GR, Hommes DW (2010) Autologous bone marrow-derived mesenchymal stromal cell treatment for refractory luminal Crohn's disease: results of a phase I study. Gut 59(12):1662–1669
24. El-Ansary M, Abdel-Aziz I, Mogawer S, Abdel-Hamid S, Hammam O, Teaema S, Wahdan M (2011) Phase II trial: undifferentiated versus differentiated autologous mesenchymal stem cells transplantation in Egyptian patients with HCV induced liver cirrhosis. Stem Cell Rev 8(3):972–981
25. Eminli S, Utikal J, Arnold K, Jaenisch R, Hochedlinger K (2008) Reprogramming of neural progenitor cells into induced pluripotent stem cells in the absence of exogenous Sox2 expression. Stem Cells 26(10):2467–2474
26. Engel MA, Neurath MF (2010) New pathophysiological insights and modern treatment of IBD. J Gastroenterol 45(6):571–583
27. Ferrara JL, Levy R, Chao NJ (1999) Pathophysiologic mechanisms of acute graft-vs.-host disease. Biol Blood Marrow Transpl 5(6):347–356
28. Fridley KM, Fernandez I, Li MT, Kettlewell RB, Roy K (2010) Unique differentiation profile of mouse embryonic stem cells in rotary and stirred tank bioreactors. Tissue Eng Part A 16(11):3285–3298
29. Friedman SL (2007) Reversibility of hepatic fibrosis and cirrhosis—is it all hype? Nat Clin Pract Gastroenterol Hepatol 4(5):236–237
30. Fusaki N, Ban H, Nishiyama A, Saeki K, Hasegawa M (2009) Efficient induction of transgene-free human pluripotent stem cells using a vector based on Sendai virus, an RNA virus that does not integrate into the host genome. Proc Jpn Acad Ser B Phys Biol Sci 85(8):348–362
31. Garcia-Bosch O, Ricart E, Panes J (2010) Review article: stem cell therapies for inflammatory bowel disease—efficacy and safety. Aliment Pharmacol Ther 32(8):939–952
32. Giuliani M, Oudrhiri N, Noman ZM, Vernochet A, Chouaib S, Azzarone B, Durrbach A, Bennaceur-Griscelli A (2011) Human mesenchymal stem cells derived from induced pluripotent stem cells down-regulate NK-cell cytolytic machinery. Blood 118(12):3254–3262
33. Gonzalez F, Boue S, Izpisua Belmonte JC (2011) Methods for making induced pluripotent stem cells: reprogramming a la carte. Nat Rev Genet 12(4):231–242
34. Gruenloh W, Kambal A, Sondergaard C, McGee J, Nacey C, Kalomoiris S, Pepper K, Olson S, Fierro F, Nolta JA (2011) Characterization and in vivo testing of mesenchymal stem cells derived from human embryonic stem cells. Tissue Eng Part A 17(11–12):1517–1525
35. Haase A, Olmer R, Schwanke K, Wunderlich S, Merkert S, Hess C, Zweigerdt R, Gruh I, Meyer J, Wagner S, Maier LS, Han DW, Glage S, Miller K, Fischer P, Scholer HR, Martin U (2009) Generation of induced pluripotent stem cells from human cord blood. Cell Stem Cell 5(4):434–441
36. Hanna J, Saha K, Pando B, van Zon J, Lengner CJ, Creyghton MP, van Oudenaarden A, Jaenisch R (2009) Direct cell reprogramming is a stochastic process amenable to acceleration. Nature 462(7273):595–601

37. Hung CN, Mar K, Chang HC, Chiang YL, Hu HY, Lai CC, Chu RM, Ma CM (2011) A comparison between adipose tissue and dental pulp as sources of MSCs for tooth regeneration. Biomaterials 32(29):6995–7005
38. Hwang NS, Varghese S, Lee HJ, Zhang Z, Ye Z, Bae J, Cheng L, Elisseeff J (2008) In vivo commitment and functional tissue regeneration using human embryonic stem cell-derived mesenchymal cells. Proc Natl Acad Sci U S A 105(52):20641–20646
39. in 't Anker PS, Noort WA, Scherjon SA, Kleijburg-van der Keur C, Kruisselbrink AB, van Bezooijen RL, Beekhuizen W, Willemze R, Kanhai HH, Fibbe WE (2003) Mesenchymal stem cells in human second-trimester bone marrow, liver, lung, and spleen exhibit a similar immunophenotype but a heterogeneous multilineage differentiation potential. Haematologica 88(8):845–852
40. Jia F, Wilson KD, Sun N, Gupta DM, Huang M, Li Z, Panetta NJ, Chen ZY, Robbins RC, Kay MA, Longaker MT, Wu JC (2010) A nonviral minicircle vector for deriving human iPS cells. Nat Methods 7(3):197–199
41. Judson RL, Babiarz JE, Venere M, Blelloch R (2009) Embryonic stem cell-specific microRNAs promote induced pluripotency. Nat Biotechnol 27(5):459–461
42. Kaji K, Norrby K, Paca A, Mileikovsky M, Mohseni P, Woltjen K (2009) Virus-free induction of pluripotency and subsequent excision of reprogramming factors. Nature 458(7239):771–775
43. Kasper G, Mao L, Geissler S, Draycheva A, Trippens J, Kuhnisch J, Tschirschmann M, Kaspar K, Perka C, Duda GN, Klose J (2009) Insights into mesenchymal stem cell aging: involvement of antioxidant defense and actin cytoskeleton. Stem Cells 27(6):1288–1297
44. Kebriaei P, Isola L, Bahceci E, Holland K, Rowley S, McGuirk J, Devetten M, Jansen J, Herzig R, Schuster M, Monroy R, Uberti J (2009) Adult human mesenchymal stem cells added to corticosteroid therapy for the treatment of acute graft-versus-host disease. Biol Blood Marrow Transpl 15(7):804–811
45. Kemp K, Gray E, Mallam E, Scolding N, Wilkins A (2010) Inflammatory cytokine induced regulation of superoxide dismutase 3 expression by human mesenchymal stem cells. Stem Cell Rev 6(4):548–559
46. Kobbe G, Schneider P, Rohr U, Fenk R, Neumann F, Aivado M, Dietze L, Kronenwett R, Hunerliturkoglu A, Haas R (2001) Treatment of severe steroid refractory acute graft-versus-host disease with infliximab, a chimeric human/mouse antiTNFalpha antibody. Bone Marrow Transpl 28(1):47–49
47. Lagasse E, Connors H, Al-Dhalimy M, Reitsma M, Dohse M, Osborne L, Wang X, Finegold M, Weissman IL, Grompe M (2000) Purified hematopoietic stem cells can differentiate into hepatocytes in vivo. Nat Med 6(11):1229–1234
48. Laurila JP, Laatikainen L, Castellone MD, Trivedi P, Heikkila J, Hinkkanen A, Hematti P, Laukkanen MO (2009) Human embryonic stem cell-derived mesenchymal stromal cell transplantation in a rat hind limb injury model. Cytotherapy 11(6):726–737
49. Le Blanc K, Frassoni F, Ball L, Locatelli F, Roelofs H, Lewis I, Lanino E, Sundberg B, Bernardo ME, Remberger M, Dini G, Egeler RM, Bacigalupo A, Fibbe W, Ringden O (2008) Mesenchymal stem cells for treatment of steroid-resistant, severe, acute graft-versus-host disease: a phase II study. Lancet 371(9624):1579–1586
50. Lian Q, Lye E, Suan Yeo K, Khia Way Tan E, Salto-Tellez M, Liu TM, Palanisamy N, El Oakley RM, Lee EH, Lim B, Lim SK (2007) Derivation of clinically compliant MSCs from CD105+, CD24− differentiated human ESCs. Stem Cells 25(2):425–436
51. Lian Q, Zhang Y, Zhang J, Zhang HK, Wu X, Lam FF, Kang S, Xia JC, Lai WH, Au KW, Chow YY, Siu CW, Lee CN, Tse HF (2010) Functional mesenchymal stem cells derived from human induced pluripotent stem cells attenuate limb ischemia in mice. Circulation 121(9):1113–1123
52. Liechty KW, MacKenzie TC, Shaaban AF, Radu A, Moseley AM, Deans R, Marshak DR, Flake AW (2000) Human mesenchymal stem cells engraft and demonstrate site-specific differentiation after in utero transplantation in sheep. Nat Med 6(11):1282–1286

53. Liu CH, Hwang SM (2005) Cytokine interactions in mesenchymal stem cells from cord blood. Cytokine 32(6):270–279
54. Liu Y, Goldberg AJ, Dennis JE, Gronowicz GA, Kuhn LT (2012) One-step derivation of mesenchymal stem cell (MSC)-like cells from human pluripotent stem cells on a fibrillar collagen coating. PLoS One 7(3):e33225
55. Locke M, Feisst V, Dunbar PR (2011) Concise review: human adipose-derived stem cells: separating promise from clinical need. Stem Cells 29(3):404–411
56. Mark AL, Sun Z, Warren DS, Lonze BE, Knabel MK, Melville Williams GM, Locke JE, Montgomery RA, Cameron AM (2010) Stem cell mobilization is life saving in an animal model of acute liver failure. Ann Surg 252(4):591–596
57. Meisel R, Zibert A, Laryea M, Gobel U, Daubener W, Dilloo D (2004) Human bone marrow stromal cells inhibit allogeneic T-cell responses by indoleamine 2,3-dioxygenase-mediated tryptophan degradation. Blood 103(12):4619–4621
58. Mohamadnejad M, Namiri M, Bagheri M, Hashemi SM, Ghanaati H, Zare Mehrjardi N, Kazemi Ashtiani S, Malekzadeh R, Baharvand H (2007) Phase 1 human trial of autologous bone marrow-hematopoietic stem cell transplantation in patients with decompensated cirrhosis. World J Gastroenterol 13(24):3359–3363
59. Mohseny AB, Szuhai K, Romeo S, Buddingh EP, Briaire-de Bruijn I, de Jong D, van Pel M, Cleton-Jansen AM, Hogendoorn PC (2009) Osteosarcoma originates from mesenchymal stem cells in consequence of aneuploidization and genomic loss of Cdkn2. J Pathol 219(3):294–305
60. Okita K, Nakagawa M, Hyenjong H, Ichisaka T, Yamanaka S (2008) Generation of mouse induced pluripotent stem cells without viral vectors. Science 322(5903):949–953
61. Olivier EN, Rybicki AC, Bouhassira EE (2006) Differentiation of human embryonic stem cells into bipotent mesenchymal stem cells. Stem Cells 24(8):1914–1922
62. Olmer R, Lange A, Selzer S, Kasper C, Haverich A, Martin U, Zweigerdt R (2012) Suspension culture of human pluripotent stem cells in controlled, stirred bioreactors. Tissue Eng Part C Methods. doi:10.1089/ten.tec.2011.0717 [Epub ahead of print]
63. Papapetrou EP, Tomishima MJ, Chambers SM, Mica Y, Reed E, Menon J, Tabar V, Mo Q, Studer L, Sadelain M (2009) Stoichiometric and temporal requirements of Oct4, Sox2, Klf4, and c-Myc expression for efficient human iPSC induction and differentiation. Proc Natl Acad Sci U S A 106(31):12759–12764
64. Penn MS, Ellis S, Gandhi S, Greenbaum A, Hodes Z, Mendelsohn FO, Strasser D, Ting AE, Sherman W (2012) Adventitial delivery of an allogeneic bone marrow-derived adherent stem cell in acute myocardial infarction: phase I clinical study. Circ Res 110(2):304–311
65. Petersen BE, Bowen WC, Patrene KD, Mars WM, Sullivan AK, Murase N, Boggs SS, Greenberger JS, Goff JP (1999) Bone marrow as a potential source of hepatic oval cells. Science 284(5417):1168–1170
66. Pfaff N, Fiedler J, Holzmann A, Schambach A, Moritz T, Cantz T, Thum T (2011) miRNA screening reveals a new miRNA family stimulating iPS cell generation via regulation of Meox2. EMBO Rep 12(11):1153–1159
67. Pfaff N, Moritz T, Thum T, Cantz T (2012) miRNAs involved in the generation, maintenance, and differentiation of pluripotent stem cells. J Mol Med 90(7):747–752
68. Poynard T, Yuen MF, Ratziu V, Lai CL (2003) Viral hepatitis C. Lancet 362(9401):2095–2100
69. Ringden O, Uzunel M, Rasmusson I, Remberger M, Sundberg B, Lonnies H, Marschall HU, Dlugosz A, Szakos A, Hassan Z, Omazic B, Aschan J, Barkholt L, Le Blanc K (2006) Mesenchymal stem cells for treatment of therapy-resistant graft-versus-host disease. Transplantation 81(10):1390–1397
70. Saito F, Nakatani T, Iwase M, Maeda Y, Hirakawa A, Murao Y, Suzuki Y, Onodera R, Fukushima M, Ide C (2008) Spinal cord injury treatment with intrathecal autologous bone marrow stromal cell transplantation: the first clinical trial case report. J Trauma 64(1):53–59
71. Salas A, Ricart E, Panes J (2009) Cell therapies for inflammatory bowel diseases. Expert Rev Gastroenterol Hepatol 3(4):321–324

72. Sanchez L, Gutierrez-Aranda I, Ligero G, Rubio R, Munoz-Lopez M, Garcia-Perez JL, Ramos V, Real PJ, Bueno C, Rodriguez R, Delgado M, Menendez P (2011) Enrichment of human ESC-derived multipotent mesenchymal stem cells with immunosuppressive and anti-inflammatory properties capable to protect against experimental inflammatory bowel disease. Stem Cells 29(2):251–262
73. Sato Y, Araki H, Kato J, Nakamura K, Kawano Y, Kobune M, Sato T, Miyanishi K, Takayama T, Takahashi M, Takimoto R, Iyama S, Matsunaga T, Ohtani S, Matsuura A, Hamada H, Niitsu Y (2005) Human mesenchymal stem cells xenografted directly to rat liver are differentiated into human hepatocytes without fusion. Blood 106(2):756–763
74. Sharma AD, Cantz T, Manns MP, Ott M (2006) The role of stem cells in physiology, pathophysiology, and therapy of the liver. Stem Cell Rev 2(1):51–58
75. Sharma AD, Cantz T, Richter R, Eckert K, Henschler R, Wilkens L, Jochheim-Richter A, Arseniev L, Ott M (2005) Human cord blood stem cells generate human cytokeratin 18-negative hepatocyte-like cells in injured mouse liver. Am J Pathol 167(2):555–564
76. Stadtfeld M, Hochedlinger K (2010) Induced pluripotency: history, mechanisms, and applications. Genes Dev 24(20):2239–2263
77. Stadtfeld M, Nagaya M, Utikal J, Weir G, Hochedlinger K (2008) Induced pluripotent stem cells generated without viral integration. Science 322(5903):945–949
78. Stockmann HB, Hiemstra CA, Marquet RL, JN IJ (2000) Extracorporeal perfusion for the treatment of acute liver failure. Ann Surg 231(4):460–470
79. Storb R, Deeg HJ, Pepe M, Appelbaum F, Anasetti C, Beatty P, Bensinger W, Berenson R, Buckner CD, Clift R et al (1989) Methotrexate and cyclosporine versus cyclosporine alone for prophylaxis of graft-versus-host disease in patients given HLA-identical marrow grafts for leukemia: long-term follow-up of a controlled trial. Blood 73(6):1729–1734
80. Storb R, Thomas ED (1985) Graft-versus-host disease in dog and man: the Seattle experience. Immunol Rev 88:215–238
81. Strober W, Fuss I, Mannon P (2007) The fundamental basis of inflammatory bowel disease. J Clin Invest 117(3):514–521
82. Taiani JT, Krawetz RJ, Zur Nieden NI, Elizabeth WuY, Kallos MS, Matyas JR, Rancourt DE (2010) Reduced differentiation efficiency of murine embryonic stem cells in stirred suspension bioreactors. Stem Cells Dev 19(7):989–998
83. Takahashi K, Yamanaka S (2006) Induction of pluripotent stem cells from mouse embryonic and adult fibroblast cultures by defined factors. Cell 126(4):663–676
84. Tan J, Wu W, Xu X, Liao L, Zheng F, Messinger S, Sun X, Chen J, Yang S, Cai J, Gao X, Pileggi A, Ricordi C (2012) Induction therapy with autologous mesenchymal stem cells in living-related kidney transplants: a randomized controlled trial. JAMA 307(11):1169–1177
85. Theise ND, Nimmakayalu M, Gardner R, Illei PB, Morgan G, Teperman L, Henegariu O, Krause DS (2000) Liver from bone marrow in humans. Hepatology 32(1):11–16
86. Thomson JA, Itskovitz-Eldor J, Shapiro SS, Waknitz MA, Swiergiel JJ, Marshall VS, Jones JM (1998) Embryonic stem cell lines derived from human blastocysts. Science 282(5391):1145–1147
87. Tiemann U, Sgodda M, Warlich E, Ballmaier M, Schöler HR, Schambach A, Cantz T (2011) Optimal reprogramming factor stoichiometry increases colony numbers and affects molecular characteristics of murine induced pluripotent stem cells. Cytom Part A 79A:426–435
88. Tolar J, Le Blanc K, Keating A, Blazar BR (2010) Concise review: hitting the right spot with mesenchymal stromal cells. Stem Cells 28(8):1446–1455
89. Trivedi P, Hematti P (2008) Derivation and immunological characterization of mesenchymal stromal cells from human embryonic stem cells. Exp Hematol 36(3):350–359
90. van Poll D, Parekkadan B, Cho CH, Berthiaume F, Nahmias Y, Tilles AW, Yarmush ML (2008) Mesenchymal stem cell-derived molecules directly modulate hepatocellular death and regeneration in vitro and in vivo. Hepatology 47(5):1634–1643
91. Villa-Diaz LG, Brown SE, Liu Y, Ross AM, Lahann J, Parent JM, Krebsbach PH (2012) Derivation of mesenchymal stem cells from human induced pluripotent stem cells cultured on synthetic substrates. Stem Cells 30(6):1174–1181

92. Voelkel C, Galla M, Maetzig T, Warlich E, Kuehle J, Zychlinski D, Bode J, Cantz T, Schambach A, Baum C (2010) Protein transduction from retroviral Gag precursors. Proc Natl Acad Sci U S A 107(17):7805–7810
93. Vosough M, Moslem M, Pournasr B, Baharvand H (2011) Cell-based therapeutics for liver disorders. Br Med Bull 100:157–172
94. Wagner W (2010) Senescence is heterogeneous in mesenchymal stromal cells: kaleidoscopes for cellular aging. Cell Cycle 9(15):2923–2924
95. Wagner W, Horn P, Castoldi M, Diehlmann A, Bork S, Saffrich R, Benes V, Blake J, Pfister S, Eckstein V, Ho AD (2008) Replicative senescence of mesenchymal stem cells: a continuous and organized process. PLoS One 3(5):e2213
96. Wang X, Willenbring H, Akkari Y, Torimaru Y, Foster M, Al-Dhalimy M, Lagasse E, Finegold M, Olson S, Grompe M (2003) Cell fusion is the principal source of bone-marrow-derived hepatocytes. Nature 422(6934):897–901
97. Warlich E, Kühle J, Cantz T, Brugman MH, Mätzig T, Galla M, Filipczyk A, Halle S, Klump H, Schöler HR, Baum C, Schroeder T, Schambach A (2011) Lentiviral vector design and imaging approaches to visualize the early stages of cellular reprogramming. Mol Ther 19(4):782–789
98. Wilmut I, Schnieke AE, McWhir J, Kind AJ, Campbell KH (1997) Viable offspring derived from fetal and adult mammalian cells. Nature 385(6619):810–813
99. Wolters FL, Russel MG, Sijbrandij J, Schouten LJ, Odes S, Riis L, Munkholm P, Langholz E, Bodini P, O'Morain C, Katsanos K, Tsianos E, Vermeire S, Van Zeijl G, Limonard C, Hoie O, Vatn M, Moum B, Stockbrugger RW (2006) Disease outcome of inflammatory bowel disease patients: general outline of a Europe-wide population-based 10-year clinical follow-up study. Scand J Gastroenterol Suppl 243:46–54
100. Wu G, Liu N, Rittelmeyer I, Sharma AD, Sgodda M, Zaehres H, Bleidissel M, Greber B, Gentile L, Han DW, Rudolph C, Steinemamm D, Schambach A, Ott M, Schöler HR, Cantz T (2011) Generation of healthy mice from gene-corrected disease-specific induced pluripotent stem cells. PLoS Biol 9(7):e1001099
101. Yen ML, Hou CH, Peng KY, Tseng PC, Jiang SS, Shun CT, Chen YC, Kuo ML (2011) Efficient derivation & concise gene expression profiling of human embryonic stem cell-derived mesenchymal progenitors (EMPs). Cell Transpl. doi:10.3727/096368910X564067 [Epub ahead of print]
102. Yu J, Hu K, Smuga-Otto K, Tian S, Stewart R, Slukvin II, Thomson JA (2009) Human induced pluripotent stem cells free of vector and transgene sequences. Science 324(5928):797–801
103. Yu J, Thomson JA (2008) Pluripotent stem cell lines. Genes Dev 22(15):1987–1997
104. Yu J, Vodyanik MA, Smuga-Otto K, Antosiewicz-Bourget J, Frane JL, Tian S, Nie J, Jonsdottir GA, Ruotti V, Stewart R, Slukvin II, Thomson JA (2007) Induced pluripotent stem cell lines derived from human somatic cells. Science 318(5858):1917–1920
105. Zaehres H, Kogler G, Arauzo-Bravo MJ, Bleidissel M, Santourlidis S, Weinhold S, Greber B, Kim JB, Buchheiser A, Liedtke S, Eilken HM, Graffmann N, Zhao X, Meyer J, Reinhardt P, Burr B, Waclawczyk S, Ortmeier C, Uhrberg M, Schöler HR, Cantz T, Wernet P (2010) Induction of pluripotency in human cord blood unrestricted somatic stem cells. Exp Hematol 38(9):809–818
106. Zhang J, Chan YC, Ho JC, Siu CW, Lian Q, Tse HF (2012) Regulation of cell proliferation of human induced pluripotent stem cell-derived mesenchymal stem cells via ether a go–go 1 (hEAG1) potassium channel. Am J Physiol Cell Physiol
107. Zhou W, Freed CR (2009) Adenoviral gene delivery can reprogram human fibroblasts to induced pluripotent stem cells. Stem Cells 27(11):2667–2674
108. Zweigerdt R, Olmer R, Singh H, Haverich A, Martin U (2010) Scalable expansion of human pluripotent stem cells in suspension culture. Nat Protoc (in press)

Adv Biochem Eng Biotechnol (2013) 130: 19–37
DOI: 10.1007/10_2012_155

Published Online: 20 December 2012

Fate of Intravenously Injected Mesenchymal Stem Cells and Significance for Clinical Application

Beate Wagner and Reinhard Henschler

Abstract Mesenchymal stromal cells (MSCs) have initially been characterized as a fibroblastlike cell population that can be expanded readily in vitro, and is able to support hematopoiesis in vitro and in vivo. By serendipity it was discovered that MSCs can also be administered into the bloodstream. This mode of application formed a major breakthrough in the clinical use of MSCs, because MSC transplantation was found to cure severe immune hyperactivation states such as graft-versus-host disease after allogeneic bone marrow transplantation, or bacterial sepsis. However, MSCs were found difficult to trace and consensus to date is lacking in the scientific community as to where transplanted MSCs end up in the body and which major principles are responsible for the therapeutic effects of MSCs. This chapter gives an overview of the current knowledge on interactions of freshly transplanted MSCs with the cells in the blood stream and the vessel wall, with major organs such as lung, liver, gut, and spleen, and discusses the limitations of the methodologies used to trace transplanted MSCs. The findings will be put into perspective on how therapeutically applied, culture-expanded MSCs may exert beneficial effects.

Keywords Homing · Mesenchymal stem cells · Mesenchymal stromal cells · Therapy · Transplantation

Contents

B. Wagner · R. Henschler (✉)
Department of Transfusion Medicine, Cell Therapeutics and Hemostaseology,
Ludwig Maximilians University München,
Marchioninistrasse 15, 81433 Munich Germany
e-mail: reinhard.henschler@med.uni-muenchen.de

1 Introduction

For more than three decades, MSCs have been found to play fundamental roles in supporting ectopic hematopoiesis [28] and replacement of tissues of mesenchymal origin [55]. The role of MSCs in the replacement of mesenchymal tissues has been revealed by important work by Arnold Caplan and colleagues, among others, who discovered and worked out the details of differentiation regimens of MSCs into a variety of mesenchymal lineages including osteocytic cells, chondroblastlike cells, adipocytic cells, ligamentlike structures, and cells with smooth muscle characteristics [9, 10]. As a consequence, a multitude of clinical trials using topical administration of MSCs has been initiated, mainly in surgical patients [11]. These applications clearly required engraftment of MSCs for their therapeutic effects, including the incorporation of the transplanted cells into tissues. To date, however, topical MSC therapy has not entered clinical routine. This is in some contrast to intravenous application. The intravenous route has gained increasing attention worldwide and has come up with remarkable therapeutic successes in severely ill patients. Starting from early milestones in the clinical development of intravenous MSCs, approaches to reveal the fate of systemically administered MSCs are discussed below.

2 Intravenous Therapeutic Use of MSCs: The Case of Osteogenesis Imperfecta

Intravenously administered MSCs have meanwhile entered clinical studies and are applied in significant numbers of patients for two major indications: to support hematopoietic regeneration or accelerate engraftment in hematopoietic stem cell transplantation, or to cope with severe disorders of immune regulation. These applications had not initially been foreseen during early clinical development of MSCs. Indeed, a ground-breaking observation that facilitated this development was the work of Horwitz and colleagues, performed in a cohort of severely ill children. In the 1990s, they treated children with an inherited monogenetic disease, named brittle bone disease or osteogenesis imperfecta, who lack correctly synthesized collagen type I thus causing formation of fragile bones in the affected children [33]. The protein defect results in retarded growth, severe malformations, and eventually premature death. Horwitz and colleagues had recognized that allogeneic bone marrow transplantation could restore normal bone formation in these children [33]. In as much as collagen synthesis is a typical function of stromal cells that are also contained in the bone marrow graft, these authors reasoned that engrafting of stromal cells could provide the missing enzymatic activity.

Six children who underwent transplantation of allogeneic culture-expanded MSCs showed significant clinical improvement, and engraftment of donor-derived osteoblasts could be demonstrated in bone specimens using microsatellite DNA markers. Next, Horwitz and colleagues replaced uncultured whole bone marrow by culture-expanded MSCs. They compared short-term cultured to longer-term cultured MSCs from the same donor, and used retroviral marking with two different reporters of the transplanted MSCs. In their seminal publication in [34], they showed that gene-marked MSCs also engraft in the diseased children's bone cavities, and that the transplanted children could principally reach growth accelerations equal to the ones transplanted with allogeneic complete bone marrow [34]. The pioneer work of Horwitz' group work provided essential data for the further successful application of MSCs through the intravenous route in other clinical settings.

These authors have in addition made another pivotal observation in the development of cellular therapies using culture-expanded cells: They found that therapeutic success was not seen in all treated patients. They showed that some patients had developed antibodies directed against bovine antigens present on the MSCs grown in media containing fetal bovine serum [34]. Altogether, these studies greatly encouraged subsequent attempts to use MSCs as an intravenously administered cellular therapeutic.

3 Tracking the Fate of Systemically Delivered MSCs

One of the next steps after these discoveries was to investigate the use of MSCs in patients with tumors along with the application of high-dose chemotherapy regimens, in an attempt to alleviate side effects of the chemotherapy. Koc et al. administered around one million MSCs/kg body weight intravenously and demonstrated that the cells were well tolerated in patients with mammary carcinoma [40], for example. Before or parallel to their first clinical uses, intravenously administered MSCs were systematically assessed in a variety of animal models, including mice, rats, baboons, and dogs. Different methods such as radioactive labeling, fluorescent dye labeling, reporter gene transduction, or detection of natural DNA markers such as microsatellite markers of the transplanted cells were used [3, 20, 30, 7, 41]. The studies in small rodents, and similarly also in non-human primates in a noninjury situation, showed that MSCs were distributed to many tissues, however, they were detectable only in minute quantities within the tissues [20, 21]. These and many other follow-up studies using similar methodologies (reviewed in [36] have revealed these major findings: (i) quantification of MSCs is difficult, and if at all, transplanted cells are found at only low absolute levels within tissues. When absolute levels were determined, quantitative recovery was mostly confined to the lungs [45, 46, 78]. (ii) Engraftment of MSCs is mostly transient, and only few studies were able to demonstrate long-term maintenance of intact MSCs. (iii) Systemically applied MSCs can accumulate in certain tissues or tissue areas, for example, in areas of hypoxia or with inflammation. The following paragraphs focus on specific aspects of MSC accumulation within a variety of tissues and under specific pathophysiological conditions. The studies include attempts to decipher the molecular mechanisms by which MSCs may interact with the vessel wall, migrate through tissues, and mediate tropism, and the question whether MSCs may display tissue-specific therapeutic effects.

4 Fate of Intravenously Administered MSCs in the Lungs

The first organs through which intravenously injected MSCs pass are the lungs. The kinetics of human MSCs injected into mice have been determined as a very rapid uptake of >80 % of injected cells within a few minutes after injection and formation of emboli of MSCs in lung vessels, and exponentially falling clearance of the injected cells with a half-life of about 24 h with practically complete elimination within 100 h [45, 46]. The *Alu* sequence assay detecting human DNA did not trace MSCs in any other tissues at significant levels; the MSCs as detected by chromosomal nucleic acid totalled less than 0.1 % of the injected cells. Similarly, the lungs are the tissue with the highest uptake of rodent-derived MSCs, as seen, for example, by the ^{99m}Tc label in rats with induced myocardial infarction [3]. In contrast to human MSCs which rapidly disappear from the (murine) lungs, murine MSCs, at least in part, have been found to colonize lung tissue. Breitbach et al. [8] demonstrated the long-term incorporation of intact murine MSCs by the later-occurring malignant conversion of the transplanted MSCs. The interaction of transplanted human or murine MSCs with lung endothelial cells has been shown to depend on the suspension medium in which intravenously applied cells are kept pre-transplantation, and occur through an interaction of MSCs with the integrin ligand VCAM-1 [19]. Long-term incorporation of murine MSCs is enhanced by pre-treatment with bleomycin, which induces damage of lung epithelia [51]. MSCs have been shown to exert an anti-inflammatory effect in the lungs by the release of interleukin 1 receptor antagonist [52]. Nemeth et al. [50] in a murine sepsis model observed that MSCs locate in close proximity to macrophages, and induce macrophages to produce anti-inflammatory IL-10 through the release of prostaglandin E. A number of other hints point to a more complex network of cellular interactions, which can influence lymphocytes, for example, by induction of regulatory T cells, or alterations in the functions of monocytic cells as well as neutrophils. More interactions of MSCs with cells of the immune cells are discussed further below. Further investigations will have to elucidate the (likely predominant) role of the lungs enabling direct interactions between intra-vasal or extra-vasated MSCs with, in particular, cells of the immune system.

5 Elucidation of Mechanistic Steps and Dynamics by which MSCs may Interact with Endothelial Cells

The MSC research field is still actively working on elucidating the fate of intravenously transplanted MSCs within tissues over time. Studies have mainly explored two major aspects: which specific adhesion molecules may be involved in interactions of MSCs with endothelial cells, and given their size and the absence of several typical adhesion molecules that are typically present on leukocytes, is the interaction of MSCs with the vessel wall rather a passive or an active process?

5.1 Expression of Major Adhesion Receptors on MSCs Normally Present on Leukocytes

Main candidate receptors for endothelial interactions that are expressed on MSCs include CD44, the alpha4 beta1 integrin, and chemokine receptors such as CXCR4 (reviewed in [36]. The latter can be modulated in their expression, for example, by cytokine stimulation or by culture of MSCs in spheroids [56, 70]. In contrast, E-selectin ligand, L-selectin, as well as beta2 integrins were mainly found to be absent on both human and murine MSCs [36, 56, 60, 62, 63, 65, 70]. Therefore, it can be expected that, not only due to their enormous cellular volume as compared to blood leukocytes, but also due to the lack of classical adhesion molecules, MSCs may show major defects in their coordinated extravasation behaviour.

5.2 Evidence for the Involvement of Specific Leukocyte like Interaction Patterns in MSC-endothelial Interaction

[68] studied the interaction of MSCs that had been pre-treated with tumor necrosis factor (TNF)-alpha to heart endothelium. They observed that blockade of VCAM-1 using function-blocking antibody resulted in a decrease in adhesion of MSCs, indicating that beta1 integrins are actively involved in this process. However, their data were mainly based on in vitro data. [35], in a homing model of murine MSCs in experimental myocardial infarction, demonstrated that alpha4 integrin was required, whereas the chemokine receptor CXCR4 was dispensable for the entry of transplanted cells into ischemic tissue. These findings propose a specific role of beta1 integrins in the homing of MSCs in infarct models. Applying a P-selectin knockout mouse and intra-vital microscopy, [60] showed that human MSCs use a P-selectin ligand that is not PSGL-1 to affect interaction of MSCs with endothelial cells in collection venules of the murine ear. [62, 63] went further and engineered human MSCs by a specific fucosylation procedure, which do not naturally bind E-selectin at high shear forces, to alter CD44 epitopes to express functional E-selectin ligand activity, and rendered them fully responsive to E-selectin expressed on HUVEC endothelial cells in vitro and on bone venules in vivo. This alteration highly promoted MSC localization into bone tissue. In further work, they suggested that these engineered MSCs mediate their arrest on endothelial cells via capturing E-selectin, followed by ligation of alpha4 beta1 integrin [75].

The ability of human MSCs to adhere reversibly to TNF-alpha pre-stimulated endothelial cells was compared to human CD34 + hematopoietic progenitor cells or freshly isolated human blood mononuclear cells in a flow chamber model employing integrins alpha4 beta1 and its ligand, vascular cell adhesion molecule (VCAM)-1 [60]. In contrast to CD34 + cells and lymphocytes which for the most part reversibly bound to HUVEC endothelial cells, MSCs were found almost unable to dissociate at increased shear stresses at any significant rate. This indicates deficiencies in the dynamics of endothelial cell binding of MSCs.

6 Direct Evidence for a Co-ordinated Extravasation of MSCs

If MSCs would successfully interact with endothelial cells, they are expected to be found in tissue sections. Although numbers of transplanted MSCs in the tissues are low (a few million cells are maximally injected into a mouse, the per kg dose for humans in phase I/II studies was mostly about one million per kg body weight), [60] described intact MSCs in interstitial areas within the mouse liver. [62, 63] found that the engineered MSCs that express functional E-selectin ligand activity accumulate in bone marrow sinusoidal vessels and remain adjacent to endothelial cells, but within 24 h slowly locate into interstitial areas outside the vessels. This confirms the principal possibility that MSCs can extravasate. However, in both models, functional readouts such as a regenerative function of the MSCs were not investigated.

7 Direct Evidence for Other Fates of MSCs within the Vasculature

Intra-vital microscopy allows us to follow the fate of injected cells more closely. It has been performed in mice, partly after intra-arterial injection [29, 76], or after intravenous administration [78]. In the studies after intra-arterial injection, reappearance of the cells given originally as a bolus has been reported, indicating that they can passage the lungs. Using the cremaster muscle model, [29] observed that blood flow and microcirculation were impaired after injection of MSCs, and that MSCs obstructed small vessels. In addition, lung emboli were also reported. Occlusion of vessels and entrapment of injected MSCs at the pre-capillary level were also reported by [76]. Within a period of three days, in situ cell death of MSCs caught in the microvessels was observed, but also the integration of transplanted cells into the vessel wall. [78] also reported a clear risk of vascular occlusion after intravenous injection in the myocardial infraction model.

In contrast to these rather discouraging findings, studies in murine experimental autoimmune encephalitis also employed intra-vital microscopy and revealed accumulation of transplanted cells in inflammatory foci using bioluminescence, and showed a role of alpha4 integrins in this process [18]. Moreover, GFP+ transplanted MSCs have been localized to lymph nodes and to spleen at increased amounts. Although absolute numbers of transplanted MSCs have not been determined, these results indicate that active inflammation may switch the fate of transplanted MSCs from unspecific entrapment to specific recruitment.

The ability of MSCs to break down interstitial matrix (e.g., by gelatinases) has been demonstrated in mouse myocardium after alpha4 beta1, VCAM-1 mediated adhesion, and transendothelial migration by in situ zymography, and involves matrix metalloproteinase (MMP)-2 [73].

Still, any specific or unspecific clearance pathway for intravenously injected circulating MSCs has so far not been determined.

8 Systemic Reactions after MSC Transplantation

The studies using intra-vital microscopy and intra-arterial delivery of MSCs have already indicated systemic hemodynamic reactions. Walczak et al. using MSCs labeled with superparamagnetic iron oxide and laser Doppler measurements along with MRI imaging, confirmed rheologic perturbations, occlusion of vessels, and an increased mortality in the cell-treated group [78].

In contrast, intravenous infusion of 1–3 × 10E6 culture expanded MSCs/kg was clinically well tolerated in a study comprising 44 patients after hematopoietic stem cell transplantation. No adverse events were detected, but a slight drop of about 15 % in platelet counts was noted as well as a fivefold increase in the coagulation marker TAT (thrombin–anti-thrombin complex) and the anaphylatoxin C3a in serum [49]. In vitro exposure of MSCs to freshly drawn human whole blood in a closed circuit system demonstrated a potential of MSCs not only to decrease platelets but also circulating granulocytes and monocytes, except lymphocytes. This points to an interaction of MSCs with circulating cells. These effects were termed "instant blood-mediated inflammatory reaction" (IBMIR) and found to be proportional to the expression of tissue factor on human MSCs and the passage number of MSCs, and varied between individual donors. IBMIR, which was initially described after pancreatic islet and hepatocyte injection, is thought to be based on the activation of the complement and coagulation systems, and may to some extent interfere with clinical efficacy and safety of MSCs as cellular therapeutics.

On the other hand, MSCs have also been found to be able to respond to complement factors, and to acquire complement factors on their cell surface [48, 67], indicating that activation of systemic inflammation and of coagulation may influence the fate of intravenously delivered MSCs and vice versa.

9 Engraftment of MSCs into Bone Marrow

Bone marrow has been the main origin of MSC preparations that have been investigated in early pre-clinical and clinical work of intravenous administration regimens. Hence, it was of interest to see whether these cells could reach their tissue of origin. Work in the course of establishment of autologous and allogeneic bone marrow transplantation protocols had already shown that stromal cell types, such as colony-forming units-fibroblast (CFU-F), generally tended not to engraft along with the hematopoietic stem cells [72]. Until recently, this finding was hardly challenged and has been confirmed also for stromal cells that have been formally qualified as MSCs [17, 58]. Rombouts et al. showed in kinetic studies of outgrowing MSCs that culture time induced an engraftment defect of MSCs into bone marrow [59]. However, the work of Horwitz et al. [34] demonstrated that MSCs engraft in bone marrow of children with osteogenesis imperfecta, because they were able to grow donor-type MSCs from bone marrow of transplanted children. On the other hand,

[39] did not find engraftment of culture-expanded MSCs in patients with Hurler syndrome or metachromatic leukodystrophy. Direct evidence of bone marrow engraftment of human MSCs was recently shown to depend on the presence of functional integrity of a CD44 epitope (hematopoietic cell E selectin ligand) that was biotechnologically expressed on MSCs [62, 63]. Normal culture-expanded MSCs display only minimal levels of E-selectin binding activity, which may explain the low or undetectable numbers of unmanipulated MSCs found in bone marrow following intravenous injection, both after bone marrow transplantation or after high-dose chemotherapy. Follenzi et al. [25] recently showed that mice suffering from hemophilia lacking coagulation factor VIII, upon transplantation of normal healthy bone marrow show engraftment not only of hematopoietic cells, but also of sub-endothelial mesenchymal stromal cells which contributed to produce functional factor VIII. This also points to engraftment of at least some functional MSCs in the course of bone marrow transplantation, at least when performed in certain enzyme deficiencies.

Still, to date, no substantial donor-type contribution of any intravenously transplanted MSCs to hematopoiesis, either within the bone marrow, or at ectopic sites such as the kidney has been established for MSCs or analogous cell types [12, 13, 28, 42, 61, 66].

10 Role of Local Cues, Including Inflammation, Ischemia, and Previous Irradiation Influencing the Fate of MSCs

In the classical concept of extravasation of circulating immune cells into tissues, local cues mediate alterations in the expression of adhesion molecules on endothelial cells, such as selectins, integrin ligands, and chemokines, This way, circulating cells that are marginalized can increase contact time with endothelia, for example, through tethering and rolling interactions, and arrest and finally transmigrate through the vessel wall into tissues. If this would hold true for MSCs, one would expect an increased accumulation of MSCs in inflamed or ischemic tissues, as observed with phagocytes or cells or lymphocytes. In several murine or rat models of myocardial infarction, MSCs have been found to accumulate preferentially in areas of ischemia (e.g., [14, 65]. Zhang [84] have demonstrated a link between expression of the chemokine, stromal-derived factor (SDF)-1 and local accumulation of MSCs. Belema-Bedada et al. [6], using a transgenic model of the monokine-CC-chemokine ligand (CCL)-2 expressed under a cardiac-specific promotor have observed that intravenously injected MSCs carrying fluorescence markers accumulate selectively in the heart. They also showed that migration to the myocardium involves certain components of the intra-cellular signaling pathway of G protein-coupled receptors, pointing to the ability of MSCs to respond in a co-ordinated way to chemokines presented on cardiac vessel endothelia. However, the model also includes chemoattraction of circulating monocytes into the myocardium, but the influence of additional signals by monocytes cannot be ruled out. Together, these data demonstrate an ability of MSCs to enter tissue in the presence of specific environmental and inflammatory cues.

Kraitchman et al. [41] have confirmed accumulation of intravenously injected MSCs into myocardial infarction areas using a radioimaging tracer and single-photon emission computer tomography in a canine model. Some studies trying to trace MSCs at later stages after infarction tend to find markers of the transplanted cells in differentiated, newly regenerated cardiomyocytes (e.g., [79]. However, it is unclear to which cell and tissue types the homed MSCs may directly contribute, whether cell fusion is also involved, or whether in some cases also only artifacts were measured. There are also studies that have failed to detect any homed MSCs in cardiac tissue in the longer term (e.g., [43]).

In favor of some engraftment and tissue incorporation of intravenously transplanted MSCs are studies in inflammatory bowel disease models. Parekkadan et al. [53] have traced MSCs in a murine chemically induced colitis model. They demonstrated the presence of the live cell label not only in the lungs, spleen, and gut of the affected animals [53]. Sasaki et al. [64] have assessed whether MSCs may differentiate into skin cells including keratinocytes, and possibly contribute to wound repair in a mouse model using intravenously injected green fluorescence protein (GFP) transgenic MSCs. They found GFP-positive cells associated with specific markers for keratinocytes, endothelial cells, and pericytes. They suggested a chemokine CCL21-mediated entry mechanism. Although there is evidence of preferential attraction of the intravenously injected MSCs into wounded versus nonwounded skin, numbers of detected cells in the wounded skin areas were low.

MSCs have been demonstrated to migrate into inflamed brain tissue. Wu et al. in a murine stroke model validated MSC migration into ischemic areas after intravenous delivery [81]. Yilmaz et al. [83] confirmed and extended these findings by providing evidence that intravenously administered MSCs that enter ischemic areas in the brain are recruited through endothelial expressed P- and E-selectin, via CD44 present on the MSCs. In a rat brain ischemia model, [80] showed that intact MSCs arrive in ischemic zones and deliver neurotrophic factors at a greater rate when they have been exposed to hypoxia before injection. This correlated with increased expression of chemokine receptor CXCR4 on the MSCs, or the flk-1 and the erythropoietin receptors, and at the same time downregulation of pro-inflammmatory regulators in the MSCs. Miroglia activity was suppressed in animals after MSC therapy, and NeuN-positive and Glut1-positive cells were increased, underscoring the beneficial effects of intravenous delivery of MSCs.

Taken together, MSCs can migrate into ischemic and pro-inflammatory regions in certain disease stages in murine models. Generally, mostly short-term actions are reported, and long-term persistence of MSCs is not reported. Also, only indirect evidence is available to indicate whether the MSCs remain intact cells in their target environments. A quantification of the amount of active MSCs within a lesion, compared with the number of originally injected MSCs, is also not available. Still, these data argue in favor of locally acting, homed MSCs in a part of the investigated pathologies.

11 Influence of Local Irradiation on the Fate of Intravenously Transplanted MSCs

Francois et al. [26] demonstrated in mice that both total body irradiation and also local irradiation (e.g., on abdomen or legs) affected the distribution of hMSCs after IV infusion in NOD/SCID mice as compared to untreated animals. Without irradiation, intravenously infused hMSCs were found only in minimal amounts exclusively in the lung, bone marrow, and muscles. Fifteen days after the above-mentioned irradiation procedures, radiation-induced damage of tissues in the irradiated regions was confirmed by histological examination. TBI-treated animals exhibited higher absolute numbers of hMSCs in the brain, heart, bone marrow, and muscles. Moreover, selective radiation of limbs or the abdomen yielded a higher hMSC engraftment in the exposed field. More hMSCs were detected in the exposed skin, quadriceps, and other muscles than with TBI alone or additional abdominal irradiation. hMSC engraftment outside the locally irradiated regions was also increased, arguing for both local and systemic effects of irradiation for MSC engraftment. Long-term engraftment was, however, not investigated. Sémont et al. [69] in an additional study specifically investigated the engraftment, but also the efficacy of MSCs in a model of radiation-induced gastrointestinal tract failure. They demonstrated accelerated recovery in the group receiving hMSC in immunodeficient mice, with decreased apoptosis of epithelial cells and increased proliferation within the small intestinal mucosa. However, the transplanted MSCs were not detected at significant amounts.

12 Homing and Engraftment of MSCs into Tumors

Tumors of different kinds inherently harbor altered microenvironments with major alterations in blood flow, blood vessel structure, immune cell activation, and accumulation of trophic factors; in these processes, endogenous local or blood-derived mesenchymal stromal cell types are thought to play a major role [24]. Therefore, tumor tissue may represent a potential target for homing of intravenously injected MSCs.

Several studies have recently highlighted the fact that MSC indeed can home into tumors and that they can have both beneficial and also unwanted effects. Direct evidence for the migration of intravenously transplanted MSCs into tumors was provided by lentivirus-transduced MSC expressing eGFP in cells isolated from human orthotopic pancreatic cancer xenografts in nude mice using intra-vital microscopy [4]. Microscopical studies confirmed the interaction of MSCs with endothelial cells of blood vessels.

A more sophisticated method to detect homing of MSCs into tumors was recently developed by insertion of a trifunctional chimeric reporter into human adipose tissue-derived MSCs and a chimeric reporter into human glioblastoma cells [1]. Because both cell types expressed luciferase reporters they could be measured noninvasively by bioluminescence both in vitro and in vivo in a SCID mouse model. Although MSCs were implanted here and not intravenously injected, the study provides evidence for a concept that treatment with ganciclovir (GCV) activates a suicide mechanism in tumor-resident MSCs, resulting in tumor regression by a factor of 10E4 relative to controls. Using a luciferase reporter regulated by an endothelial-specific (PECAM, platelet/endothelial cell adhesion molecule) promoter and in vivo BLI to detect MSC differentiation, a mechanism was elucidated: implanted MSCs homed to tumor vessels, where they differentiated to endothelial cells [1].

12.1 Pro-Tumorigenic Effects of Tumor Engrafted MSCs

Enhanced angiogenesis as a mechanism of tumor promotion by MSCs was confirmed in vitro and in vivo using a murine BM-derived MSC line against B16 melanoma cells expressing LacZ (B16-LacZ) and Lewis lung carcinoma (LLC) [74]. Both co-culture with MSCs and treatment with MSC-conditioned media led to enhanced growth of tumor cells, although the magnitude of growth stimulation in co-cultured cells was greater than that of cells treated with conditioned media. Co-injection of tumor cells and MSCs into syngeneic mice led to increased tumor size compared with injection of tumor cells alone. Consistent with a role for neovascularization in MSC-mediated tumor growth, tumor vessel area was greater in tumors resulting from co-injection of tumor cells with MSCs than in tumors induced by injection of cancer cells alone. Co-injected MSCs directly supported the tumor vasculature by localizing close to vascular walls and by expressing the endothelial marker CD31. CCL25 was identified as a major chemoattractant for MSCs which was produced by multiple myeloma cells and has been made responsible for growth support of multiple myeloma cells through MSCs [82], providing a rationale to engineer chemokine receptors on MSCs, tailoring them towards an anti-tumor response profile.

12.2 Anti-Tumor Efficacy of MSCs

In a nude mice model, interleukin-12 (IL-12) expressing MSCs was injected intravenously to treat established Ewing sarcomas [22]. Although transplanted MSCs were not directly identified, the ongoing secretion of IL-12 in the tumor microenvironment strongly suggested homing of the injected MSCs to the tumor sites. In addition, growth suppression of the Ewing sarcomas was observed. [37] investigated potential phenotypes of tumor-associated MSCs by multicolored tissue transplant procedures in mice [37]. In syngeneic ovarian and breast cancer subpopulations they showed tumor-associated fibroblasts (TAFs) originated from MSCs located in the bone marrow, whereas most vascular and fibrovascular stroma (pericytes, α-SMA(+) myofibroblasts, and endothelial cells) were recruited from neighboring adipose tissue. These data form a basis that intravenously injected MSCs follow a path that is already established for endogenous MSCs, circulating to the tumor through the bloodstream. Grisendi et al. [32] demonstrated that this process implies epithelial/endothelial mesenchymal transitions in situ, or occurs through circulating pools of fibroblasts deriving from mesenchymal progenitors. These findings explain the apparent tumor tropism of MSCs, making them an attractive tool in tumor therapy.

There are some promising approaches to using MSCs as a tumor-suppressing cell therapy after incorporation of Paclitaxel, a widely used anti-cancer drug that also inhibits endothelial cell proliferation. Co-injection of tumor cells and a Paclitaxel-loaded MSCs cell line showed the anti-tumor efficacy of this approach [54].

13 Tracing of MSCs In Vivo that may Induce Alterations of the Immune System

MSCs have been shown to affect the immune system and also affect systemic administration in many ways (reviewed in [44]. However, tracing of MSCs has been cumbersome. As already indicated by Horwitz et al. for patients, development of antibodies against allogeneic MSCs, mostly against bovine antigens after ex vivo culture in fetal bovine serum-containing media, has given a good explanation of why MSCs are eliminated from transplanted humans or baboons (Horwitz et al. [5, 34]. The concept of locally acting MSCs to control immune reactions has been visualized in the lungs in the sepsis model of Nemeth et al. [50], but otherwise few MSCs have been found both in mice and in patients; Studies in mice [57, 31] found little evidence of the transplanted MSCs, as did Lee et al. in the TSG-6-mediated myocardial infarction mouse model [45, 46]. The same applied to patients who had received MSCs in the course of severe graft-versus-host disease, in whom it was very rarely possible to trace transplanted MSCs [77].

14 Interactions of i.v. Transplanted MSCs with Other Cell Types

Recently, the first studies have been published that describe responses in distinct immune cell types after intravenous application of MSCs. Although not all studies provide direct evidence for a cellular contact between the MSCs and other cells of the immune system, direct interactions between these cell types could be visualized in vitro, pointing to a realistic possibility that the interactions also occur in vivo. One study by Chiesa et al. [15] has visualized the almost complete cessation of migrating dendritic cells (DCs) in mice after intravenous delivery of MSCs. Using an in vitro system they show that murine MSCs inhibit DCs through toll-like receptor (TLR) 4. Mechanistic studies using mixed lymphocyte reactions co-culture studies with monocytes and hMSCs revealed a unique immunophenotype of alternatively activated human monocytes being CD206-high, IL-10-high, IL-6-high, IL12-low, and TNF-alpha-low [38]. The immunosuppressive potential of MSCs has been shown to depend on the inducibility of indoleamine 2,3-dioxygenase (IDO) [27]. Other work has identified MSC-secreted prostaglandin E2 as a main mediator of inflammation [47]; Nemeth et al. [50] in their sepsis model have revealed that intravenously administered MSCs use PGE2 as a mediator. This work put into the middle of its central hypothesis a direct influence of MSCs and macrophages in the lung. However, overall, direct interactions have rarely been demonstrated between i.v. injected MSCs and monocytes/macrophages [71].

Recent work by Akiyama et al. [2] has elucidated a role for fas ligand expressed on MSCs, by transiently inducing apoptosis in T cells. It has been shown to involve the secretion of MCP-1 by MSCs, which recruits T cells to apoptosis. The apoptosing T cells activate macrophages to produce TGF-beta, increasing regulatory T cells and thus promoting immune tolerance. More or less direct actions of MSCs on immune cells also involve the secretion of anti-inflammatory protein TSG-6 by activated MSCs which in a zymosan-induced mouse peritonitis model decreased TLR2/NF-κB signaling in resident macrophages [16].

15 Other Fates of MSCs and Outlook

Surprisingly, ectopic tissue formation was not found after systemic administration of MSCs. Also, fusion of MSCs, as has been shown to occur during tissue culture (e.g., in the presence of epithelial cells; [23], has not been regularly observed after MSC administration through the intravenous route. In summary, the terminal fate of the bulk of intravenously injected MSCs therefore remains elusive, inasmuch as studies have generally only been able to detect small amounts of injected cells. An exception is the approximately 80 % of MSCs that have been found transiently in the lungs of mice by [45, 46]. A natural clearance pathway for circulating MSCs has not yet been established. Future work will have to continue to trace transplanted MSCs, involving more quantitative assessments, in order to reveal what we do not really know about these scientifically and clinically fascinating cells.

References

1. Alieva M, Bagó JR, Aguilar E, Soler-Botija C, Vila OF, Molet J, Gambhir SS, Rubio N, Blanco J (2012) Glioblastoma therapy with cytotoxic mesenchymal stromal cells optimized by bioluminescence imaging of tumor and therapeutic cell response. PLoS One 7(4):e35148 (Epub 2012 Apr 17)
2. Akiyama K, Chen C, Wang D, Xu X, Qu C, Yamaza T, Cai T, Chen W, Sun L, Shi S (2012) Mesenchymal-stem-cell-induced immunoregulation involves FAS-ligand-/FAS-mediated T cell apoptosis. Cell Stem Cell 10(5):544–555 (Epub 2012 Apr 26)
3. Barbash IM, Chouraqui P, Baron J, Feinberg MS, Etzion S, Tessone A, Miller L, Guetta E, Zipori D, Kedes LH et al (2003) Systemic delivery of bone marrow-derived mesenchymal stem cells to the infarcted myocardium: feasibility, cell migration, and body distribution. Circulation 108:863–868
4. Beckermann BM, Kallifatidis G, Groth A, Frommhold D, Apel A, Mattern J, Salnikov AV, Moldenhauer G, Wagner W, Diehlmann A, Saffrich R, Schubert M, Ho AD, Giese N, Büchler MW, Friess H, Büchler P, Herr I (2008) VEGF expression by mesenchymal stem cells contributes to angiogenesis in pancreatic carcinoma. Br J Cancer 99(4):622–631 (Epub 2008 July 29)
5. Beggs KJ, Lyubimov A, Borneman JN, Bartholomew A, Moseley A, Dodds R, Archambault MP, Smith AK, McIntosh KR (2006) Immunologic consequences of multiple, high-dose administration of allogeneic mesenchymal stem cells to baboons. Cell Transplant 15(8–9):711–721
6. Belema-Bedada F, Uchida S, Martire A, Kostin S, Braun T (2008) Efficient homing of multipotent adult MSCs depends on FROUNT-mediated clustering of CCR2. Cell Stem Cell 2:566–575
7. Bensidhoum M, Chapel A, Francois S, Demarquay C, Mazurier C, Fouil- lard L, Bouchet S, Bertho JM, Gourmelon P, Aigueperse J, et al (2004) Homing of in vitro expanded Stro-1- or Stro-1 + human mesenchymal stem cells into the NOD/SCID mouse and their role in supporting human CD34 cell engraftment. Blood 103:3313–3319
8. Breitbach M, Bostoni T, Roell W, et al (2007) Potential risk of bone marrow cell transplantation into infarcted hearts. Blood 110:1362–1369
9. Caplan AI, Dennis JE (2006) Mesenchymal stem cells as trophic mediators. J Cell Biochem 98:1076–1084
10. Caplan AI (2009) Why are MSCs therapeutic? new data: new insight. J Pathol 217:318–324
11. Caplan A, Correa D (2011) The MSC: an injury drugstore. Cell Stem Cell 9:11–15
12. Calvi LM, Adams GB, Weibrecht KW, Weber JM, Olson DP, Knight MC, Martin RP, Schipani E, Divieti P, Bringhurst FR, Milner LA, Kronenberg HM, Scadden DT (2003) Osteoblastic cells regulate the haematopoietic stem cell niche. Nature 425:841–846
13. Chen Y, Jacamo R, Shi YX, Wang RY, Battula VL, Konoplev S, Strunk D, Hofmann NA, Reinisch A, Konopleva M, Andreeff M (2012) Human extramedullary bone marrow in mice: a novel in vivo model of genetically controlled hematopoietic microenvironment. Blood 119(21):4971–4980 (Epub 2012 Apr 5)
14. Chen J, Li Y, Wang L, Zhang Z, Lu D, Lu M, Chopp M (2001) Therapeutic benefit of intravenous administration of bone marrow stromal cells after cerebral ischemia in rats. Stroke 32:1005–1011
15. Chiesa S, Morbelli S, Morando S, et al (2011) Mesenchymal stem cells impair in vivo T-cell priming by dentritic cells. Proc Natl Acad Sci USA 108:17384–17389
16. Choi H, Lee RH, Bazhanov N, Oh JY, Prockop DJ (2011) Anti-inflammatory protein TSG-6 secreted by activated MSCs attenuates zymosan- induced mouse peritonitis by decreasing TLR2/NF-KB signaling in resident macrophages. Blood 118:330–338
17. Cilloni D et al (2000) Limited engraftment capacity of bone marrow-derived mesenchymal cells following T-cell-depleted hematopoietic stem cell transplantation. Blood 96:3637–3643

18. Constantin G, Marconi S, Rossi B, Angiari S, Calderan L, Anghileri E, Gini B, Bach SD, Martinello M, Bifari F, Galiè M, Turano E, Budui S, Sbarbati A, Krampera M, Bonetti B (2009) Adipose-derived mesenchymal stem cells ameliorate chronic experimental autoimmune encephalomyelitis. Stem Cells. 27(10):2624–2635
19. Deak E, Rüster B, Keller L, Eckert K, Fichtner I, Seifried E, Henschler R (2010) Suspension medium influences interaction of mesenchymal stromal cells with endothelium and pulmonary toxicity after transplantation in mice. Cytotherapy 12(2):260–264
20. Devine SM, Bartholomew A, Mahmud N, Nelson M, Patil S, Hardy W, Cord S, Terry H, Chung T, Stock W et al (2001) Mesenchymal stem cells are capable of homing to the bone marrow of non-human primates following systemic infusion. Exp Hematol 29:244–255
21. Devine SM, Cobbs C, Jennings M, Bartholomew A, Hoffman R (2003) Mesenchymal stem cells distribute to a wide range of tissues following systemic infusion into nonhuman primates. Blood 101:2999–3001
22. Duan X, Guan H, Cao Y, Kleinerman ES (2009) Murine bone marrow-derived mesenchymal stem cells as vehicles for interleukin-12 gene delivery into Ewing sarcoma tumors. Cancer 115(1):13–22
23. Ferrand J, Noël D, Lehours P, Prochazkova-Carlotti M, Chambonnier L, Ménard A, Mégraud F, Varon C (2011) Human bone marrow-derived stem cells acquire epithelial characteristics through fusion with gastrointestinal epithelial cells. PLoS One 6(5):e19569
24. Fakhrejahani E, Toi M (2012) Tumor angiogenesis: pericytes and maturation are not to be ignored. J Oncol 2012:261750 (Epub 2011 Oct 9)
25. Follenzi A, Raut S, Merlin S, Sarkar R, Gupta S (2012) Role of bone marrow transplantation for correcting hemophilia A in mice. Blood 119(23):5532–5542 (Epub 2012 Feb 24)
26. Francois S, Bensidhoum M, Mouiseddine M, Mazurier C, Allenet B, Semont A, Frick J, Sache A, Bouchet S, Thierry D et al (2006) Local irradiation not only induces homing of human mesenchymal stem cells at exposed sites but promotes their widespread engraftment to multiple organs: a study of their quantitative distribution after irradiation damage. Stem Cells 24:1020–1029
27. Francois M, Romieu-Mourez R, Li M, Galipeau J (2011) Human MSC suppression correlates with cytokine induction of indoleamine 2,3-dioxygenase and bystander M2 macrophage differentiation. Mol Ther 20:187–195
28. Friedenstein AJ, Chailakhyan RK, Latsinik NV, Panasyuk AF, Keiliss-Borok IV (1974) Stromal cells responsible for transferring the micro- environment of the hemopoietic tissues. Cloning in vitro and retransplantation in vivo. Transplantation 17:331–340
29. Furlani D, Ugurlucan M, Ong L, Bieback K, Pittermann E, Westien I, Wang W, Yerebakan C, Li W, Gaebel R, Li RK, Vollmar B, Steinhoff G, Ma N (2009) Is the intravascular administration of mesenchymal stem cells safe? mesenchymal stem cells and intravital microscopy. Microvasc Res 77(3):370–376 (Epub 2009 Feb 26)
30. Gao J, Dennis JE, Muzic RF, Lundberg M, Caplan AI (2001) The dynamic in vivo distribution of bone marrow-derived mesenchymal stem cells after infusion. Cells Tissues Organs 169:12–20
31. Gonzalez-Rey E, Anderson P, González MA, Rico L, Büscher D, Delgado M (2009) Human adult stem cells derived from adipose tissue protect against experimental colitis and sepsis. Gut 58:929–939
32. Grisendi G, Bussolari R, Veronesi E, Piccinno S, Burns JS, De Santis G, Loschi P, Pignatti M, Di Benedetto F, Ballarin R, Di Gregorio C, Guarneri V, Piccinini L, Horwitz EM, Paolucci P, Conte P, Dominici M (2011) Understanding tumor-stroma interplays for targeted therapies by armed mesenchymal stromal progenitors: the mesenkillers. Am J Cancer Res 1(6):787–805 (Epub 2011 May 28)
33. Horwitz EM, Prockop DJ, Fitzpatrick LA, et al (1999) Transplantability and therapeutic effects of bone marrow-derived mesenchymal cells in children with osteogenesis imperfecta. Nat Med 5:309–313
34. Horwitz EM, Gordon PL, Koo WK, Marx JC, Neel MD, McNall RY, Muul L, Hofmann T (2002 Jun 25) Isolated allogeneic bone marrow-derived mesenchymal cells engraft and

stimulate growth in children with osteogenesis imperfecta: implications for cell therapy of bone. Proc Natl Acad Sci USA 99(13):8932–8937
35. Ip JE, Wu Y, Huang J, Zhang L, Pratt RE, Dzau VJ (2007) Mesen- chymal stem cells use integrin beta1 not CXC chemokine receptor 4 for myocardial migration and engraftment. Mol Biol Cell 18:2873–2882
36. Karp JM, Leng Teo GS (2009) Mesenchymal stem cell homing: the devil is in the details. Cell Stem Cell 4:206–216
37. Kidd S, Spaeth E, Watson K, Burks J, Lu H, Klopp A, Andreeff M, Marini FC (2012) Origins of the tumor microenvironment: quantitative assessment of adipose-derived and bone marrow-derived stroma. PLoS One 7(2):e30563 (Epub 2012 Feb 20)
38. Kim J, Hematti P (2009) Mesenchymal stem cell-educated macrophages: a novel type of alternatively activated macrophages. Exp Hematol 37:1445–1453
39. Koç ON, Gerson SL, Cooper BW, Dyhouse SM, Haynesworth SE, Caplan AI, Lazarus HM (2000) Rapid hematopoietic recovery after coinfusion of autologous-blood stem cells and culture- expanded marrow mesenchymal stem cells in advanced breast cancer patients receiving high-dose chemotherapy. J Clin Oncol 18:307–316
40. Koç ON, Day J, Nieder M, Gerson SL, Lazarus HM, Krivit W (2002) Allogeneic mesenchymal stem cell infusion for treatment of metachro- matic leukodystrophy (MLD) and Hurler syndrome (MPS-IH). Bone Marrow Transplant 30:215–222
41. Kraitchman DL, Tatsumi M, Gilson WD, Ishimori T, Kedziorek D, Walczak P, Segars WP, Chen HH, Fritzges D, Izbudak I et al (2005) Dynamic imaging of allogeneic mesenchymal stem cells trafficking to myocardial infarction. Circulation 112:1451–1461
42. Lange C, Brunswig-Spickenheier B, Cappallo-Obermann H, Eggert K, Gehling UM, Rudolph C, Schlegelberger B, Cornils K, Zustin J, Spiess AN, Zander AR (2011) Radiation rescue: mesenchymal stromal cells protect from letal irradiation. PLoS One 6(1):e14486
43. Laurila JP, Laatikainen L, Castellone MD, Trivedi P, Heikkila J, Hinkkanen A, Hematti P, Laukkanen MO (2009) Human embryonic stem cell-derived mesenchymal stromal cell transplantation in a rat hind limb injury model. Cytotherapy 11(6):726–737
44. Le Blanc K, Mougiakanos D (2012) Multipotent mesenchymal stromal cells and the innate immune system. Nat Rev Immunol 12:383–396
45. Lee JW, Fang X, Gupta N, Serikov V, Matthay MA (2009) Allogeneic human mesenchymal stem cells for treatment of E. coli endotoxin-induced acute lung injury in the ex vivo perfused human lung. Proc Natl Acad Sci USA 106(38):16357–16362 (Epub 2009 Aug 31. (a))
46. Lee RH, Pulin AA, Seo MJ, Kota DJ, Ylostalo J, Larson BL, Semprun-Prieto L, Delafontaine P, Prockop DJ (2009) Intravenous hMSCs improve myocardial infarction in mice because cells embolized in lung are activated to secrete the anti-inflammatory protein TSG-6. Cell Stem Cell 5(1):54–63 (b)
47. Maggini J, Mirkin G, Bognanni I, Holmberg J, Piazzón IM, Nepomnaschy I, Costa H, Cañones C, Raiden S, Vermeulen M, Geffner JR (2010) Mouse bone marrow-derived mesenchymal stromal cells turn activated macrophages into a regulatory-like profile. PLoS One 5(2):e9252
48. Moll G, Jitschin R, von Bahr L, Rasmusson-Duprez I, Sundberg B, Lönnies L, Elgue G, Nilsson-Ekdahl K, Mougiakakos D, Lambris JD, Ringdén O, Le Blanc K, Nilsson B (2011) Mesenchymal stromal cells engage complement and complement receptor bearing innate effector cells to modulate immune responses. PLoS One 6(7):e21703 (Epub 2011 Jul 1)
49. Moll G, Rasmusson-Duprez I, von Bahr L, Connolly-Andersen AM, Elgue G, Funke L, Hamad OA, Lönnies H, Magnusson PU, Sanchez J, Teramura Y, Nilsson-Ekdahl K, Ringdén O, Korsgren O, Nilsson B, Le Blanc K (2012) Are therapeutic human mesenchymal stromal cells compatible with human blood? Stem Cells 30(7):1565–1574
50. Nemeth K, Leelahavanichkul A, Yuen PS, Mayer B, Parmelee A, Doi K, Robey PG, Leelahavanichkul K, Koller BH, Brown JM et al (2009) Bone marrow stromal cells attenuate sepsis via prostaglandin E(2)-dependent reprogramming of host macrophages to increase their interleukin-10 production. Nat Med 15:42–49

51. Ortiz LA, Gambelli F, McBride C, Gaupp D, Baddoo M, Kaminski N, Phinney DG (2003) Mesenchymal stem cell engraftment in lung is enhanced in response to bleomycin exposure and ameliorates its fibrotic effects. Proc Natl Acad Sci USA 100:8407–8411
52. Ortiz LA, Dutreil M, Fattman C, Pandey AC, Torres G, Go K, Phinney DG (2007) Interleukin 1 receptor antagonist mediates the antiinflammatory and antifibrotic effect of mesenchymal stem cells during lung injury. Proc Natl Acad Sci USA 104(26):11002–11007 (Epub 2007 Jun 14)
53. Parekkadan B, Upadhyay R, Dunham J, Iwamoto Y, Mizoguchi E, Mizoguchi A, Weissleder R, Yarmush ML (2011) Bone marrow stromal cell transplants prevent experimental enterocolitis and require host CD11b + splenocytes. Gastroenterology 140(3):966–975 (Epub 2010 Oct 16)
54. Pessina A, Bonomi A, Coccè V, Invernici G, Navone S, Cavicchini L, Sisto F, Ferrari M, Viganò L, Locatelli A, Ciusani E, Cappelletti G, Cartelli D, Arnaldo C, Parati E, Marfia G, Pallini R, Falchetti ML, Alessandri G (2011) Mesenchymal stromal cells primed with paclitaxel provide a new approach for cancer therapy. PLoS One 6(12):e28321 (Epub 2011 Dec 20)
55. Pittenger MF, Mackay AM, Beck SC, et al (1999) Multilineage potential of adult human mesenchymal stem cells. Science 284:143–147
56. Potapova IA, Brink PR, Cohen IS, Doronin SV (2008) Culturing of human mesenchymal stem cells as three-dimensional aggregates induces functional expression of CXCR4 that regulates adhesion to endothelial cells. J Biol Chem 283:13100–13107
57. Ren C, Kumar S, Chanda D, et al (2008) Therapeutic potential of mesenchymal stem cells producing interferon-alpha in a mouse melanoma lung metastasis model. Stem Cells 26:2332–2338
58. Rieger K et al (2005) Mesenchymal stem cells remain of host origin even a long time after allogeneic peripheral blood stem cell or bone marrow transplantation. Exp Hematol 33:605–611
59. Rombouts WJ, Ploemacher RE (2003) Primary murine MSC show highly efficient homing to the bone marrow but lose homing ability following culture. Leukemia 17(1):160–170
60. Ruster B, Gottig S, Ludwig RJ, Bistrian R, Muller S, Seifried E, Gille J, Henschler R (2006) Mesenchymal stem cells display coordinated rolling and adhesion behavior on endothelial cells. Blood 108:3938–3944
61. Sacchetti B, Funari A, Michienzi S, Di Cesare S, Piersanti S, Saggio I, Tagliafico E, Ferrari S, Robey PG, Riminucci M, Bianco P (2007) Self-renewing osteoprogenitors in bone marrow sinusoids can organize a hematopoietic microenvironment. Cell 131:324–336
62. Sackstein R, Merzaban JS, Cain DW, Dagia NM, Spencer JA, Lin CP, Wohlgemuth R (2008) Ex vivo glycan engineering of CD44 programs human multipotent mesenchymal stromal cell trafficking to bone. Nat Med 14:181–187
63. Sackstein R, Merzaban JS, Cain DW, Dagia NM, Spencer JA, Lin CP, Wohlgemuth R (2008) Ex vivo glycan engineering of CD44 programs human multipotent mesenchymal stromal cell trafficking to bone. Nat Med 14:181–187
64. Sasaki M, Abe R, Fujita Y, Ando S, Inokuma D, Shimizu H (2008) Mesenchymal stem cells are recruited into wounded skin and contribute to wound repair by transdifferentiation into multiple skin cell type. J Immunol 180(4):2581–2587
65. Schenk S, Mal N, Finan A, Zhang M, Kiedrowski M, Popovic Z, McCarthy PM, Penn MS (2007) Monocyte chemotactic protein-3 is a myocar-dial MSC homing factor. Stem Cells 25:245–251
66. Schofield R (1983) The stem cell system. Biomed Pharmacother 37:375–380
67. Schraufstatter IU, Discipio RG, Zhao M, Khaldoyanidi SK (2009) C3a and C5a are chemotactic factors for human mesenchymal stem cells, which cause prolonged ERK1/2 phosphorylation. J Immunol 182:3827–3836
68. Segers VFM, Van Riet I, Andries LJ, Lemmens K, Demolder MJ, De Becker AJML, Kockx MM, De Keulenaer GW (2006) Mesenchymal stem cell adhesion to cardiac microvascular endothelium: activators and mechanisms. Am J Physiol Heart Circ Physiol 290:H1370–H1377
69. Sémont A, Mouiseddine M, François A, Demarquay C, Mathieu N, Chapel A, Saché A, Thierry D, Laloi P, Gourmelon P (2010) Mesenchymal stem cells improve small intestinal

integrity through regulation of endogenous epithelial cell homeostasis. Cell Death Differ 17(6):952–961 (Epub 2009 Dec 18)
70. Shi M, Li J, Liao L, Chen B, Li B, Chen L, Jia H, Zhao RC (2007) Regulation of CXCR4 expression in human mesenchymal stem cells by cyto-kine treatment: role in homing efficiency in NOD/SCID mice. Haematologica 92:897–904
71. Shi C, Jia T, Mendez-Ferrer S, Hohl TM, Serbina NV, Lipuma L, Leiner I, Li MO, Frenette PS, Pamer EG (2011) Bone marrow mesenchymal stem and progenitor cells induce monocyte emigration in response to circulating toll-like receptor ligands. Immunity 34:590–601
72. Simmons PJ, Przepiorka D, Thomas ED, Torok-Storb B (1987) Host origin of marrow stromal cells following allogeneic bone marrow transplantation. Nature 328:429–432
73. Steingen C, Brenig F, Baumgartner L, Schmidt J, Schmidt A, Bloch W (2008). Characterization of key mechanisms in transmigration and invasion of mesenchymal stem cells. J Mol Cell Cardiol 44(6):1072–1084 (Epub 2008 Mar 19)
74. Suzuki K, Sun R, Origuchi M, Kanehira M, Takahata T, Itoh J, Umezawa A, Kijima H, Fukuda S, Saijo Y (2011) Mesenchymal stromal cells promote tumor growth through the enhancement of neovascularization. Mol Med 17(7–8):579–587 10.2119/molmed.2010.00157. Epub 2011 Mar 11
75. Thankamony SP, Sackstein R (2011) Enforced hematopoietic cell E- and L-selectin ligand (HCELL) expression primes transendothelial migration of human mesenchymal stem cells. Proc Natl Acad Sci USA 108(6):2258–2263 (Epub 2011 Jan 21)
76. Toma C, Wagner WR, Bowry S, Schwartz A, Villanueva F (2009) Fate of culture-expanded mesenchymal stem cells in the microvasculature: in vivo observations of cell kinetics. Circ Res 104(3):398–402 (Epub 2008 Dec 18)
77. von Bahr L, Batasis I, Moll G, et al (2012) Analysis of tissues following mesenchymal stromal cell therapy in human indicates limited long-term engraftment and no ectopic tissue formation. Stem Cells 30:1575–1578
78. Walczak P, Zhang J, Gilad AA, Kedziorek DA, Ruiz-Cabello J, Young RG, Pittenger MF, van Zijl PCM, Huang J, Bulte JWM (2008) Dual-modality monitoring of targeted intraarterial delivery of mesenchymal stem cells after transient ischemia. Stroke 39:1569–1574
79. Wang Y, Zhang D, Ashraf M, Zhao T, Huang W, Ashraf A, Balasubramaniam A (2010) Combining neuropeptide Y and mesenchymal stem cells reverses remodeling after myocardial infarction. Am J Physiol Heart Circ Physiol 298(1):H275–H286 (Epub 2009 Nov 6)
80. Wei L, Fraser JL, Lu ZY, Hu X, Yu SP (2012) Transplantation of hypoxia preconditioned bone marrow mesenchymal stem cells enhances angiogenesis and neurogenesis after cerebral ischemia in rats. Neurobiol Dis 46(3):635–645 (Epub 2012 Mar 9)
81. Wu J, Sun Z, Sun H-S, Wu J, Weisel RD, Keating A, Li Z-H, Feng Z-P, Li R-K (2008) Intravenously administered bone marrow cells migrate to damaged brain tissue and improve neural fnction in ichemic rats. Cell Transplant 16:993–1005
82. Xu S, Menu E, De Becker A, Van Camp B, Vanderkerken K, Van Riet I (2012) Bone marrow-derived mesenchymal stromal cells are attracted by multiple myeloma cell-produced chemokine CCL25 and favor myeloma cell growth in vitro and in vivo. Stem Cells 30(2):266–279
83. Yilmaz G, Vital S, Yilmaz CE, Stokes KY, Alexander JS, Granger DN (2011) Selectin-mediated recruitment of bone marrow stromal cells in the postischemic cerebral microvasculature. Stroke 42(3):806–811 (Epub 2011 Jan 21)
84. Zhang M, Mal N, Kiedrowski M, Chacko M, Askari AT, Popovic ZB, Koc ON, Penn MS (2007) SDF-1 expression by mesenchymal stem cells results in trophic support of cardiac myocytes after myocardial infarction. FASEB J 21:3197–3207

Adv Biochem Eng Biotechnol (2013) 130: 39–54
DOI: 10.1007/10_2012_153

Published Online: 23 August 2012

The Implications of Stem Cell Applications for Diseases of the Respiratory System

Mei Ling Lim, Philipp Jungebluth and Paolo Macchiarini

Abstract Stem cells possess the unique properties of unlimited self-renewal capability and a broad differentiation spectrum to produce multiple different cell types. This provides many platforms to explore novel multidisciplinary approaches to create and/or restore functional three-dimensional tissues or organs for the treatment of a range of diseases. In this chapter, in the context of respiratory diseases, we review the unique properties of stem cells, and how they have been studied for their therapeutic potential in cell therapy and tissue engineering. In addition, we give a brief overview of the current clinical studies on the use of stem cells for both acute and chronic respiratory diseases.

Keywords Stem cells · Induced pluripotent stem cells · Embryonic stem cells · Adult progenitor cells · Celltherapy · Respiratory diseases

Abbreviations

hESCs	Human embryonic stem cells
ASCs	Adult stem cells
MSCs	Mesenchymal stem cells
iPSCs	Induced pluripotent stem cells
BMNCs	Bone marrow mononuclear cells
EPCs	Epithelial progenitor cells
COPD	Chronic obstructive pulmonary disease
PH	Pulmonary hypertension
IPAH	Idiopathic pulmonary arterial hypertension
RILI	Radiation-induced lung injury

M. L. Lim · P. Jungebluth · P. Macchiarini (✉)
Advanced Center for Translational Regenerative Medicine (ACTREM),
European Airway Institute, Division of Ear, Nose and Throat (CLINTEC),
Karolinska Institutet, Alfred Nobels Allé 8 Huddinge,
141 86 Stockholm, Sweden
e-mail: paolo.macchiarini@ki.se

Contents

1 Introduction

Respiratory diseases are a major cause of mortality and morbidity. It has been estimated that more than 1 billion people suffer from chronic respiratory diseases worldwide, with approximately 4 million deaths occurring every year. It is expected that by 2030, chronic respiratory disease will become the third leading cause of death in the world [1].

There is a need for novel therapeutic strategies for the treatment of chronic and acute lung diseases such as lung cancer, pulmonary hypertension, chronic obstructive pulmonary disease (COPD), and asthma because their current treatment options are limited and are mainly aimed at controlling or delaying the progression of the disease. The only plausible option with curative intent would be organ transplantation. However, organ transplantation has its limitations: shortage of donor organs, as well as the life-long need for immunosuppressant therapy after transplantation.

This chapter reviews the promising field of regenerative medicine for respiratory diseases. There are two different ways that regenerative medicine can be applied: (1) it can be a multidisciplinary approach to create and/or restore functional three-dimensional tissues or organs that utilizes a complex combination of stem cells (SCs), scaffolds, and signaling molecules (tissue engineering), or (2) direct application of SCs to the site of injury (cell therapy). In recent years, novel approaches to the treatment of previously incurable diseases by repairing, replacing, or regenerating damaged tissues or organs using SCs have been extensively studied. Hence, the focus of stem cell research is to elucidate their potential for specific diseases and explore methods of controlling differentiation into specific progenitor cells such as endothelial progenitor cells, neural crest cells, or differentiated cell types including type II pneumocytes, cardiomyocytes, and

Table 1 Comparative analysis of different types of stem cells

	Cell sources		
	ESCs	iPSCs	Progenitor cells
Cell characteristics			
Cell of origin	Embryonic	Adult	Adult
Potency	Pluripotent	Pluripotent	Multipotent
Differentiation potential	Unlimited	Unlimited	Limited
Self-renewal	Unlimited	Unlimited	Limited
Karyotype	Stable	Stable	Stable
Homogeneity	Low	Low	High
Immunogenicity	High	Undefined	Low
Limitations			
Technical difficulty	High	Low	Low
Risk of teratoma formation	High	High	Low
Risk of infectious disease	No	Yes	No
Ethical issues	High	Low	Low
Logistical issues	High	Low	Low
Therapeutic benefits			
Cell therapy	Allogenic	Autologous	Autologous
Gene modification	No	Yes	No
Immunomodulation	Undefined	Undefined	Yes

neuronal cells. Stem cells can be applied by direct administration, namely cell therapy or by combining SCs with different components via tissue engineering. If successful, stem cell applications can potentially provide a successful alternative therapeutic option for acute and chronic respiratory diseases.

2 Cell Therapy

Cell therapy (using local or systemic administration) is a well-recognized treatment modality, for example, hematopoietic stem cell transplants for leukemia and epithelial stem cell-based treatments for burns and corneal disorders. The advantages are:

1. Cells can be isolated, expanded, frozen (banked), and retrieved when needed. This enables cells to be harvested before degeneration occurs.
2. Cells can be quality controlled by screening for the presence of pathogens. This ensures safe application for future use.
3. Cells are receptive to gene modification by DNA recombinant technology, for example, gene therapy. This provides the added benefit of genetic manipulation of cells when required.
4. Site-specific cell transplantation can also increase the efficacy and safety of therapy by localized delivery of therapeutic substances at the target site. This would reduce toxicity as a lower dose of cells can be administered directly and be contained at the site of a lesion. In addition, implanted cells can release

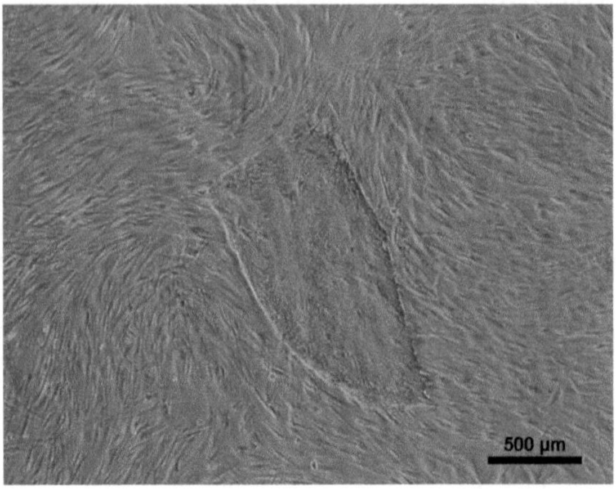

Fig. 1 Human embryonic stem cells (HESCs) on gamma-irradiated human fibroblasts

therapeutic substances at an adaptive dynamic rate determined by the cellular feedback mechanism, which prevents inappropriate dosages.

Nevertheless, cell therapy has its challenges. The ability to achieve cell homogeneity, ensure cell survival and engraftment efficiency, and maintain low immunogenicity during local or systemic administration of cells varies with the type of SCs used. Hence, it is important to discuss the benefits and clinical risks of utilizing different sources of SCs, that is, embryonic stem cells (ESCs), adult stem cells (ASCs), progenitor cells, or genetically modified cells, that is, induced pluripotent stem cells (iPSCs; Table 1).

2.1 Human Embryonic Stem Cells

Human embryonic stem cells (hESCs) can be derived from the inner cell mass of possibly discarded, day 5 healthy, nontransferred embryos (Fig. 1). hESCs have the advantage of being the most therapeutically versatile source of cells. They can proliferate indefinitely to generate daughter cells of identical characteristics without senescence (self-renewing) and are able to maintain their ability to differentiate into almost all tissue-specific cell lineages upon receiving appropriate signals (pluripotency) [2].

Recent experimental studies have examined the reparative and engraftment ability of differentiated cells, that is, type II pneumocytes from hESCs in mouse models. Mouse lung injury models using either bleomycin [3] or silica [4] have demonstrated a reduced inflammation and improved lung function following hESC treatment. The successful engraftment of the differentiated cells from hESC at the site of injury suggests that cell therapy can be useful for the treatment of fibrotic lung diseases. Toya and colleagues (2011) have also shown that mesoderm-induced cell aggregates (embryoid bodies) significantly reduced lung inflammation and edema in a mouse sepsis-induced lung injury model [5].

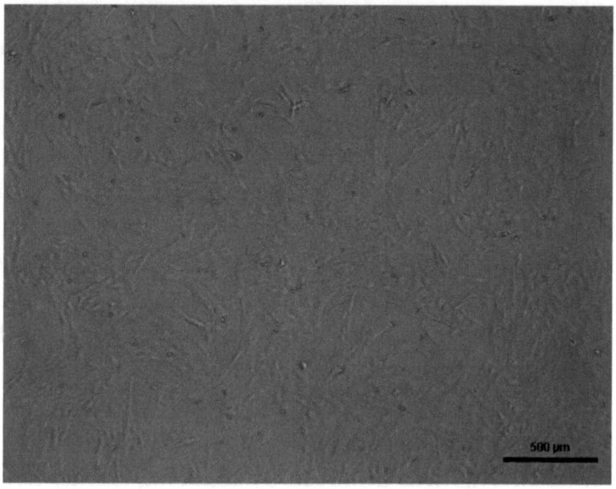

Fig. 2 Mesenchymal stromal cells

Despite the numerous studies that have examined the hESC pluripotency, there are inherent challenges with the use of hESCs: (1) hESCs are derived from an allogenic source and the inability to remove the immunogenic barriers suggests that patients would require long-term immunosuppression; (2) the clinical use of hESCs may be limited by the inability to control differentiation to derive a homogeneous differentiated population of cells; (3) the residing SCs may also pose a risk of teratoma formation in patients; and lastly (4) the use of hESCs in medical research is riddled with ethical and logistical issues because oocytes are needed for ESC isolation. To circumvent these challenges, alternative sources of SCs are required for therapeutic use.

2.2 Adult (Progenitor) Stem Cells

Adult organs set aside reservoirs of stem cells for replenishing cells that are lost in either tissue injury or homeostasis [6]. These stem cells are known as ASCs. ASCs are considered multipotent as they can produce a whole spectrum of cell types within a single cell lineage. They can either be found residing in organs with high cell turnover, such as skin and intestinal tract where the cells' lifespans are measured in days or weeks [7], or in stem cell "factory" organs, that is, adipose tissue and the bone marrow.

In the body, the bone marrow has the largest reservoir of SCs with two distinct residing populations of SCs: hematopoietic and nonhematopoietic. These cells can also be isolated from peripheral and umbilical cord blood. Hematopoietic stem cells (HSCs) can form all blood cells in the body [8] whereas nonhematopoietic stem cells, now known as mesenchymal stromal cells (MSCs) can form cell types mainly associated with skeletal tissue, that is, bone, cartilage, and fat [9].

In recent years, there has been a growing interest in understanding MSCs in terms of their immunomodulatory capability and regenerative potential. MSCs show many advantages. They are readily available, proliferative, display multi-lineage potential [10], and are more immune-privileged (Fig. 2). MSCs can release several growth factors and anti-inflammatory cytokines, which regulate endothelial and epithelial permeability and reduce the severity of inflammation.

Numerous experimental studies have explored the effects of MSC therapy in the context of acute lung injury and chronic lung disorders, that is, asthma and COPD. Pati and colleagues (2011) showed in a rat hemorrhagic shock-induced acute lung injury model that human bone marrow-derived MSCs suppress lung edema and inflammatory cells [11]. Sun and colleagues (2011) demonstrated that the intravenous administration of autologous adipose-derived MSCs could attenuate the inflammatory response and oxidative stress in an acute rat ischemia–reperfusion lung injury model [12]. In experiments using asthma as a model for chronic lung disease, it was shown that the decline in lung function was inhibited by rat bone marrow MSCs [13, 14]. This was achieved by reducing airway hyperactivity, inflammation, and remodeling. These murine experimental studies clearly highlight the immunomodulatory role of MSCs. Ingenito and colleagues (2011) on the other hand have demonstrated the regenerative properties of MSCs in an experimental emphysema sheep model [10]. They showed that transplantation of autologous lung-derived MSCs attached to scaffolds induced the regeneration of functional lung tissue in emphysematous regions of the lungs. From these studies, there is compelling evidence that MSCs have beneficial effects on lung development, repair, and remodeling.

Although the use of human MSCs has been successfully used in several cases, there are still hurdles that scientists and clinicians must overcome before incorporating MSC transplantation into routine clinical practice. The invasive procedures on the patient, the low MSC levels present in the marrow (approximately 1 in 100,000 to 500,000 cells), diminished expansion and differentiation ability, the resistance to trypsinization during passaging in vitro, morphological changes in culture, and the requirement of serum-containing media are some of the problems faced with the clinical use of MSCs [8, 15].

2.3 Induced Pluripotent Stem Cells

One potential source of alternative cell types for transplantation is SCs derived from cell reprogramming. Cell reprogramming is described as resetting the developmental clock [16]. It is a process that reverts the genetic status in a somatic cell nucleus to a state of developmental pluripotency to produce an autologous multipotent population of cells [17]. This would potentially be the best solution for cell therapy.

In 2006, Takahashi and Yamanaka developed a novel strategy to derive pluripotent cells from somatic cells called direct reprogramming [18]. iPSCs were successfully derived from both mouse and human somatic cells by ectopic retroviral or lentiviral expression [19] with four transcription factors. The overall estimated efficiency of establishing iPSCs from somatic cells was reported to be less than 0.1 % [18, 20]. However, iPSCs exhibited the essential characteristics of

ESCs: normal karyotype, ESC-like morphology, express cell surface markers and genes that characterize ESC, teratoma formation that showed contribution to all three germ layers (i.e., endoderm, mesoderm, ectoderm), and the contribution to viable chimeras. Numerous studies have suggested that iPSC are almost ESC-like and have a significantly lower level of immunological and ethical concern as compared to ESCs. There are still many ongoing refinements for the method of generating iPSCs. Although Takahashi and Yamanaka (2006) have demonstrated that iPSCs generation is simple, the use of the retroviral reprogramming system may not be safe because the somatic cells could be at risk of permanent genetic alteration and the retroviral vectors could become reactivated. There are other approaches that were explored and the aim of the different methodologies allows one to understand the homing mechanism: this includes cell survival, proliferation, differentiation, and reprogramming.

It is clear from the current literature that the different methodologies of generating iPSCs have demonstrated a successful upregulation of the expression of specific pluripotency genes without transferring potentially harmful genes into the somatic genome. However, despite all the similarities reported between iPSCs and ESCs, microarray gene expression data have been reported to show differences between the two stem cell groups. When these data were reanalyzed in seven different laboratories, it was revealed that nearly one-third of the genes with lab-specific expression signatures were differently expressed between ESCs and iPSCs. This suggests that the ESCs and iPSCs gene expressions differ and the in vitro microenvironment may partially contribute to these differences [21]. Another common problem with the different reprogramming strategies is the extremely low reprogramming efficiencies, ranging from less than 0.0001 % to 0.001 % [22, 23]. The low frequency of reprogramming may be attributed to one or more of these possibilities: (1) the heterogeneous fibroblast population could prevent a subpopulation of cells from reprogramming (2) a small minority population of primitive multipotent cells rather than fully differentiated cells may actually be the source of reprogrammed cells, and (3) the retro- or lentivirus integration used to deliver the reprogramming factors may have modified a small fraction of cells [24].

It is certainly very attractive to know that iPSCs can offer a promising platform for generating patient-specific SCs of any lineage without the need for embryonic materials. Studies performed on endotoxin-induced acute lung injury (ALI) in rodents have also shown that the intravenous delivery of iPSCs can provide a beneficial effect to attenuate the severity of ALI and improve the physiological impairment, which is partly attributed to NF-kB and neutrophil accumulation [25, 26].

3 Tissue Engineering

The goals of tissue engineering are to repair, regenerate, and replace diseased or dysfunctional tissue to restore organ function. Tissue engineering can be conducted either in vivo or ex vivo. In vivo tissue engineering involves the body's own regenerative capability to generate cells on an appropriate biomaterial. The ex vivo technique involves culturing cells on a scaffold and reimplanting it into the host [27].

In scaffold-based tissue engineering, there are a number of important factors to consider: route of delivery, pre-conditioning using different growth factors or cytokines, immunological function, and accessibility and availability of cell sources. Cells may be allogeneic (same species, different individual) or autologous (same individual) [28]. Although autologous cells are preferred because an immunologic response is not evoked, there might be problems with achieving an adequate cell yield for expansion and transplantation, especially in patients with end-stage organ disease. Furthermore, some primary autologous human cells cannot be expanded from particular organs (i.e., brain, pancreas, and lung).

4 Clinical Applications of Cell Therapy

Animal models have been used to examine its engraftment during transplantation and airway reconstitution in animal models with experimentally induced tracheal and lung defects. Current trends in tracheal and lung transplantation include the use of autologous cells, development of bioactive cell-free scaffolds that are capable of supporting activation, and differentiation of host SCs at the site of injury.

4.1 Chronic Obstructive Pulmonary Disease

Chronic obstructive pulmonary disease (COPD) is a preventable and treatable disease that is characterized by a progressive airflow limitation that is not fully reversible. Cigarette smoking is one of the most common risk factors in COPD although inhalational exposures to certain occupational dusts and chemicals have also been implicated. A genetic risk factor for COPD is alpha1-antitrypsin deficiency.

Small airway disease and lung parenchymal destruction are two components that contribute to airflow limitation, and these are caused by an abnormally amplified inflammatory response of the lungs to noxious particles and gases. An increased level of pro-inflammatory mediators and oxidative stress in the respiratory tract characterizes this inflammatory response. This results in airflow limitation and air trapping (small airway disease). It also impairs gaseous exchange, causes mucus hypersecretion, and parenchymal destruction [26, 29–31].

Several animal models have demonstrated that MSCs have its "protective" mechanism towards an inflammatory response [31–33]. To evaluate the immunomodulatory role of MSCs in humans, a clinical trial was carried out on allogenic-derived bone marrow MSCs transfusion to patients with moderate to severe COPD (NCT00683722). The study showed that at 6 months the intravenous infusion of MSCs significantly reduced the inflammation as measured by C-reactive protein (CRP). No adverse events or increased incidence of infections were seen. This reinforces the immunomodulatory capability of MSCs. However, after 6 months of treatment with MSC, pulmonary function tests were not significantly improved over placebo control, although it did reveal positive trends in functional endpoints such as the six-minute walk test, especially in patients with less severe COPD. Nevertheless, a longer follow-up is necessary to evaluate the MSC efficacy on improving the pulmonary function and quality of life.

Another study with a longer follow-up period of 12 months showed that patients with stage-4 COPD, who underwent stem cell infusion using bone marrow mononuclear cells (BMNCs), reported significant improvement in quality of life. Their clinical condition also stabilized, which can be inferred that the natural progression of the disease was altered. More important, no adverse events were reported (NCT01110252). Nevertheless this study had a small sample size of four patients, but has been expanded and to date is still ongoing. Studies with a larger sample size and longer follow-up period would be helpful to assess the efficacy of stem cell therapy in COPD patients. To date, there have been no other clinical trials that have utilized stem cell therapy in COPD patients.

4.2 Pulmonary Hypertension

Pulmonary hypertension (PH) is defined as elevated pulmonary arterial pressure (mean pulmonary artery pressure > 25 mmHg at rest) and right ventricular failure. Based on the different mechanisms of PH, the World Health Organization (WHO) has classified it into five groups; pulmonary arterial hypertension, pulmonary hypertension owing to left heart disease, pulmonary hypertension owing to lung diseases or hypoxemia, chronic thromboembolic pulmonary hypertension, and pulmonary hypertension with unclear multifactorial mechanisms [34, 35].

The pathogenesis of PH was studied largely in the context of idiopathic pulmonary arterial hypertension (IPAH), in which vascular endothelial dysfunction appears to be a key element. This dysfunction manifests as a proliferative vasculopathy, characterized by deregulated cell proliferation leading to intimal hyperplasia and smooth muscle hypertrophy, fibrosis, vasoconstriction, and thrombosis. This form of remodeling is a “disordered angiogenesis” process and it increases pulmonary vascular resistance, causing pulmonary hypertension, and with increasing severity leads to right ventricular failure. Hence, IPAH is a slowly progressive disease with poor prognosis, which ultimately leads to death.

Based on the concept of vascular endothelial dysfunction, it has been postulated that endothelial progenitor cells (EPCs) could support angiogenesis via a paracrine mechanism as it was shown to secrete crucial pro-angiogenic factors such as VEGF-A, CXCL12, and insulinlike growth factor-1. Hence, in supporting angiogenesis, EPC promotes repair and reduces the pathological morphology seen in PH. This hypothesis was supported by a number of murine studies [34]. In addition, because adenoviral overexpression of endothelial nitric oxide synthase (eNOS) in the lung is known to reduce PH, Kanki-Horimoto et al. (2006) showed in a rat study that implanting MSCs that overexpressed eNOS could reduce the effects on PH-related right ventricular impairment and increase survival time [36]. Jungebluth et al. (2011) have also demonstrated the restoration of lung function at a proteomic level in a PH rat model when allogenic MSCs were administered [37].

Based on encouraging pre-clinical data, two clinical trials conducted in Zhejiang University, Hangzhou, China (NCT00257413, NCT00641836) investigated the feasibility, safety, and clinical outcome of intravenous infusion of autologous EPCs in patients with IPAH. This was compared with conventional therapy. At 12 weeks of follow-up, the cell infusion group reported a significant improvement in the six-minute walk distance compared with the conventional therapy group. There was also significant improvement in mean pulmonary artery pressure, pulmonary vascular resistance, and cardiac output. No severe adverse events with cell infusion were reported. In lieu of this, a 12-week clinical trial of autologous EPCs transplantation in an open-label pilot study was conducted, involving 13 pediatric patients with IPAH [38]. The pilot study showed that EPCs intravenous infusion was associated with significant improvements in the six-minute walk distance, New York Heart Association (NYHA) functional class, and pulmonary hemodynamics. No adverse events with cell infusion were reported. The results from the clinical trials suggest EPC infusion has its potential benefits in both adult and pediatric age groups.

In 2005, autologous progenitor cell-based gene therapy of heNOS was intravenously infused in patients with severe IPAH that was refractory to conventional treatment (NCT00469027). This was a landmark trial involving the use of EPCs combined with a therapeutic gene therapy (heNOS) to treat IPAH. To date, the six patients show significant reduction in total pulmonary vascular resistance, and there is no safety concerns reported thus far. Therefore, stem cell applications for IPAH show promising results and may offer an alternative therapeutic solution.

4.3 Radiation-Induced Lung Injury

Radiation-induced lung injury comprises radiation pneumonitis and fibrosis, in which pneumonitis tends to present in the subacute stage whereas fibrosis tends to present late. It is largely observed in patients who have undergone chest wall irradiation for the treatment of lung, breast, and hematological malignancies.

The pathogenesis of radiation-induced lung injury is a combination of radiation-induced cytotoxicity and inflammatory responses. Radiation causes DNA damage resulting in cellular death. It also induces cellular apoptosis. Moreover, radiation is known to upregulate a milieu of inflammatory cytokines (e.g., TGF-beta, TNF-a, IL-1a, IL-6, PDGF, bFGF) [39]. Currently, there are no standard guidelines in the treatment of radiation-induced lung injury. However, the general consensus is the use of glucocorticoids or other immunosuppressants such as

Table 2 Current clinical trials of stem cell therapy in respiratory diseases

Diseases	NCT	Stem cell type	Origin	Mode of administration	Findings
COPD	00683722	BM-MSC	Allogenic	Intravenous	No significant improvement in pulmonary function but possible improvement in functional end points (e.g., 6-minute walk test
	01110252	BM-MNCs	Autologous	Intravenous	Significant improvement in quality of life
IPAH	00257413 00641836	EPCs	Autologous	Intravenous	Significant improvement in mean pulmonary artery pressure, pulmonary vascular resistance, cardiac output, and 6-minute walk distance
	00469027	EPCs + gene therapy (heNOS)	Autologous	Intravenous	Significant improvement in pulmonary vascular resistance
Radiation-induced lung injury	Nil	MNCs	Autologous	Intravenous	No progression of lung injury at 1 year post-treatment

azathioprine and cyclosporine, which appear to benefit in the pneumonitis stage but not at the fibrosis stage. Experimental agents including pentoxifylline and inhibitors of collagen synthesis (e.g., colchicine, penicillamine) were suggested for use in the fibrosis stage, but to date there is still no concrete evidence to support this. Therapeutic effects from stem cell therapy were explored in a rat model and in a pilot clinical trial [40]. It was reported that a single transplantation of autologous MSCs was associated with a decrease in mortality rate in mice that underwent radiation-induced lung injury. The pilot clinical trial also demonstrated that when standard pharmacotherapy was combined with intravenous autologous MSCs administration in 11 patients with radiation-induced lung injury, there was no progression of the lung injury at 1 year post-treatment. This suggests that stem cell therapy can play a role in the treatment of radiation-induced lung injury. However, the results of this clinical trial could be confounded by the use of pharmacotherapy. In addition, prospective placebo controlled trials would be more appropriate to assess the therapeutic effect of SCs accurately.

Table 3 Current clinical trials of tissue engineering in respiratory diseases

	Tissue engineered	Tissue scaffold	Findings
Omori et al.	Trachea	Marlex mesh tube	Good epithelialization to cover the implant
	Larynx	Marlex mesh tube	Good epithelialization and no airway obstruction
Zhang et al.	Pharynx and larynx	Alloderm, ADM	Satisfactory wound healing but some degree of stenosis reported
Macchiarini et al.	Trachea	Decellularized donor trachea	Airway remained vascularized and quality of life improved. At long-term, limited collapse occurred in 30 % of the patients
Jungebluth et al.	Trachea and bronchi	Synthetic micro- and nanofibers	In progress

5 Clinical Applications of SCs in Tissue Engineering

With the ongoing search for SCs that can be applied to human treatment, precise delivery and homing to the disease site must be ensured for successful therapy. SCs were shown in clinical studies to be safely inoculated into the trachea (Table 2). Most tracheal tissue-engineering approaches use biodegradable three-dimensional scaffolds, which are important for neotracheal formation by promoting cell attachment, cell redifferentiation, and production of the extracellular matrix (Table 3). An important milestone in applying regenerative medicine techniques to the respiratory system is using artificial tissue scaffolds to reconstruct organs that could later be used for implantation.

Omori and colleagues (2005) applied this concept by using a tissue scaffold (Marlex mesh tube) and covering it with collagen sponge to form part of a trachea [41]. A two-year follow-up reported that the reconstructed trachea showed good epithelialization to cover the implant with no complications. In view of this success, Omori and colleagues (2008) used the same technique to repair the larynx and trachea in another four patients (one with subglottic stenosis, three with thyroid cancer) [42]. Within 8–34 months follow-up, post-operative endoscopy showed that the lumen of the implants was well epithelialized and no obstruction was observed.

Recently, significant advances were made in the field of tissue bioengineering. Zhang and colleagues (2010) performed a study on patients with hypopharyngeal carcinoma [43]. They used tissue patches made from artificial biological material (i.e., acellular dermal matrix; Alloderm, ADM), combined with pectoralis major myocutaneous flaps (PMMFs), to reconstruct the surgical defect created from total laryngectomy and total hypopharyngectomy. ADM tissue patches were also used to reconstruct the defect in the posterior pharyngeal wall from patients who only underwent tumor resection. This study reported satisfactory wound healing with good coverage of the defect by the growing epithelium 18–37 days post surgery. There was also no pharyngeal fistula. However, they reported some degree of stenosis in the pharyngeal cavity occurred, but following dilatation of the stenosis, patients could have a regular diet. This study demonstrated that stem cell technology could be incorporated into tissue engineering, to create "biological plasters" that could be used to mend surgical defects.

A major milestone was achieved with the transplantation of the first bioengineered trachea: a combination of SCs and tissue-engineering techniques. In 2008, a bioengineered human trachea was constructed from a human donor trachea and was transplanted to a patient [44]. Its MHC antigens and cells were removed, and the trachea was recolonized with the recipient's epithelial cells and MSC-derived chondrocytes that had been cultured ex vivo. This was surgically transplanted to replace a stenosed left main bronchus in a patient with end-stage bronchomalacia. Post-surgery, the patient did not require immunosuppressants and the blood had no traces of anti-donor antibodies. At 4 months post-surgery, the patient's airway remained functional and quality of life was improved dramatically. This landmark case highlights the tremendous potential of regenerative medicine that combines tissue regeneration techniques using SCs to create functional organs that can be transplanted in humans. It also offers a paradigm shift with the concept that human organ transplantation does not require long-term immunosuppressant therapy to ensure organ viability.

6 Conclusions

In the foreseeable future, stem cell and tissue-engineering technology can offer a therapeutic solution in the treatment of previously incurable and possibly certain genetic, diseases. It has been shown in many animal studies and a few clinical trials to have an immense potential in treating diseases by repairing, replacing, or regenerating tissues and restoring the function of an organ. Nonetheless, the field of regenerative medicine is still in its infancy and continuing research into genomics and bioinformatics technologies will continue to offer new insights into the understanding of SCs growth, differentiation, and their application to engineering tissues in the future.

References

1. World Health Organization (2011) Living with chronic lung diseases. http://www.whoint/features/2007/copd/en/. Accessed 13 Dec 2011
2. Doss MX, Koehler CI, Gissel C, Hescheler J, Sachinidis A (2004) Embryonic stem cells: a promising tool for cell replacement therapy. J Cell Mol Med 8(4):465–473. doi:008.004.06 [pii]
3. Wang D, Morales JE, Calame DG, Alcorn JL, Wetsel RA (2012) Transplantation of human embryonic stem cell-derived alveolar epithelial type II cells abrogates acute lung injury in mice. Mol Ther 18(3):625–634. doi:mt2009317 [pii]. 10.1038/mt.2009.317

4. Spitalieri P, Quitadamo MC, Orlandi A, Guerra L, Giardina E, Casavola V, Novelli G, Saltini C, Sangiuolo F (2012) Rescue of murine silica-induced lung injury and fibrosis by human embryonic stem cells. Eur Respir J doi:09031936.00005511 [pii]. 10.1183/09031936.00005511
5. Toya SP, Li F, Bonini MG, Gomez I, Mao M, Bachmaier KW, Malik AB (2011) Interaction of a specific population of human embryonic stem cell-derived progenitor cells with CD11b+ cells ameliorates sepsis-induced lung inflammatory injury. Am J Pathol 178(1): 313–324. doi:S0002-9440(10)00046-5 [pii]. 10.1016/j.ajpath.2010.09.041
6. Guasch G, Fuchs E (2005) Mice in the world of stem cell biology. Nat Genet 37(11): 1201–1206. doi:ng1667 [pii]. 10.1038/ng1667
7. Gudjonsson T, Magnusson MK (2005) Stem cell biology and the cellular pathways of carcinogenesis. APMIS 113(11–12):922–929. doi:APMapm_371 [pii]. 10.1111/j.1600-0463. 2005.apm_371.x
8. Ulloa-Montoya F, Verfaillie CM, Hu WS (2005) Culture systems for pluripotent stem cells. J Biosci Bioeng 100(1):12–27. doi:S1389-1723(05)70423-0 [pii]. 10.1263/jbb.100.12
9. Caplan AI (2005) Review: mesenchymal stem cells: cell-based reconstructive therapy in orthopedics. Tissue Eng 11(7–8):1198–1211. doi:10.1089/ten.2005.11.1198
10. Ingenito EP, Tsai L, Murthy S, Tyagi S, Mazan M, Hoffman A (2011) Autologous lung-derived mesenchymal stem cell transplantation in experimental emphysema. Cell Transplant. doi:ct0078ingenito [pii]. 10.3727/096368910X550233
11. Pati S, Gerber MH, Menge TD, Wataha KA, Zhao Y, Baumgartner JA, Zhao J, Letourneau PA, Huby MP, Baer LA, Salsbury JR, Kozar RA, Wade CE, Walker PA, Dash PK, Cox CS, Jr., Doursout MF, Holcomb JB (2011) Bone marrow derived mesenchymal stem cells inhibit inflammation and preserve vascular endothelial integrity in the lungs after hemorrhagic shock. PLoS One 6(9):e25171. doi:10.1371/journal.pone.0025171. PONE-D-11-11403 [pii]
12. Sun CK, Yen CH, Lin YC, Tsai TH, Chang LT, Kao YH, Chua S, Fu M, Ko SF, Leu S, Yip HK (2011) Autologous transplantation of adipose-derived mesenchymal stem cells markedly reduced acute ischemia-reperfusion lung injury in a rodent model. J Transl Med 9:118. doi:1479-5876-9-118 [pii]. 10.1186/1479-5876-9-118
13. Firinci F, Karaman M, Baran Y, Bagriyanik A, Ayyildiz ZA, Kiray M, Kozanoglu I, Yilmaz O, Uzuner N, Karaman O (2011) Mesenchymal stem cells ameliorate the histopathological changes in a murine model of chronic asthma. Int Immunopharmacol 11(8):1120–1126. doi:S1567-5769(11)00148-2 [pii]. 10.1016/j.intimp.2011.03.009
14. Lee SH, Jang AS, Kwon JH, Park SK, Won JH, Park CS (2011) Mesenchymal stem cell transfer suppresses airway remodeling in a toluene diisocyanate-induced murine asthma model. Allergy Asthma Immunol Res 3(3):205–211. doi:10.4168/aair.2011.3.3.205
15. Brown SE, Tong W, Krebsbach PH (2009) The derivation of mesenchymal stem cells from human embryonic stem cells. Cells Tissues Organs 189(1–4):256–260. doi:000151746 [pii]. 10.1159/000151746
16. Collas P, Hakelien AM (2003) Teaching cells new tricks. Trends Biotechnol 21 (8):354–361. doi:S0167779903001471 [pii]
17. Ambrosi DJ, Rasmussen TP (2005) Reprogramming mediated by stem cell fusion. J Cell Mol Med 9(2):320–330. doi:009.002.08 [pii]
18. Takahashi K, Yamanaka S (2006) Induction of pluripotent stem cells from mouse embryonic and adult fibroblast cultures by defined factors. Cell 126(4):663–676. doi:S0092-8674(06)00976-7 [pii]. 10.1016/j.cell.2006.07.024
19. Yu J, Vodyanik MA, Smuga-Otto K, Antosiewicz-Bourget J, Frane JL, Tian S, Nie J, Jonsdottir GA, Ruotti V, Stewart R, Slukvin, II, Thomson JA (2007) Induced pluripotent stem cell lines derived from human somatic cells. Science 318(5858):1917–1920. doi:1151526 [pii]. 10.1126/science.1151526
20. Wernig M, Meissner A, Foreman R, Brambrink T, Ku M, Hochedlinger K, Bernstein BE, Jaenisch R (2007) In vitro reprogramming of fibroblasts into a pluripotent ES-cell-like state. Nature 448(7151):318–324. doi:nature05944 [pii]. 10.1038/nature05944
21. Newman AM, Cooper JB Lab-specific gene expression signatures in pluripotent stem cells. Cell Stem Cell 7(2):258–262. doi:S1934-5909(10)00331-0 [pii]. 10.1016/j.stem.2010.06.016

22. Okita K, Nakagawa M, Hyenjong H, Ichisaka T, Yamanaka S (2008) Generation of mouse induced pluripotent stem cells without viral vectors. Science 322(5903):949–953. doi:1164270 [pii]. 10.1126/science.1164270
23. Stadtfeld M, Nagaya M, Utikal J, Weir G, Hochedlinger K (2008) Induced pluripotent stem cells generated without viral integration. Science 322(5903):945–949. doi:1162494 [pii]. 10.1126/science.1162494
24. Sridharan R, Plath K (2008) Illuminating the black box of reprogramming. Cell Stem Cell 2(4):295–297. doi:S1934-5909(08)00129-X [pii]. 10.1016/j.stem.2008.03.015
25. Yang KY, Shih HC, How CK, Chen CY, Hsu HS, Yang CW, Lee YC, Perng RP, Peng CH, Li HY, Chang CM, Mou CY, Chiou SH (2011) IV delivery of induced pluripotent stem cells attenuates endotoxin-induced acute lung injury in mice. Chest 140(5):1243–1253. doi:chest.11-0539 [pii]. 10.1378/chest.11-0539
26. Gupta N, Su X, Popov B, Lee JW, Serikov V, Matthay MA (2007) Intrapulmonary delivery of bone marrow-derived mesenchymal stem cells improves survival and attenuates endotoxin-induced acute lung injury in mice. J Immunol 179(3):1855-1863. doi:179/3/1855 [pii]
27. Jungebluth P, Alici E, Baiguera S, Blanc KL, Blomberg P, Bozoky B, Crowley C, Einarsson O, Grinnemo KH, Gudbjartsson T, Guyader SL, Henriksson G, Hermanson O, Juto JE, Leidner B, Lilja T, Liska J, Luedde T, Lundin V, Moll G, Nilsson B, Roderburg C, Stromblad S, Sutlu T, Teixeira AI, Watz E, Seifalian A, Macchiarini P (2011) Tracheobronchial transplantation with a stem-cell-seeded bioartificial nanocomposite: a proof-of-concept study. Lancet. doi:S0140-6736(11)61715-7 [pii]. 10.1016/S0140-6736(11)61715-7
28. Hipp J, Atala A (2004) Tissue engineering, stem cells, cloning, and parthenogenesis: new paradigms for therapy. J Exp Clin Assist Reprod 1(1):3. doi:1743 1050 1-3 [pii]. 10.1186/1743-1050-1-3
29. Adcock IM, Caramori G, Barnes PJ Chronic obstructive pulmonary disease and lung cancer: new molecular insights. Respiration 81(4):265–284. doi:000324601 [pii]. 10.1159/000324601
30. D'Agostino B, Sullo N, Siniscalco D, De Angelis A, Rossi F Mesenchymal stem cell therapy for the treatment of chronic obstructive pulmonary disease. Expert Opin Biol Ther 10(5):681–687. doi:10.1517/14712591003610614
31. Bonfield TL, Caplan AI Adult mesenchymal stem cells: an innovative therapeutic for lung diseases. Discov Med 9(47):337–345
32. Liang ZX, Sun JP, Wang P, Tian Q, Yang Z, Chen LA Bone marrow-derived mesenchymal stem cells protect rats from endotoxin-induced acute lung injury. Chin Med J (Engl) 124(17):2715–2722
33. Zhao F, Zhang YF, Liu YG, Zhou JJ, Li ZK, Wu CG, Qi HW (2008) Therapeutic effects of bone marrow-derived mesenchymal stem cells engraftment on bleomycin-induced lung injury in rats. Transplant Proc 40(5):1700–1705. doi:S0041-1345(08)00471-5 [pii]. 10.1016/j.transproceed.2008.01.080
34. Majka S, Burnham E, Stenmark KR (2011) Cell-based therapies in pulmonary hypertension: who, what, and when? Am J Physiol Lung Cell Mol Physiol 301(1):L9–L11. doi:ajplung.00118.2011 [pii]. 10.1152/ajplung.00118.2011
35. Simonneau G, Robbins IM, Beghetti M, Channick RN, Delcroix M, Denton CP, Elliott CG, Gaine SP, Gladwin MT, Jing ZC, Krowka MJ, Langleben D, Nakanishi N, Souza R (2009) Updated clinical classification of pulmonary hypertension. J Am Coll Cardiol 54(1 Suppl): S43–54. doi:S0735-1097(09)01216-9 [pii]. 10.1016/j.jacc.2009.04.012
36. Kanki-Horimoto S, Horimoto H, Mieno S, Kishida K, Watanabe F, Furuya E, Katsumata T (2006) Implantation of mesenchymal stem cells overexpressing endothelial nitric oxide synthase improves right ventricular impairments caused by pulmonary hypertension. Circulation 114(1 Suppl): I181–185. doi:114/1_suppl/I-181 [pii]. 10.1161/CIRCULATIONAHA.105.001487
37. Jungebluth P, Luedde M, Ferrer E, Luedde T, Vucur M, Peinado VI, Go T, Schreiber C, Richthofen MV, Bader A, Haag J, Darsow KH, Bartel SJ, Lange HA, Furlani D, Steinhoff G, Macchiarini P (2011) Mesenchymal stem cells restore lung function by recruiting resident and non-resident proteins. Cell Transplant. doi:ct0201jungebluthetalpdf [pii]. 10.3727/096368910X557254

38. Zhu JH, Wang XX, Zhang FR, Shang YP, Tao QM, Chen JZ (2008) Safety and efficacy of autologous endothelial progenitor cells transplantation in children with idiopathic pulmonary arterial hypertension: open-label pilot study. Pediatr Transplant 12(6):650–655. doi:PTR863 [pii]. 10.1111/j.1399-3046.2007.00863.x
39. Graves PR, Siddiqui F, Anscher MS, Movsas B (2010) Radiation pulmonary toxicity: from mechanisms to management. Semin Radiat Oncol 20(3):201–207. doi:S1053-4296(10)00018-4 [pii]. 10.1016/j.semradonc.2010.01.010
40. Kursova LV, Konoplyannikov AG, Pasov VV, Ivanova IN, Poluektova MV, Konoplyannikova OA (2009) Possibilities for the use of autologous mesenchymal stem cells in the therapy of radiation-induced lung injuries. Bull Exp Biol Med 147(4):542–546
41. Omori K, Nakamura T, Kanemaru S, Asato R, Yamashita M, Tanaka S, Magrufov A, Ito J, Shimizu Y (2005) Regenerative medicine of the trachea: the first human case. Ann Otol Rhinol Laryngol 114(6):429–433
42. Omori K, Tada Y, Suzuki T, Nomoto Y, Matsuzuka T, Kobayashi K, Nakamura T, Kanemaru S, Yamashita M, Asato R (2008) Clinical application of in situ tissue engineering using a scaffolding technique for reconstruction of the larynx and trachea. Ann Otol Rhinol Laryngol 117(9):673–678
43. Zhang L, Cheng JP, Zhang WC, Li W, Anniko M Reconstruction of defects following surgery for hypopharyngeal carcinoma using artificial biological material. Acta Otolaryngol 130(11):1293–1299. doi:10.3109/00016489.2010.486798
44. Macchiarini P, Jungebluth P, Go T, Asnaghi MA, Rees LE, Cogan TA, Dodson A, Martorell J, Bellini S, Parnigotto PP, Dickinson SC, Hollander AP, Mantero S, Conconi MT, Birchall MA (2008) Clinical transplantation of a tissue-engineered airway. Lancet 372(9655):2023–2030. doi:S0140-6736(08)61598-6 [pii]. 10.1016/S0140-6736(08)61598-6

Adv Biochem Eng Biotechnol (2013) 130: 55–67
DOI: 10.1007/10_2012_162

Published Online: 6 November 2012

Potential of Mesenchymal Stem Cell Applications in Plastic and Reconstructive Surgery

Birgit Weyand and Peter M. Vogt

Abstract Novel therapy with mesenchymal stem cells from bone marrow, adipose tissue, or other sources has raised high hopes for treatment of a variety of diseases. For plastic and reconstructive surgery, first pilot studies and clinical trials using stem cells for treatment of chronic wounds, radiation injury, or soft tissue augmentation have furnished encouraging results compared with the limitations of standard therapy, for example autologous fat grafting. Further research must be conducted to reveal the complex physiological interactions between activated stem cells and the host environment. Long-term effects and safety aspects of these novel treatment options also require randomized controlled studies. For future clinical applications, guidelines and standardized procedures for stem cell isolation and preparation, and techniques for application must be established.

Keywords Allotransplantation · Burns · Fat grafting · Lipofilling · Mesenchymal stem cells · Radiation injury · Scar repair · Tissue engineering · Wound healing

Contents

B. Weyand (✉) · P. M. Vogt
Department of Plastic, Hand and Reconstructive Surgery, Hannover Medical School, Carl-Neubergstr. 1, 30625 Hannover, Germany
e-mail: Weyand.birgit@mh-hannover.de

1 The Researcher's View of the Discovery of Mesenchymal Stem Cells

When, in 2001, Patricia Zuk and colleagues reported the characterization of a mesenchymal stem cell source isolated from lipoaspirates of adipose tissue, the scientific community realized the implications of her findings for tissue engineering and reconstructive medicine [1]. The term "stem cell" was introduced by the German biologist Ernst Haeckel in the late nineteenth century with two meanings—the fertilized egg and the ancestor unicellular organism from which all multicellular organisms evolved [2]. Side by side, in the early twentieth century, Artur Pappenheim, Alexander Maximow, Ernst Neumann, and others proposed use of this term for a common progenitor cell of the haematopoietic system [2]. In the 1960s, McCulloch and Till reported the presence of self-renewing cells in mouse bone marrow [3–5] and Friedenstein described osteogenesis arising from transplanted bone marrow cells of non-haematopoietic origin [6]. Further studies on the multi-lineage differentiation potential of bone marrow cells from the non-haematopoietic fraction led to coinage of the name "mesenchymal stem cells (MSCs)" by Arnold Caplan in the early 1990s [7–10].

Nowadays several terms are used to characterize stem cells from fatty tissue, for example adipose-derived stromal cells (ADSC), also abbreviated to adipose stromal cells (ASC), adipose tissue-derived stromal cells (ATSC or ATDSC), adipose-derived adult stromal/stem (ADAS) cells, also, as a result of their isolation technique, known as processed lipoaspirate cells (PLAs), or simply as MSCs [11]. Sometimes, the term ADSC has also been used to describe the stromal vascular fraction (SVF), which is defined as the heterogeneous cell portion obtained after collagenase digestion and centrifugation of fat tissue containing ADSC, endothelial cells, and haematopoietic cells.

Because these stem cells have the potential to differentiate not only into cells developing from mesenchyme, for example adipocytes, chondrocytes, and osteocytes, but also because differentiation into neuronal cells or vascular cells has been described, the term "mesenchymal" stem cells has been under ongoing debate in the scientific community [12, 13]. Besides, bone and muscle tissue have distinct progenitors, not a common ancestor, and craniofacial bone develops from neuroectodermal progenitors [12]. Caplan himself, in a recent publication, suggested re-naming "mesenchymal stem cell" as "medicinal stem cell" [14]. Alternatively, the term multipotent (mesenchymal) stromal cell (MSC) is also used.

2 The Introduction of Mesenchymal Stem Cell Therapy to Plastic Surgery

The origins of plastic and reconstructive surgery can be dated back approximately 1000 BC, or even further. Reports about complex nose reconstruction in India or ear reconstructive procedures on mummies in Egypt tells us about efforts to restore and form body's outer shape and function. Nowadays, plastic and reconstructive surgery is an independent surgical speciality at the intersection of other surgical specialities, for example vascular surgery, orthopaedic surgery, traumatology, ear, nose and throat, gynaecology, urology, or visceral surgery.

Plastic surgery as a surgical speciality can be classified into four disciplines. The first column is the field of reconstructive surgery which includes treatment of functional and formal defects caused by traumatic events, tumour lesions, or impairment of wound healing. The second consists in the treatment of acute burn injuries and sequelae. The third covers the field of hand surgery for primary or secondary injuries. Finally, the fourth entails aesthetic surgery to adjust form, contour and shape of body parts according to aesthetic ideals.

The introduction of fat grafting to correct contour deformities can be regarded as the basis of today's stem cell therapy, because fat contains MSCs besides adipocytes, preadipocytes, fibroblasts, and haemaopoietic and vascular cells [15]. In the literature, the term "fat grafting" is used as often as "stem cell therapy", because these procedures are related to each other. However, true stem cell therapy approaches require steps to isolate and purify the stem cell fraction, and some procedures also use expansion of MSCs under good manufacturing practice (GMP) conditions.

Free fat grafting was described by Neuber in 1893, at the 22nd conference of the German National Surgery Association, for corrections of facial defects and by Czerny for breast reconstruction procedures. Because of complications experienced with fat grafting, for example formation of oil cysts, infection, involution, and death of the graft, this procedure was subsequently discontinued. Fat grafting underwent a revival on introduction of the subcutaneous aspiration–liposuction technique by Illouz in 1977 [16]; this was further modified by Fournier and also by Gasparotti and Toledo in the 1980s [17, 18]. Furthermore, in the early 1990s the novel ultrasound-assisted liposuction technique introduced by the Italian Zocchi and the French Fitoussi, each with a different system, enabled minimally invasive removal of excessive fat deposits and body contouring [19, 20]. From there it was a small step to use the lipoaspirate to correct contour deformities in the subcutaneous tissues. Colemann developed a special syringe system which enabled easy preparation and re-injection of fat grafts after centrifugation. He also studied, very thoroughly, the effects of suction technique, suction force, tumescent solution, and concentration and technique of fat grafting, on adipocyte survival and content [21]. By centrifugation of the syringe content it was possible to separate the lipoaspirate from the tumescent fluid. Nowadays, several syringe systems for lipofilling are commercially available, and the lipoaspirate can be separated via a filter device from the oil and water fraction before injection [22, 23]. This procedure stands in contrast to the technique of preparation of the SVF for stem cell isolation, in which the lipoaspirate (or alternatively fat tissue "en bloc") is treated by collagenase digestion before centrifugation and followed by several washing steps.

On the basis of this procedure, a novel technique called "cell-assisted lipotransfer" was introduced by Yoshimura in 2006 with the underlying idea of enhancing the potential of fat grafts by supplementation of the lipoaspirate with stem cells before transplantation [24]. Study results in recent years have demonstrated superior survival of stem cell-enhanced fat grafts compared with conventional fat grafts [24, 25]. Other clinical applications have used expanded isolated bone marrow stromal cells or isolated adipose-derived stem cells obtained by use of the Celution system for topical application to treat chronic radiation wounds [26–28].

3 Characteristics of Mesenchymal Stem Cells and Implications for Medical Application

3.1 Definition

The multipotent MSCs is defined by (1) its plastic-adherence for in vitro culture with a fibroblast-like morphology; (2) its potential for differentiation into the adipogenic, osteogenic, and chondrogenic lineage; and (3) the presence of specific antigen surface markers viewed by fluorescence-assisted cell sorting (FACS) analysis as defined by the International Society for Cellular Therapy in 2006 [13]. The term "mesenchymal stem cell" is problematic because of inconsistency of described surface markers in the scientific community and the question of true self-renewal as main property of a "stem cell" [12]. Furthermore, the nomenclature uses the terms mesenchymal stem or stromal cells, multipotent progenitor cells or stromal progenitor cells or multipotent precursor cell, as already stated above. In the clinical literature, the use of MSCs is even less defined often without proper assessment of surface markers and the differentiation potential of the cells. A variety of terms, for example preadipocytes, processed lipoaspirate, stromal vascular fraction, ADSC, adipose stromal vascular cell fraction, or cell-assisted lipotransfer, has been associated as "MSCs" often without testing for stem cell criteria or purity of the cell fraction used.

3.2 Sources

MSCs can be isolated from bone marrow aspirates or from fat tissue. Bone marrow stromal cells are the cells most commonly used for clinical studies so far. Because the estimated number of stem cells which can be derived from adipose tissue is 100 to 500-fold higher than that from bone marrow aspirates [29], this source has become an interesting option, especially for plastic surgeons, and several techniques are available for isolation of cells from lipoaspirate or fat tissue. Novel stem cell sources are umbilical cord blood or the umbilical cord Wharton's jelly itself [30–33]. Nowadays, stem cell banking has become a well-established procedure in perinatal medicine to preserve stem cells of newborns for future possible medical applications for themselves or their siblings [33, 34]. Stem cells have been also been identified and isolated from a variety of other tissues and locations, for example synovium, tooth, gum, perivascular niche, keloid scars, perinodular fat in Morbus Dupuytren, or even burn eschar [35–39]. However, these sources do not yet provide sufficient numbers of cells for treatment options, but are certainly of interest for research into possible future options.

3.3 Isolation and Tracing Techniques

Techniques for isolation and preparation of MSCs from adipose tissue lack proper standardization and structured procedures [40, 41]. Because MSCs, in contrast with haematopoietic stem cells, do not require enrichment and immunophenotypic sorting before ex vivo expansion, cell fractions are often heterogeneous and most of the isolated cells will not form so-called "colony-forming units—fibroblasts". In addition, the current application of stem cell-containing tissue transplants in plastic surgery for example fat, cell-enriched lipoaspirate or the so-called "SVF" is usually conducted in one surgical procedure without leaving the operating theatre. This, in fact, often prevents further characterization of MSCs in the cell or tissue graft, for example immunophenotyping, differentiation potential, and senescence, or assessment of the proportional contribution of stem cells to the whole graft.

Several studies have compared the effects of isolation variables on adipose-derived MSCs viability and stem cell ratio. Although data are already available for different sampling systems on the effect of donor site, liposuction devices, and pressures applied, for liposuction solutions and composition, and for optimized procedures for centrifugation and further processing and purification of cell aspirates [21, 23, 42–47], there is still a need for further research toward standardized clinical procedures [48].

Another important aspect to consider is the difficulty in tracking the implanted MSCs during homing, tissue migration in the recipient site, organ engraftment, and cell expansion and apoptosis in vivo [49, 50]. This problem has been well recognized in measurement of the survival and volume of autologous fat grafts. In a rodent model, Rieck et al. were able to trace autologous adipocytes transplanted to various body regions (subcutaneously, inguinal fat pad, epididymal fat pad, and intramuscularly) by use of the lipophilic rhodamine-like dye PKH26 [51]. Human adipocyte viability can be assessed by immunohistochemical staining with the novel marker perilipin [52]. Newer techniques for stem cell labelling use direct labelling of cells with super paramagnetic iron oxides (SPIO) or perfluoropolyether or perfluorocarbon nano-beacons for magnetic resonance imaging or radionuclides for positron emission tomography or gamma scintigraphic imaging [50]. Other indirect labelling techniques rely on the expression of reporter genes which are transduced in the cells before transplantation and are visualized by application of suitable substrates or probes, enabling real-time monitoring of dynamic biological processes [50]. For clinical use, MRT volumetric measurement has furnished feasible results in monitoring the retention of fat grafts, e.g. after breast reconstruction [53, 54].

3.4 Biological Properties

After an injury, MSCs receive signals which activate them to migrate towards the site of injury, either after being released from the bone marrow into the circulation or by migration of MSCs resting within the local tissue [55, 56]. Knowing this fact, MSCs are possibly activated after every single surgical procedure, and their role during scar formation and wound healing still needs further elucidation. Because we can assume that MSCs per se are already involved into the healing process, the interesting question to be asked is whether additionally added stem cells are beneficial (or detrimental) to the tissue-regeneration process. Experimental studies have shown that MSCs at the site of injury interact through paracrine effects with surrounding cells by secreting growth factors, for example IGF, VEGF, HGF, or TGF-β [48]. These paracrine actions stimulate local cell proliferation, migration, assembly and differentiation leading to angiogenesis and tissue repair [48, 57]. Some studies also suggest that MSCs themselves participate in new tissue formation by transdifferentiation into vascular cells, connective tissue cells, or even keratinocytes or glandular cells for skin appendices [58]. We have learned from cell-tracking studies that locally administered MSCs decrease markedly in number even within the first week. Also systemically applied stem cells from bone marrow which migrate toward the site of injury are not detectable within the circulation after a few days. It would, therefore, be interesting to further investigate the whereabouts of those stem cells during the healing process and the role of these stem cells in the activation and course of the healing cascade. Yoshimura recently described a three-zone concept for cell survival after lipotransfer [52]. In the outer core, adipocytes and adipose progenitor cells can both survive, because nutrition by diffusion is sufficient for the first hours and days until blood vessels have entered the graft. In the inner zone, both cell types die, because of a lack of nutrient and oxygen supply. The middle zone contains dying adipocytes but surviving adipose-derived stem cells, because this cell type can withstand severe hypoxia for up to several days. The stem cells then initiate a period of regeneration of the tissue, leading to replacement of dead adipocytes [52].

Experimental studies of animal models have shown an age-dependent effect of exposure of stem cells to ischemia before transplantation, leading to a pronounced paracrine response in the old donor group. Effects of senescence on stem cell markers and gene expression are also currently being investigated [59].

Safety concerns about stem cell therapy, especially for patients after cancer treatment, must be addressed. Although studies of fat transfer for breast reconstruction after mastectomy with large numbers of patients have so far not shown any alarming results regarding cancer recurrence, more research in the form of prospective randomized studies are required, and also for novel technology, for example cell-assisted lipotransfer [47, 60–66]. There are case reports of cancer, for example sarcoma, recurrence after local stem cell therapy [67]. Furthermore, in vitro studies of MSCs with different cancer cell lines have furnished contradictory results, and further investigation is needed [68–70].

The antitumourigenic effects observed in some studies are thought to relate also to the unique immunomodulatory properties of MSCs [69–71]. It will be interesting to learn if these effects also contribute to the wound or graft healing process and how they can be guided to support cellular interaction between donor and host cells, and graft take and survival.

4 Current Clinical Studies and Potential of Mesenchymal Stem Cells in Plastic Surgery

4.1 Clinical Studies

The clinical trials website http://clinicaltrials.gov provides a summary of registered completed or ongoing studies using stem cells for therapeutic use. Most of the trials are phase I or II studies. In a recent review paper Casteilla differentiates among studies using autologous fat, autologous stromal vascular fraction, or ADSc. Here ADSc are restricted to purified and in culture expanded ASCs and, by definition, as separate from the term "heterogeneous crude SVF" which is derived directly from digested fat extracts [29]. A concise review of current clinical studies using MSCs in plastic and reconstructive surgery applications has recently been published by Rohrich's group [72]. The main fields are wound healing and soft tissue augmentation, followed by tissue engineering. Because of the lack of standardized procedures for cell harvesting, purification, processing, and culture, results from different centres are difficult to compare. Furthermore, because of the study design, two treatment options are usually compared, where in studies of soft tissue augmentation the control group as standard of care often includes treatment by fat grafting. In pilot studies with small numbers, especially, problems with drop outs from the designed control groups, because of patient comorbidity, deteriorating health condition, or even death, may contribute to a complete loss of the control group, which impairs statistical analysis and study strength [73].

4.2 Therapeutic Potential

Novel forthcoming therapy in plastic surgery includes use of MSCs derived either from bone marrow or from adipose tissue, which are currently being introduced for a huge variety of clinical problems. For plastic and reconstructive surgery, several aspects of mesenchymal stem cell therapy in support of established surgical techniques are very promising:

(1) MSCs can serve as adjuvants to improve local angiogenesis [74, 75]. For this function, the main effect is assumed by paracrine factors which stimulate endothelial cells and mural cells in the wound bed [75, 76]. This preparation of the host graft bed can improve wound healing, graft take, and graft survival, e.g. of fat or skin grafts. Furthermore, the angiogenic effects might be part of the beneficial factors explaining improved healing in difficult wound beds, for example burn wounds or radiation injury after local adipose-derived stromal cell transplantation.
(2) MSCs can also improve survival of fat grafts in dynamic remodelling processes. In mice experiments, survival of transplanted human progenitor cells was better than that of differentiated adipocytes, because of better resistance to ischaemic cell death; the cells subsequently became activated and participated in the remodelling and regeneration process of the fat tissue [52].
(3) MSCs have the potential for differentiation into adipogenic, osteogenic, or chondrogenic lineage and possible transdifferentiation into vasculogenic and neurogenic cells [1, 7, 77]. This potential suggests the future use of MSCs in tissue-engineering applications to repair or replace tissues or organs [48, 78–80].
(4) MSCs have immunomodulatory properties which suggests possible use as a medical additive in composite allograft transplantation to suppress adverse immune reaction or as treatment option for graft versus host disease [81–84]. A recent animal study of allogenic hindlimb transplantation in a rodent model demonstrated changes in anti-inflammatory cytokine expression and T cell function when transient immunosuppression was combined with donor adipose stem cell application [85].
(5) MSCs can be locally activated for guided new tissue formation in situ. This method was shown by Yoshimura by local injection of hyaluronic acid fillers under and immediately around the periosteum [15].

(6) MSCs are systemically activated by factors released after wounding and migrate to the site of injury either after being released from the bone marrow into the blood circulation or by migration from local or adjacent tissue sources, for example by activation of perivascular cells in fat tissue [86–88]. Circulating bone marrow progenitor cells may also contribute to an increase in adipocyte number in fat tissue when activated by a high-fat diet or by substances such as thizolinedione (TZD) and rosiglitazone, which are PPARγ ligands used, e.g. in the treatment of type 2 diabetes mellitus [89]. Further progress in understanding this regulatory mechanism and defining the factors involved may inspire development of medical solutions to support this process.

References

1. Zuk PA, Zhu M, Mizuno H, Huang J, Futrell JW, Katz AJ et al (2001) Multilineage cells from human adipose tissue: implications for cell-based therapies. Tissue Eng 7(2):211–228
2. Ramalho-Santos M, Willenbring H (2007) On the origin of the term “stem cell”. Cell Stem Cell 1(1):35–38
3. Till JE, McCulloch EA (1961) A direct measurement of the radiation sensitivity of normal mouse bone marrow cells. Radiat Res 14:213–222
4. Becker AJ, McCulloch EA, Till JE (1963) Cytological demonstration of the clonal nature of spleen colonies derived from transplanted mouse marrow cells. Nature 197:452–454
5. Siminovitch L, McCulloch EA, Till JE (1963) The distribution of colony-forming cells among spleen colonies. J Cell Physiol 62:327–336
6. Friedenstein AJ, Piatetzky-Shapiro II, Petrakova KV (1966) Osteogenesis in transplants of bone marrow cells. J Embryol Exp Morphol 16(3):381–390
7. Caplan AI (1991) Mesenchymal stem cells. J Orthop Res. 9(5):641–650
8. Caplan AI (1990) Stem cell delivery vehicle. Biomaterials 11:44–46
9. Nakahara H, Bruder SP, Haynesworth SE, Holecek JJ, Baber MA, Goldberg VM et al (1990) Bone and cartilage formation in diffusion chambers by subcultured cells derived from the periosteum. Bone 11(3):181–188
10. Bruder SP, Gazit D, Passi-Even L, Bab I, Caplan AI (1990) Osteochondral differentiation and the emergence of stage-specific osteogenic cell-surface molecules by bone marrow cells in diffusion chambers. Bone Miner 11(2):141–151
11. Cawthorn WP, Scheller EL, MacDougald OA (2012) Adipose tissue stem cells meet preadipocyte commitment: going back to the future. J Lipid Res 53(2):227–246
12. Bianco P, Robey PG, Simmons PJ (2008) Mesenchymal stem cells: revisiting history, concepts, and assays. Cell Stem Cell. 2(4):313–319
13. Dominici M, Le Blanc K, Mueller I, Slaper-Cortenbach I, Marini F, Krause D et al (2006) Minimal criteria for defining multipotent mesenchymal stromal cells. the international society for cellular therapy position statement. Cytotherapy 8(4):
14. Caplan AI (2010) What’s in a name? Tissue Eng Part A 16(8):2415–2417.
15. Yoshimura K, Eto H, Kato H, Doi K, Aoi N (2011) In vivo manipulation of stem cells for adipose tissue repair/reconstruction. Regen Med 6(6):33–41
16. Illouz YG (1983) Body contouring by lipolysis: a 5-year experience with over 3000 cases. Plast Reconstr Surg 72(5):591–597
17. Fournier PF, Otteni FM (1983) Lipodissection in body sculpturing: the dry procedure. Plast Reconstr Surg 72(5):598–609

18. Gasparotti M (1992) Superficial liposuction: a new application of the technique for aged and flaccid skin. Aesthetic Plast Surg Spring 16(2):141–153
19. Zocchi M (1992) Ultrasonic liposculpturing. Aesthetic Plast Surg Fall 16(4):287–298
20. Fitoussi H (1993) Ultrasonic liposculpture: the French technique. Extrait de la Revue de Chirurgie Esthétique
21. Coleman SR (2006) Structural fat grafting: more than a permanent filler. Plast Reconstr Surg 118(3 Suppl):108–120
22. Rennekampff HO, Reimers K, Gabka CJ, Germann G, Giunta RE, Knobloch K et al (2010) Current perspective and limitations of autologous fat transplantation–"consensus meeting" of the German society of plastic, reconstructive and aesthetic surgeons at Hannover; September 2009. Handchir Mikrochir Plast Chir 42(2):137–142
23. Kuhbier JW, Weyand B, Radtke C, Vogt PM, Kasper C, Reimers K (2010) Isolation, characterization, differentiation, and application of adipose-derived stem cells. Adv Biochem Eng Biotechnol 123:105–155
24. Matsumoto D, Sato K, Gonda K, Takaki Y, Shigeura T, Sato T et al (2006) Cell-assisted lipotransfer: supportive use of human adipose-derived cells for soft tissue augmentation with lipoinjection. Tissue Eng 12(12):3375–3382
25. Yoshimura K, Sato K, Aoi N, Kurita M, Hirohi T, Harii K (2008) Cell-assisted lipotransfer for cosmetic breast augmentation: supportive use of adipose-derived stem/stromal cells. Aesthetic Plast Surg 32(1):48–55, discussion 56–57
26. Akita S, Akino K, Hirano A, Ohtsuru A, Yamashita S (2010) Mesenchymal stem cell therapy for cutaneous radiation syndrome. Health Phys 98(6):858–862
27. Bey E, Prat M, Duhamel P, Benderitter M, Brachet M, Trompier F et al (2010) Emerging therapy for improving wound repair of severe radiation burns using local bone marrow-derived stem cell administrations. Wound Repair Regen 18(1):50–58
28. Herold C, Rennekampff HO, Kramer R, Hillmer A, Knobloch K, Vogt P (2011) Stem cell-enhanced fat transplantation—a potential therapeutic option for radiation ulcus? Zentralbl Chir. doi:10.1055/s-0030-1247476. [Epub ahead of print]
29. Casteilla L, Planat-Benard V, Laharrague P, Cousin B (2011) Adipose-derived stromal cells: Their identity and uses in clinical trials, an update. World J Stem Cells 3(4):25–33
30. Ilic D, Miere C, Lazic E (2012) Umbilical cord blood stem cells: clinical trials in non-hematological disorders. Br Med Bull 102:43–57
31. Yang S, Huang S, Feng C, Fu X (2012) Umbilical cord-derived mesenchymal stem cells: strategies, challenges, and potential for cutaneous regeneration. Front Med 6(1):41–47
32. Pelosi E, Castelli G, Testa U (2012) Human umbilical cord is a unique and safe source of various types of stem cells suitable for treatment of hematological diseases and for regenerative medicine. Blood Cells Mol Dis 49(1):8–20
33. Ballen K (2010) Challenges in umbilical cord blood stem cell banking for stem cell reviews and reports. Stem Cell Rev 6(1):14–18
34. Bieback K, Brinkmann I (2010) Mesenchymal stromal cells from human perinatal tissues: from biology to cell therapy. World J Stem Cells 2(4):81–92
35. Iqbal SA, Manning C, Syed F, Kolluru V, Hayton M, Watson S et al (2012) Identification of mesenchymal stem cells in perinodular fat and skin in Dupuytren's disease: a potential source of myofibroblasts with implications for pathogenesis and therapy. Stem Cells Dev 21(4): 609–622
36. Iqbal SA, Syed F, McGrouther DA, Paus R, Bayat A (2010) Differential distribution of haematopoietic and nonhaematopoietic progenitor cells in intralesional and extralesional keloid: do keloid scars provide a niche for nonhaematopoietic mesenchymal stem cells? Br J Dermatol 162(6):1377–1383
37. Hindocha S, Iqbal SA, Farhatullah S, Paus R, Bayat A (2011) Characterization of stem cells in Dupuytren's disease. Br J Surg 98(2):308–315
38. van der Veen VC, Vlig M, van Milligen FJ, de Vries SI, Middelkoop E, Ulrich MM (2011) Stem cells in burn eschar. Cell Transpl 21(5):933–942

39. Kerkis I, Caplan AI (2012) Stem cells in dental pulp of deciduous teeth. Tissue Eng Part B Rev 18(2):129–138
40. Gimble JM, Bunnell BA, Guilak F (2012) Human adipose-derived cells: an update on the transition to clinical translation. Regen Med 7(2):225–235
41. Bieback K, Kinzebach S, Karagianni M (2011) Translating research into clinical scale manufacturing of mesenchymal stromal cells. Stem Cells Int 20(2010):193519
42. Pu LL (2012) Towards more rationalized approach to autologous fat grafting. J Plast Reconstr Aesthet Surg 65(4):413–419
43. Kurita M, Matsumoto D, Shigeura T, Sato K, Gonda K, Harii K, et al (2008) Influences of centrifugation on cells and tissues in liposuction aspirates: optimized centrifugation for lipotransfer and cell isolation. Plast Reconstr Surg 121(3):1033–41, discussion 1042–1043
44. Sterodimas A, de Faria J, Nicaretta B, Papadopoulos O, Papalambros E, Illouz YG (2010) Cell-assisted lipotransfer. Aesthet Surg J 30(1):78–81
45. Herold C, Pflaum M, Utz P, Wilhelmi M, Rennekampff HO, Vogt PM (2011) Viability of autologous fat grafts harvested with the Coleman technique and the tissue trans system (Shippert method): a comparative study. Handchir Mikrochir Plast Chir 43(6):361–367
46. Herold C, Utz P, Pflaum M, Wilhelmi M, Vogt PM (2012) Rennekampff HO. Negative pressure of manual liposuction with Coleman technique is highly dependant on the position of plunger of the syringe, J Plast Reconstr Aesthet Surg
47. Doi K, Tanaka S, Iida H, Eto H, Kato H, Aoi N, Kuno S, Hirohi T, Yoshimura K (2012) Stromal vascular fraction isolated from lipo-aspirates using an automated processing system: bench and bed analysis. J Tissue Eng Regen Med. doi:10.1002/term.1478. [Epub ahead of print]
48. Philips BJ, Marra KG, Rubin JP (2012) Adipose stem cell-based soft tissue regeneration. Expert Opin Biol Ther 12(2):155–163
49. Yan L, Han Y, He Y, Xie H, Liu J, Zhao L et al (2007) Cell tracing techniques in stem cell transplantation. Stem Cell Rev 3(4):265–269
50. Lee Z, Dennis JE, Gerson SL (2008) Imaging stem cell implant for cellular-based therapies. Exp Biol Med (Maywood) 233(8):930–940
51. Rieck B, Schlaak S (2003) In vivo tracking of rat preadipocytes after autologous transplantation. Ann Plast Surg 51(3):294–300
52. Eto H, Kato H, Suga H, Aoi N, Doi K, Kuno S et al (2012) The fate of adipocytes after nonvascularized fat grafting: evidence of early death and replacement of adipocytes. Plast Reconstr Surg 129(5):1081–1092
53. Herold C, Knobloch K, Rennekampff HO, Ueberreiter K, Vogt PM (2010) Magnetic resonance imaging-based progress control after autologous fat transplantation. Plast Reconstr Surg 126(5):260–261
54. Herold C, Reichelt A, Stieglitz LH, Dettmer S, Knobloch K, Lotz J et al (2010) MRI-based breast volumetry-evaluation of three different software solutions. J Digit Imaging 23(5): 603–610
55. Ceradini DJ, Kulkarni AR, Callaghan MJ, Tepper OM, Bastidas N, Kleinman ME et al (2004) Progenitor cell trafficking is regulated by hypoxic gradients through HIF-1 induction of SDF-1. Nat Med 10(8):858–864
56. Gao J, Dennis JE, Muzic RF, Lundberg M, Caplan AI (2001) The dynamic in vivo distribution of bone marrow-derived mesenchymal stem cells after infusion. Cells Tissues Organs 169(1):12–20
57. Eto H, Suga H, Inoue K, Aoi N, Kato H, Araki J et al (2011) Adipose injury-associated factors mitigate hypoxia in ischemic tissues through activation of adipose-derived stem/progenitor/stromal cells and induction of angiogenesis. Am J Pathol 178(5):2322–2332
58. Tabit CJ, Slack GC, Fan K, Wan DC, Bradley JP (2012) Fat grafting versus adipose-derived stem cell therapy: distinguishing indications, techniques, and outcomes. Aesthetic Plast Surg 36(3):704–713
59. Lepperdinger G, Brunauer R, Jamnig A, Laschober G, Kassem M (2008) Controversial issue: is it safe to employ mesenchymal stem cells in cell-based therapies? Exp Gerontol 43(11):1018–1023

60. Pearl RA, Leedham SJ, Pacifico MD (2012) The safety of autologous fat transfer in breast cancer: Lessons from stem cell biology. J Plast Reconstr Aesthet Surg 65(3):283–288
61. Zimmerlin L, Donnenberg AD, Rubin JP, Basse P, Landreneau RJ, Donnenberg VS (2011) Regenerative therapy and cancer: in vitro and in vivo studies of the interaction between adipose-derived stem cells and breast cancer cells from clinical isolates. Tissue Eng Part A 17(1–2):93–106
62. Eaves FF, Haeck PC, Rohrich RJ (2012) ASAPS/ASPS position statement on stem cells and fat grafting. Plast Reconstr Surg 129(1):285–287
63. Mizuno H, Hyakusoku H (2010) Fat grafting to the breast and adipose-derived stem cells: recent scientific consensus and controversy. Aesthet Surg J 30(3):381–387
64. MacIsaac ZM, Shang H, Agrawal H, Yang N, Parker A, Katz AJ (2012) Long-term in vivo tumorigenic assessment of human culture-expanded adipose stromal/stem cells. Exp Cell Res 318(4):416–423
65. Kanchwala SK, Glatt BS, Conant EF, Bucky LP (2009) Autologous fat grafting to the reconstructed breast: the management of acquired contour deformities. Plast Reconstr Surg 124(2):409–418
66. Rigotti G, Marchi A, Stringhini P, Baroni G, Galie M, Molino AM et al (2010) Determining the oncological risk of autologous lipoaspirate grafting for post-mastectomy breast reconstruction. Aesthetic Plast Surg 34(4):475–480
67. Perrot P, Rousseau J, Bouffaut AL, Redini F, Cassagnau E, Deschaseaux F et al (2010) Safety concern between autologous fat graft, mesenchymal stem cell and osteosarcoma recurrence. PLoS One 5(6):10999
68. Burns JS, Abdallah BM, Schroder HD, Kassem M (2008) The histopathology of a human mesenchymal stem cell experimental tumor model: support for an hMSC origin for Ewing's sarcoma? Histol Histopathol 23(10):1229–1240
69. Grisendi G, Bussolari R, Veronesi E, Piccinno S, Burns JS, De Santis G et al (2011) Understanding tumor-stroma interplays for targeted therapies by armed mesenchymal stromal progenitors: the mesenkillers. Am J Cancer Res 1(6):787–805
70. Singer NG, Caplan AI (2011) Mesenchymal stem cells: mechanisms of inflammation. Annu Rev Pathol 6:457–478
71. Puissant B, Barreau C, Bourin P, Clavel C, Corre J, Bousquet C et al (2005) Immunomodulatory effect of human adipose tissue-derived adult stem cells: comparison with bone marrow mesenchymal stem cells. Br J Haematol 129(1):118–129
72. Gir P, Oni G, Brown SA, Mojallal A, Rohrich RJ (2012) Human adipose stem cells: current clinical applications. Plast Reconstr Surg 129(6):1277–1290
73. Kirana S, Stratmann B, Prante C, Prohaska W, Koerperich H, Lammers D et al (2012) Autologous stem cell therapy in the treatment of limb ischaemia induced chronic tissue ulcers of diabetic foot patients. Int J Clin Pract 66(4):384–393
74. Meirelles Lda S, Fontes AM, Covas DT, Caplan AI (2009) Mechanisms involved in the therapeutic properties of mesenchymal stem cells. Cytokine Growth Factor Rev 20(5–6):419–427
75. Sorrell JM, Baber MA, Caplan AI (2009) Influence of adult mesenchymal stem cells on in vitro vascular formation. Tissue Eng Part A 15(7):1751–1761
76. Sorrell JM, Caplan AI (2010) Topical delivery of mesenchymal stem cells and their function in wounds. Stem Cell Res Ther 1(4):30
77. Mizuno H (2009) Adipose-derived stem cells for tissue repair and regeneration: ten years of research and a literature review. J Nihon Med Sch 76(2):56–66
78. Levi B, Longaker MT (2011) Concise review: adipose-derived stromal cells for skeletal regenerative medicine. Stem Cells 29(4):576–582
79. Sterodimas A, de Faria J, Nicaretta B, Pitanguy I (2010) Tissue engineering with adipose-derived stem cells (ADSCs): current and future applications. J Plast Reconstr Aesthet Surg 63(11):1886–1892
80. Arthur A, Zannettino A, Gronthos S (2009) The therapeutic applications of multipotential mesenchymal/stromal stem cells in skeletal tissue repair. J Cell Physiol 218(2):237–245

81. Caplan AI, The Correa D (2011) The MSC An injury drugstore. Cell Stem Cell 9(1):5–11
82. Ringden O, Le Blanc K (2011) Mesenchymal stem cells for treatment of acute and chronic graft-versus-host disease, tissue toxicity and hemorrhages. Best Pract Res Clin Haematol 24(1):65–72
83. Tholpady SS, Ogle RC, Katz AJ (2009) Adipose stem cells and solid organ transplantation. Curr Opin Organ Transplant 14(1):51–55
84. Aggarwal S, Pittenger MF (2005) Human mesenchymal stem cells modulate allogeneic immune cell responses. Blood 105(4):1815–1822
85. Kuo YR, Chen CC, Goto S, Lee IT, Huang CW, Tsai CC et al (2011) Modulation of immune response and T-cell regulation by donor adipose-derived stem cells in a rodent hind-limb allotransplant model. Plast Reconstr Surg 128(6):661–672
86. Tang W, Zeve D, Suh JM, Bosnakovski D, Kyba M, Hammer RE et al (2008) White fat progenitor cells reside in the adipose vasculature. Science 322(5901):583–586
87. Traktuev DO, Merfeld-Clauss S, Li J, Kolonin M, Arap W, Pasqualini R et al (2008) A population of multipotent CD34-positive adipose stromal cells share pericyte and mesenchymal surface markers, reside in a periendothelial location, and stabilize endothelial networks. Circ Res 102(1):77–85
88. Cherubino M, Rubin JP, Miljkovic N, Kelmendi-Doko A, Marra KG (2011) Adipose-derived stem cells for wound healing applications. Ann Plast Surg 66(2):210–215
89. Crossno JT Jr, Majka SM, Grazia T, Gill RG, Klemm DJ (2006) Rosiglitazone promotes development of a novel adipocyte population from bone marrow-derived circulating progenitor cells. J Clin Invest 116(12):3220–3228

Adv Biochem Eng Biotechnol (2013) 130: 69–88
DOI: 10.1007/10_2012_145

Published Online: 18 July 2012

General Principles for the Regeneration of Bone and Cartilage

Michael Jagodzinski and C. Haasper

Abstract For the regeneration of bone and cartilage, mesenchymal stem cells are currently used invitro and in-vivo. For bone, the existence of viable cells, scaffolds, mechanical environment, growth factors and vascularization are of paramount importance. Mesenchymal stem cells can be harvested from the bone marrow using minimally invasive techniques. Centrifugation can increase the number of transplanted cells per volume. The use of cell therapy is under current clinical investigation and the benefit from these systems has to be proven in level I studies. For cartilage, current techniques recruiting stem cells from the subchondral bone have been demonstrated to be nearly as effective as autologous chondrocyte transplantation, requiring less invasive surgery. The efficacy of mesenchymal stem cell concentrates remains to be proven. There is high potential for tissue engineered joint surfaces to become an option for joint surface defects and degeneration.

Contents

M. Jagodzinski (✉)
Department of Orthopedic Trauma, Hannover Medical School (MHH),
Carl-Neuberg-Str. 1, 30625 Hannover, Germany
e-mail: Jagodzinski.Michael@MH-Hannover.de

C. Haasper
ENDO Clinic, Orthopaedic Surgery, Holstenstr. 2,
22767 Hamburg, Germany
e-mail: Carl.Haasper@damp.de

Even though there is a strong link between cartilage and bone in embryonic development, [1] there are fundamental differences between the regeneration of bone and cartilage. Hyalin cartilage is found within a nonunion (Fig. 1), demonstrating that a natural environment exists where precursor cells from the cortical bone or bone marrow can transform into chondrocytes. This phenomenon may explain why there is a high healing potential found for chondral injuries, [2] especially in the adolescent population. This healing potential is used when marrow-stimulating techniques such as microfracturing or drilling are applied [3]. Likewise, these techniques work best for young individuals with intact corresponding cartilage, stable knee joints, and proper joint alignment [4]. However, the clinical use of mesenchymal stem cells amplified in vitro has not yet been achieved for use in clinical trials [5].

There are several established ways to regenerate small and large volumes of bone: spongiosa transplantation, segment transportation, and lately, bone marrow concentrate implantation [6]. All of these options take advantage of the regenerative capacities of precursor cells.

For the successful reconstruction of bone, the "diamond concept" (Figure 2) addresses the needs for successful treatment. It consists of five components: [7, 8]

- Viable osteogenic cells
- Osteoconductive scaffolds
- Mechanical environment
- Growth factors
  Recently, another important factor was added to the concept: [9]
- Vascularization

1 Viable Osteogenic Cells

In a fracture with potential to heal, there is an orchestra of different viable and damaged cells, regulating cytokines, and precursor cells with variable functions. Areas of high cellular density are created, which trigger transformation of MSCs into an osteoblastic phenotype and stimulate angiogenesis [10]. Likewise, if cells are expanded in vitro, the implantation of different cell phenotypes (e.g.,

Table 1 Scaffolds and characteristics

Category	Brand name (example)	Advantage	Disadvantage
Organic (allogeneic, xenogeneic)		Similar structure as human spongiosa, pore size and interconnectivity	Foreign body reaction if protein structure is preserved
Decellularized allograft/ xenograft limited heat treated	Tutobone®	Higher mechanical stability compared with heat-treated material	Foreign body reaction Higher risk of infectious disease (CJD)
Decellularized allograft/ xenograft, sintered	Orthoss® Endobon®	Limited foreign body reaction	Reduced mechanical stability Residual risk of infectious disease
Demineralized bone matrix	Grafton®	Osteoinductivity: contains bone morphogenic protein (BMP)	No porosity, has shown inferior results in animal studies compared with porous scaffolds [85]
Synthetic		Large variability of structure and composition	No material is identical with the structure and composition of human bone
Hydroxyapatite	Ostim®	Limited foreign body reaction	No porosity if used as a paste
Ca_2SO_4	Vitoss®	High porosity and interconnectivity	Limited mechanical stability
Glass ceramics	Skelite®	Good mechanical properties	Variation of degradation; possible degradation without bone formation [18, 86]
Polymers/ composites	Trufit®	Good mechanical properties High variability of the architecture	Bone formation in pure Ca_2SO_4-scaffolds higher than in some composites [22, 87]

endothelial cells mixed with mesenchymal stem cells) or cells expressing both osteogenic and angiogenic growth factors is more effective than MSC transplantation alone [11, 12].

2 Scaffolds

During fracture healing, there are fragments of cortical and spongious bone that act as a scaffold. This scaffold contains and attracts both osteoclasts and osteoblasts to regenerate the architecture of bone. The ideal pore diameter for artificial scaffolds has been described ranging from 150 to 500 μm.[13, 14] Numerous variations of

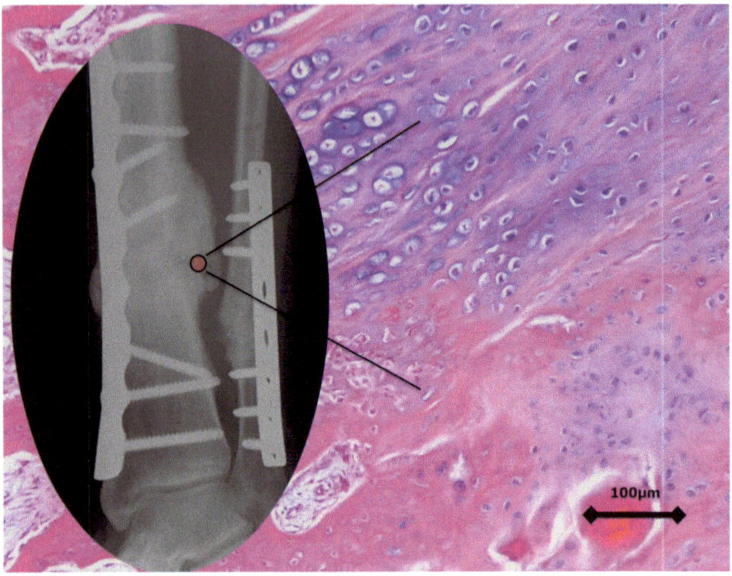

Fig. 1 A hypertrophic nonunion illustrates the pluripotency of progenitor cells that form clusters of chondrocytes as a result of insufficient stability. Courtesy of Dr. Länger, Department of Pathology, Hannover Medical School (MHH)

scaffolds have been suggested for tissue engineering of bone, and a variety are applicable for clinical use (Table 1).

Organic scaffolds may be derived from allogeneic or xenogeneic bone, mostly spongious bone. They come in a variety of shapes (blocks, wedges, cylinders, chips). Due to their origin, they are somewhat limited in size. The main differences in their biologic behaviour result from heat treatment: If proteins are preserved during processing (demineralized bone matrix, DBM) osteoinductive properties may be preserved [15]. Growth factors such as BMP-7 have been eluted out of DBM [16]. However, the preservation of the protein structure makes an immune response and implant weakening more likely [17].

If heat treatment is applied, the biomaterial becomes more brittle but less immunogenic [18]. In addition, the likelihood of transmission of infectious disease is reduced [19].

Synthetic scaffolds have virtually unlimited variations in form and structure. The most prominent candidates are calcium sulphate (Ca_2SO_4) and hydroxyapatite ($Ca_5(PO_4)_3(OH)$). Whereas beta-tricalciumphosphate degrades rapidly 3-6 months after implantation in the human tibia, [20] hydroxyapatite remains visible and mostly unchanged on plain radiographs, even after 10 years [21].

Composites such as silica-bound hydroxyapatite were thought to be osteoinductive, but they failed to demonstrate equity to bone marrow in an ovine long-bone defect [22]. In the future, the optimum biomaterial should allow for the

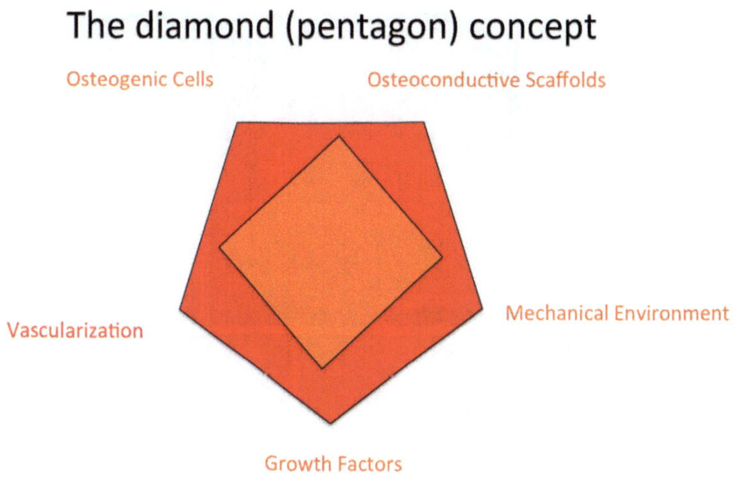

Fig. 2 The diamond concept originally described the following key factors for successful treatment of bone defects: Osteogenic cells, osteoconductive scaffold, mechanical environment, and growth factors [8, 88]. Later on, vascularization was added to the concept [9]

transplanted or locally available osteogenic cells to rapidly propagate through the scaffold and be replaced within 3–12 months [23].

Depending on the volume of bone that needs to be reconstructed, there is an increasing need to implant cells on a scaffold. Even though many scaffolds are applicable for bone defect repair, the ideal material has yet to be found [5, 24]. It is hoped that the results of cell scaffolds in combination with growth factors may exceed the healing times of critical long-bone defects in the future (Fig. 1).

3 Mechanical Environment

The mechanical environment of naturally healing bone is characterized by high motion in the beginning, leading to a soft callus that contains collagen fibers that are structured according to the direction of motion [25, 26]. This process closely resembles endomembranous ossification [27]. Later, the callus starts mineralization, which leads to hardening of the regenerate. Therefore, high strain rates are tolerated in the beginning but not in the last cycle of fracture healing [28]. For most tissue engineering applications in the long bones, external fixation devices have been used [29, 30]. However, the switch to dynamic internal fixation has been described for the treatment of nonunions. Nails can be inserted in a static and dynamic fashion to allow for controlled micro-motion of the regenerate [31].

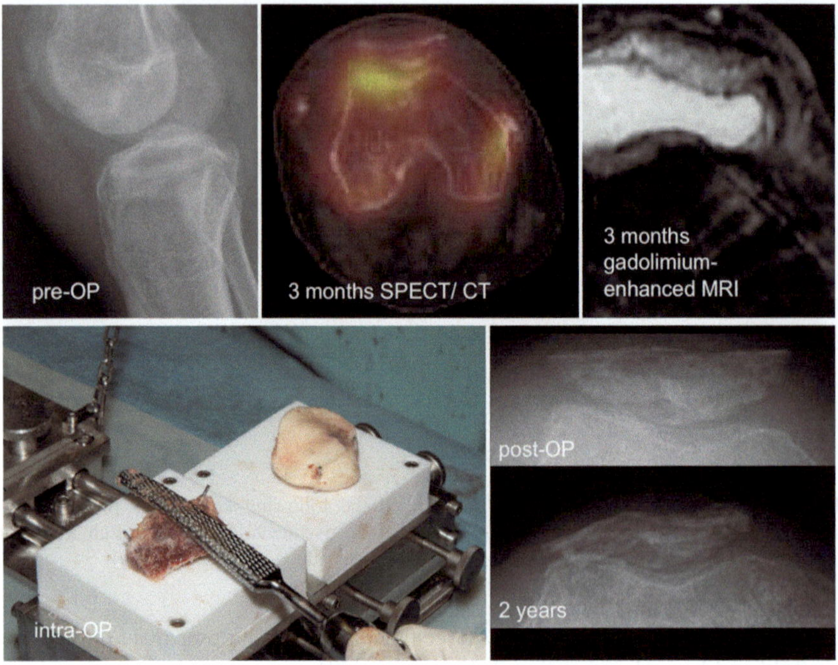

Fig. 3 Patellar defect after grade III open fracture: The compromised vascularity leads to insufficient viability of the transplanted customized iliac crest bone graft, seeded with autologous chondrocytes

4 Growth Factors

In a fracture site, the hematoma has been proven to induce the release of several signalling molecules (interleukin [IL]-1, IL-6, TNF-a, fibroblast growth factor [FGF], insulin-like growth factor [IGF], platelet-derived growth factor [PDGF], vascular endothelial growth factor [VEGF], and the transforming growth factor [TGF]-β superfamily members). Each of these growth factors triggers and interacts with other cells and signalling molecules. Endothelial cells, macrophages, platelets, monocytes, mesenchymal stem cells, chondrocytes, osteocytes, osteoblasts, and osteoclasts are involved in this complex process (Fig. 2).

The growth factors that have been identified to be beneficial for callus formation and enhance fracture healing are members of the TGF-β-superfamily. BMP-2 and BMP-7 have been approved for clinical application in nonunions of the tibia [32, 33]. Several other growth factors are currently under investigation in clinical trials, including parathyroid hormone (PTH) and antibodies to sclerostin and dickkopf1 [34].

Table 2 Comparison of bone marrow aspirate concentrates

System	BMAC™ (Harvest technologies)	MarrowStim® (Biomet)
Volume of aspirate (mL)	60	30/60
Centrifuge Time/Speed	14 min/3,200 RPM	15 min/3,200 RPM
Final volume (mL)	6	3/6
Number of MNC/mL	$18 \pm 7 \times 10^6$	$26.6 \pm 8.3 \times 10^3/\mu l$
Number of CFUs	33 ± 8	742 ± 66/ml
Progenitors in aspirate/mL	612 ± 134	151.2^
Progenitors in concentrate/mL	2579 ± 1121	894^
Concentration factor	4.2 ± 1.8	6.0 (Progenitors) 6.9 (MNC)*^
Costs per application (disposables)	≈800,- Euro	550,- Euro
Distributor	Harvest technologies corp., Plymouth, MA	Biomet biologics, LLC, Warsaw, Indiana
Reference	[41]	*

^-Progenitors were considered cells, which produced a colony-forming unit in either of the two assays performed (CFU-F and CFU-EPC)
*Specification according to the manufacturer

5 Vascular Supply

Lately, the high demand for appropriate vascularization has been stressed for bone reconstruction [9]. Clinically, a lack of vascularization can occur as a result of trauma, surgical exposure, or infection [35]. In cases of insufficient oxygen and nutrient supply, both transplanted and regional cells suffer from apoptosis and bone resorption occurs (Fig. 3). There are two ways to improve local blood supply for the preparation of an MSC transplantation:

1. A foreign membrane can be induced to provide both vascularization and growth factors [36, 37].
2. Regional or free flaps can be used to improve the local blood supply.

6 Available Cell Sources and Application Systems

6.1 Harvesting

For the harvesting of bone marrow cells and/or spongiosa, three different approaches can be used:

1. Bone marrow aspiration can be performed on various locations of the iliac crest by means of a trocar system, such as a Yamcidi needle.
2. Minimally invasive harvesting of spongiosa or/and cortical bone is facilitated by drilling or cutting systems (e.g., Diamond Bone Cutting System [DBCS], Storz AG, Tuttlingen, Germany).

3. Open bone marrow harvesting is the most invasive means, with several drawbacks and the highest complication rate [38].

6.2 Concentration

The concentration of bone marrow stromal cells by gradient centrifugation has been used since the beginning of tissue engineering [39]. Within the last decade, several companies have developed kits that facilitate onsite concentration of bone mesenchymal stromal cells (Table 2). There have been various reports of the benefits of using these cells for bone reconstruction, [40] pseudarthrosis treatment, [41, 42] cartilage, [43] and tendon repair [44]. However, there is an obvious lack of prospective randomized clinical studies with adequate control groups.

6.3 Stem Cell Banks

The most appealing option for the future is the use of a mesenchymal stem cell bank [45]. Such institutions already exist in most countries for the use of umbilical cord stem cells (UCSCs). Few reports exist about the successful use of these cells for regenerative medicine [46] and no published data for their clinical application for bone defects.

However, in vitro expanded cells seeded on hydroxyapatite carriers have been clinically applied in critically sized bone defects [47]. A 7-year follow-up for these cases has been reported [30]. In vitro cultivation of bone mesenchymal cells seeded on spongiosa disks was used to treat a 73-mm defect of the distal tibia [35]. The positron-emission tomography (PET)/computed tomography (CT) scan after 3 months showed that there was callus formation around the implants with highly enhanced bone and perfusion signals. Cell-seeded carriers may be an alternative for both segment transport and spongiosa transplantation, especially if the vascularity of the recipient site is enhanced [36, 37] and the soft tissues have been conditioned.

6.4 Differentiation In Vitro

If the differentiation of bone mesenchymal stromal cells is supposed to be enhanced in vitro (and it is questionable if this is necessary for successful bone repair), [48] this can be achieved by either growth factors such as dexamethasone [49] and bone morphogenic protein (BMP)-2 or BMP-7, [50] or by using mechanical stimulation, [51] ultrasound, [52] magnetic fields, [52] or laser stimulation [53]. However, each of these protocols has to be approved by a country's legislation for clinical use, so there is a high barrier for these approaches to be used in controlled clinical trials.

Bioreactors for this purpose have been suggested by our group [54] and other groups [55] and have been used in clinical case studies [35].

6.5 *Differentiation In Vivo*

Several growth factors from the TGF-β-superfamily have been or are currently under investigation to be used in the clinical setting. BMP-2 and BMP-7 have been approved for clinical application for nonhealing of the tibia [32, 33]. Further studies include systemic or local administration of PTH and antibodies to sclerostin and dickkopf1 [34].

7 Clinical Application of MSC for Bone Repair

7.1 *Benign Tumors*

Small defects such as osteoidosteomas can be successfully treated by CT guided drilling and can be left empty. Currently, there is no evidence that the results of treating larger volumes of cavities with a scaffold achieves better results than treating these defects with spongiosa. Likely, the center of the defects will remain unvascularized or will vascularize within a long time period. PET/CT studies have shown that after 3 months, only 10–12-mm of porous scaffold is vascularized [56].

7.2 *Long-Bone Defects*

For small circumferential long-bone defects in humans, shortening and segment transportation remain the criterion standard [57]. If cell-loaded scaffolds are used in the future, healing must be compared in a large-scale animal model. Even the latest animal models have not demonstrated a benefit of a scaffold-based healing approach over spongiosa transplantation [22].

For large circumferential long-bone defects, segment transportation remains the procedure of choice. The same restrictions apply for bone mesenchymal cells seeded on scaffolds. Treatment may be considered for defects with good adjacent vascularity and preconditioned defects after osteitis (Fig. 4a, b). However, level I evidence for clinical superiority is needed.

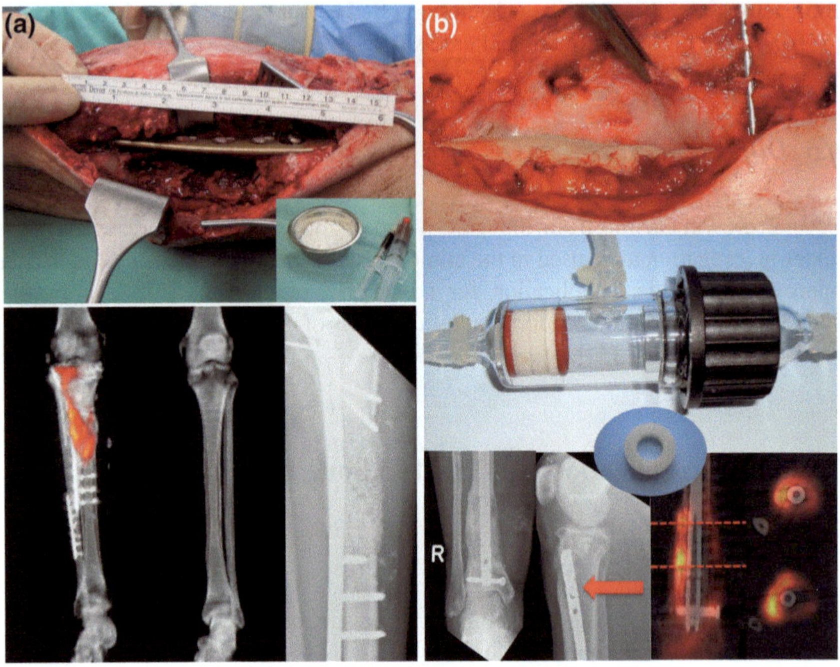

Fig. 4 **a** A 65-year-old man with a 70-mm circumferential defect of the proximal tibia: Treatment with medial and lateral gastrocnemius flaps and 4.5-mm locking plate (LCP, Synthes AG, Umkirch, Switzerland). **b** A 60-year-old woman with a 73-mm segmental defect after osteitis of the distal tibia following an AO 43C3.2 fracture. After insertion of a spacer for 6 weeks, a foreign body membrane was used to insert xenogeneic, cell-seeded disks. The construct was covered by a latissimus dorsi flap (Department of Plastic and Reconstructive Surgery, Prof. Vogt, Hannover Medical School (MHH)). The inserted nail was locked in a dynamic fashion (*arrow*) to enhance mechanical stimulation. After 12 weeks of partial weight bearing and solid callus formation, the patient rapidly progressed to full weight bearing [35]

8 Clinical Application for Cartilage Repair

8.1 Preclinical Trials

Progenitor cells such as bone marrow stromal cells (BMSC) have been differentiated into chondrocytes [58, 59]. Thus, stem cells—especially those that are derived from the mesenchyme—are a potentially attractive alternative to the use of chondrocytes, such as in autologous chondrocyte transplantation (ACT). They can be subsummized as mesenchymal stem cells (MSCs); however, cells out of the marrow compartment are more often specified in the literature. Next to the proliferation of adult differentiated cell lines of bone and cartilage, human bone marrow stromal cells are considered to be the premier source for tissue engineering. This subgroup of MSCs can be obtained easily from various locations and amplified in vitro [39]. Recent studies also involve mesenchymal stem cell transplantation from the umbilical cord.

This source of MSCs needs to be further studied and requires more attention in cartilage repair. Approximately 80 registered clinical trial sites are evaluating MSC therapy throughout the world, although there is still a long way to go before these cells are used as a routinely applied therapy in clinics [60].

These pluripotent stromal stem cells can be isolated and cultured ex vivo, and then their histogenic differentiation can be induced by external factors, such as TGF and dexamethasone, to maintain high-density aggregates towards a chondrogenic phenotype [61]. Restoration of the diseased articular cartilage is a challenge but has considerable appeal for researchers and clinicians. Techniques that cause multipotent adult MSCs to differentiate into cells of the chondrogenic lineage have led to a variety of experimental strategies to investigate whether MSCs instead of chondrocytes can be used for the regeneration and maintenance of articular cartilage [62].

Theoretically, scaffolds, membranes, gels, and many other biological and synthetic materials could serve as a matrix for the construct. The criterion standard for a matrix is autologous tissue, such as spongious bone from the iliac crest as in bone surgery. Next, periosteum could serve as a membrane; it was often used in the first-generation procedures, such as ACT. The design of joint surfaces, such as in cartilage repair, needs a specific strategy. The implant needs to mimic the contours of the articulating surface, show adequate mechanical properties, and have early functional load-bearing abilities [63].

Chondrogenic differentiation of MSCs can be performed in the micromass culture model or pellet culture, adding recombinant growth factors such as TGF-β, dexamethasone, β-glycerophosphate and ascorbic acid [64] to the media for different time periods and points. Additional serum of the donor species supports proper nutrition of cells in the culture. The transforming growth factor (TGF) family is the most potent inducer of chondrogenesis in MSC [65]. The bone morphogenetic proteins (BMP) present in bone are members of the TGF-ß superfamily, and some have chondrogenic properties [66]. Chemokines, including some types of IGF, FGF, and PDGF, are known to promote various physiological parameters, such as proliferation; they work synergistically together with the TGFs [67]. However, currently they have no relevance in clinically applied systems for cartilage repair.

Culture conditions could be optimized using bioreactors. The basic concept of a bioreactor is to provide an environment that is advantageous to the creation of a desired product. Nutrients, waste, temperature, and gas levels must be carefully controlled. If these conditions are kept at an optimal level, then the reactors can be run successfully for long periods of time. In addition to media supply, mechanical stimuli could be applied to enhance the production process [63].

9 Therapeutical Approaches

At present, there is no standard clinical application of in vitro generated tissue engineering techniques using the promising approach of progenitor cells in the field of a cartilage-like or osteochondral transplant to repair articular surfaces. In general, size-dependent repair is recommended using tissue response (e.g. microfracturing) for defects up to 2 cm^2. Osteochondral transplantation (OCT), a diamond bone cutting system (DBCS), or another mosaic plasty device (e.g., Osteo Articular Transplantation System (OATS)) can be used for defects larger than 2 cm^2 but are limited by the donor site at approximately 4 cm^2. Modern procedures such as ACT (following a two-step protocol with in vitro multiplication) or matrix-induced ACT are suitable for larger defects [63]. Because of regulatory problems, cell-free procedures using only a matrix have become more popular for larger defects, with evidence showing that they achieve similar clinical scores [68].

10 *Tissue Response and ACT*

Marrow-stimulation procedures, such as microfracturing for cartilage repair, rely on the formation of a primitive mesenchymal blood clot that forms fibrous tissue with a variable outcome. Limitations of osteochondral grafts include limited donor sites, morbidity, questionable cell viability, and fibrocartilage formation in between osteochondral plugs [69]. An important aspect of bone marrow–stimulating techniques is the real amount of MSCs that is derived from the microfractured areas of the subchondral bone. A wide variability between individual patients or surgeons in terms of the presence of MSCs has been reported. This variability radically affects the clinical outcome. In addition, as in most techniques, the cell mixture is often not well defined and shows a very high interindividual variance. Inconsistent clinical results have been reported when comparing ACT and microfracturing for the treatment of cartilage defects of the knee [70–72]. Microfracturing is a technically simple and cheap classical marrow stimulation technique that causes hardly any patient morbidity [73]. It results in good medium- to long-term outcomes in all compartments of the knee for defects up to about 4 cm^2 and is an accepted first-line procedure in younger patients for treating small, isolated defects, not only on the femoral condyles but also in the patellofemoral joint [74]. More invasive and expensive cell transplantation techniques should yield clinical outcomes that exceed those of microfracturing. Possible shortcomings of microfracturing include limited hyaline repair tissue and variable repair cartilage volume that may result in a possible functional deterioration over time. The Cartilage Regeneration System (CaReS®, Arthrokinetics) is another matrix-associated ACT technique for the treatment of chondral and osteochondral lesions. Using this technique, no expansion of chondrocytes in a monolayer culture is needed, and a homogeneous cell distribution within the gel is guaranteed. Recently, it was reported as being a safe and clinically effective treatment that yields significant functional improvement and improvement of pain level [75]. However, there are a variety of surgical techniques in use and many products available. Furthermore, study designs are often diverse and include a limited number of patients; therefore, evidence regarding the best surgical technique is inconclusive.

11 *Clinically Applicable Cell Sources*

Other potential clinical cell resources for tissue-engineered chondral grafts are cartilage autograft, cartilage allograft (banked), induced skin fibroblasts, induced fat cells, bone marrow stroma, and progenitors/precursors, e.g. umbilical cord cells (stem cells). Different biological preparations of cells have been proposed for use in cartilage defect treatment. Primary cells that reside within bone or cartilage and maintain and remodel the tissue are initially ideal candidates, but they are limited in number and expansion leads, often in cases of chondrogenic cells with de-differentiation or a loss of the cell phenotype. However, the potential use of stem cells derived from bone marrow relies on their well-known regenerative effects during the spontaneous repair of articular defects or when marrow-stimulating techniques are performed during surgery. The so-called super-clot that spontaneously forms during these procedures is proposed to have an enriched content of such cells [76]. In contrast, some authors have reported microfracturing that yields to repair tissue based on the super-clot, which contains a mixture of cells with a relatively small portion of MSCs [77]. However, experimental data indicate that application of an enriched cell population might be beneficial for chondrogenesis. The number of cells available can be increased in vitro by 10^6 times; moreover, they can be frozen and thawed without losing their ability to proliferate and differentiate into bone, [78] cartilage, [79] and fibrous tissue [55]. The process is enhanced by individual cell cultivation protocols, of which a variety of methods were published during the last two decades. Thus, they are applicable for multistep procedures in cartilage repair comparable to ACT.

12 Applying MSC in Cartilage Repair

Theoretically, two types of clinical application could be considered: MSCs could be applied in conjunction or below a supportive matrix or by intraarticular injection. However, their relative contribution under conditions in an articular environment with regard to other cells from the synovial membrane or the underlying bone marrow has not yet been evaluated.

Experimental data generated so far have shown that genetically modified chondrocytes and MSCs allow for sustained transgene expression when transplanted into articular cartilage defects [80]. Overexpression of selected factors should enhance the structural features of the repair tissue. Significant benefits have been also observed in preclinical animal models. There is a phase I clinical gene therapy study in which transduced fibroblasts were injected into the metatarsophalangeal joints of patients without adverse events [81]. The key issues are the development of a safe and highly efficient gene delivery system with sustained duration of transgene expression, identification of optimal therapeutic genes, and a combination of genetically modified articular chondrocytes and/or MSCs with scaffolds that better support chondrogenesis in vivo. The techniques of applying gene therapy/modification may provide an important step toward the unanswered question of cartilage regeneration.

The autologous matrix-induced chondrogenesis technique (AMIC) combines the microfracturing technique with the use of a type I/III porcine collagen matrix (Chondro-Gide®, Geistlich Pharma), which is able to stabilize and protect the blood clot coming from the subchondral bone. Although recently introduced, satisfactory results have been already reported. A pilot study treated patients with the AMIC technique, enhanced by autologous concentrated bone marrow [77]. A small fraction of bone marrow samples, both from the iliac crest and from the created microfractures, were analyzed by FACS and then cultured to verify their proliferative and differentiation potentials. An average of only 0.04 % of concentrated bone marrow cells harvested from the iliac crest presented with a mesenchymal stem cell phenotype, whereas just 0.02 % of these cells were identified from the samples harvested during the creation of microfracturing of the knee. After two passages in culture, cells expressed a peculiar profile for MSCs. Only MSCs from bone marrow could be propagated in the long-term and were able to efficiently differentiate in the cultures. Although the AMIC approach has many advantages, the surgical technique in the application of microfracturing remains essential and affects the final result. Interestingly, such study results are somehow damasking the magical effects of such therapies, focusing on the small portion of stem cells within the clot.

Another matrix-associated transplantation of MSCs beside many other cell lines is platelet-rich plasma combined with scaffolds such as fibrin glue. These can be used clinically as a scaffold to deliver autologous culture-expanded MSCs for cartilage repair. In one report of clinical results after implantation, [60] all patients' symptoms improved during the follow-up period. Arthroscopic scores were nearly normal. MRI revealed complete defect fill and complete surface congruity with native cartilage, whereas one-fifth of patients showed incomplete congruity. Autologous MSC transplantation on a scaffold may be an effective approach to promote the repair of articular cartilage defects of the knee in humans. However, this was only a small case series without any control group. Several other studies applied Platelet Rich Plasma (PRP) in case series with good results, although they focused on cytokine storms in this enriched environment rather than MSC applications [82].

Clinical studies using MSCs only in cartilage repair currently exist only as case reports or technical notes [43] in the literature. Most often, bone marrow aspirates from the iliac crest of patients were taken as cell sources in these studies. High cell density was achieved by different protocols prior to re-implantation, and it seems as if this method was translated from in vitro tests as described previously. For example, Wakitani et al. published two cases using autologous bone marrow aspirates [83]. At 1 and 2 years after transplantation into patellar cartilage defects, arthroscopy demonstrated fibrocartilaginous defect repair. In another publication, the same group reported improved arthroscopic and histologic findings without clinical improvement in patients after leg axis correction combined with transplantation of MSCs bedded in a collagen gel into defects of the femoral condyle [84].

13 *Osteochondral Defects*

For an osteochondral lesion, repair should followed a stepwise protocol [63]. Reconstruction or stimulation of biological regeneration of the joint′s surface formed by cartilage is the superior task. Most efforts have focused on correcting the two main shortcomings of the articular cartilage: bringing in new cells capable of chondrogenesis and facilitating access to the vascular system. Many methods have been used, with varying success in animal studies and clinical trials. These include shaving or debridement of damaged articular cartilage; perforation of the subchondral plate by multiple drill holes or abrasion subsumed under tissue response; and transplantation of osteochondral grafts, perichondrium, periosteum, chondrocytes, and (mesenchymal) stem cells. Synthetic gels and implants, such as carbon fibber pads, biodegradable matrices, and collagen gels, have been used by themselves or as carriers for chondrocytes or growth-stimulating factors. As with most orthopaedic research, these approaches have been studied with diverse methods and frequently disparate grading systems. To demonstrate the superiority of a certain technique, it must be proven to achieve better long-term clinical results than natural history and the current methods of treatment. Obtaining long-term results for these procedures may require as much as 20–30 years of follow-up. An alternative standard may be to examine how closely the regenerative tissue resembles articular cartilage—structurally, biochemically, and biomechanically. The more closely the native structure of articular cartilage is replicated, the more likely it is that the tissue will result in normal, symptom-free joint function and will not deteriorate prematurely over time. A critical examination of each of these approaches will enable us to better evaluate the promise of chondrocyte cell transfer, as well as future modalities.

14 *Future Perspective*

The progressively aging population has an increasing demand for therapies to regenerate or replace musculoskeletal tissues. Designing cartilage and osteochondral tissue needs a successful strategy. The natural contours of the articulating surface, the mechanical properties, and functional load-bearing ability have to be achieved as optimally as possible. Last but not least, integration to the host tissue must be provided. There have been numerous attempts to repair articular injuries as described in this chapter. However, results have demonstrated that articular cartilage has limited potential to repair itself. Biology, gene therapy, and tissue engineering may provide a breakthrough to treat those injuries [63]. A combination of cells, scaffolds, and bioactive factors is a promising approach. Pluripotent stromal stem cells from bone marrow could be isolated and cultured ex vivo; then their histogenic differentiation could be induced by external factors [63]. The proof of histotypical function after implantation in vivo is essential.

Composite-engineered tissues, such as an engineered joint, represent a future goal. Tissue engineering is clinically established for small volumes of tissue that are of limited three-dimensional complexity. The integration of mechanical stimulation in the tissue-engineering process may lead to progress in the structural and biomechanical properties of osteochondral tissues and offer new possibilities for managing joint injuries and degenerative diseases. Cell selection, density, scaffold design, and biological stimulation remain the key challenges of functional tissue engineering. Advances in materials design may generate smart scaffolds that will control tissue topology and have surface modifications to stimulate cell attachment, differentiation, and growth. Last but not least, ethical problems have to be considered and prior industrial product certification (e.g., good manufacturing practices) has to be obtained. Country-dependent health economic capacities and regulations have to be taken into account.

References

1. Choi YA, Kim DK, Bang OS, Kang SS, Jin EJ (2010) Secretory phospholipase A2 promotes MMP-9-mediated cell death by degrading type I collagen via the ERK pathway at an early stage of chondrogenesis. Biol Cell/under Auspices Europ Cell Biol Organ 102:107–19
2. Widuchowski W, Widuchowski J, Faltus R et al (2011) Long-term clinical and radiological assessment of untreated severe cartilage damage in the knee: a natural history study. Scand J Med Sci Sports 21:106–10
3. Cain EL, Clancy WG (2001) Treatment algorithm for osteochondral injuries of the knee. Clin Sports Med 20:321–42
4. Salzmann GM, Niemeyer P, Steinwachs M, Kreuz PC, Sudkamp NP, Mayr HO (2011) Cartilage repair approach and treatment characteristics across the knee joint: a European survey. Arch Orthop Trauma Surg 131:283–91
5. Hollister SJ (2009) Scaffold engineering: a bridge to where? Biofabrication 1:012001
6. Jager M, Herten M, Fochtmann U et al (2011) Bridging the gap: bone marrow aspiration concentrate reduces autologous bone grafting in osseous defects. J Orthop Res 29:173–80
7. Giannoudis PV, Einhorn TA, Schmidmaier G, Marsh D (2008) The diamond concept–open questions. Injury 39(Suppl 2):S5–8
8. Giannoudis PV, Einhorn TA, Marsh D (2007) Fracture healing: the diamond concept. Injury 38(Suppl 4):S3–6
9. Schmidmaier G, Schwabe P, Strobel C, Wildemann B (2008) Carrier systems and application of growth factors in orthopaedics. Injury 39(Suppl 2):S37–43
10. Brighton CT, Hunt RM (1991) Early histological and ultrastructural changes in medullary fracture callus. J Bone Joint Surg Am Vol 73:832–47
11. Koob S, Torio-Padron N, Stark GB, Hannig C, Stankovic Z, Finkenzeller G (2011) Bone formation and neovascularization mediated by mesenchymal stem cells and endothelial cells in critical-sized calvarial defects. Tissue Eng Part A 17:311–21
12. Kumar S, Wan C, Ramaswamy G, Clemens TL, Ponnazhagan S (2010) Mesenchymal stem cells expressing osteogenic and angiogenic factors synergistically enhance bone formation in a mouse model of segmental bone defect. Molec Ther J Am Soc Gene Ther 18:1026–34
13. Gil-Albarova J, Salinas AJ, Bueno-Lozano AL et al (2005) The in vivo behaviour of a sol-gel glass and a glass-ceramic during critical diaphyseal bone defects healing. Biomaterials 26:4374–82
14. Vaccaro AR (2002) The role of the osteoconductive scaffold in synthetic bone graft. Orthopedics 25:s571–8

15. Schwartz Z, Hyzy SL, Moore MA et al (2011) Osteoinductivity of demineralized bone matrix is independent of donor bisphosphonate use. J Bone Joint Surg Am Vol 93:2278–86
16. Pietrzak WS, Dow M, Gomez J, Soulvie M, Tsiagalis G (2011) The in vitro elution of BMP-7 from demineralized bone matrix. Cell and tissue banking 2011
17. Gorschewsky O, Browa A, Vogel U, Stauffer E (2002) Klinisch-histologischer Vergleich des allogenen und autologen Patellarsehnendrittels (Bone-Tendon-Bone) zur Rekonstruktion des vorderen Kreuzbands. Unfallchirurg 105:703–14
18. Tadic D, Epple M (2004) A thorough physicochemical characterisation of 14 calcium phosphate-based bone substitution materials in comparison to natural bone. Biomaterials 25:987–94
19. Sutton JM, Dickinson J, Walker JT, Raven ND (2006) Methods to minimize the risks of Creutzfeldt-Jakob disease transmission by surgical procedures: where to set the standard? Clin Infect Dis Official Publ Infect Dis Soc America 43:757–64
20. Dehoux E, Madi K, Fourati E, Mensa C, Segal P (2005) Valgisation tibiale par ouverture mediale utilisant un coin de ceramique de phosphate tricalcique: a propos de 70 cas revus avec un recul moyen de 18 mois. Revue de chirurgie orthopedique et reparatrice de l'appareil moteur 91:143–8
21. Bucholz RW (2002) Nonallograft osteoconductive bone graft substitutes. Clin Orthop Relat Res 395:44–52
22. Giannoni P, Mastrogiacomo M, Alini M et al (2008) Regeneration of large bone defects in sheep using bone marrow stromal cells. J Tissue Eng Regen Med 2:253–62
23. Cancedda R, Giannoni P, Mastrogiacomo M (2007) A tissue engineering approach to bone repair in large animal models and in clinical practice. Biomaterials 28:4240–50
24. Hollister SJ, Murphy WL (2011) Scaffold translation: barriers between concept and clinic. Tissue Eng Part B Rev 17:459–74
25. Augat P, Merk J, Ignatius A, et al. (1996) Early, full weightbearing with flexible fixation delays fracture healing. Clin Orthop Relat Res 328:194–202
26. Claes L, Reusch M, Gockelmann M et al (2011) Metaphyseal fracture healing follows similar biomechanical rules as diaphyseal healing. J Orthop Res 29:425–32
27. Stavropoulos A, Kostopoulos L, Mardas N, Karring T (2003) Influence of demineralized bone matrix's embryonic origin on bone formation: an experimental study in rats. Clin Implant Dent Relat Res 5:184–92
28. Hente R, Fuchtmeier B, Schlegel U, Ernstberger A, Perren SM (2004) The influence of cyclic compression and distraction on the healing of experimental tibial fractures. J Orthop Res 22:709–15
29. Quarto R, Mastrogiacomo M, Cancedda R et al (2001) Repair of large bone defects with the use of autologous bone marrow stromal cells. NEnglJMed 344:385–6
30. Marcacci M, Kon E, Moukhachev V et al (2007) Stem cells associated with macroporous bioceramics for long bone repair: 6- to 7-year outcome of a pilot clinical study. Tissue Eng 13:947–55
31. Jagodzinski M, Krettek C (2007) Effect of mechanical stability on fracture healing–an update. Injury 38(Suppl 1):S3–10
32. Friedlaender GE (2001) OP-1 clinical studies. J Bone Joint Surg Am 83(A Suppl 1):S160–161
33. Friedlaender GE, Perry CR, Cole JD, et al. (2001) Osteogenic protein-1 (bone morphogenetic protein-7) in the treatment of tibial nonunions. J Bone Joint Surg Am 83(A Suppl 1):S151–158
34. Baron R, Hesse E (2012) Update on bone anabolics in osteoporosis treatment: rationale, current status, and perspectives. The journal of clinical endocrinology and metabolism 2012
35. Hesse E, Kluge G, Atfl A et al (2010) Repair of a segmental long bone defect in human by implantation of a novel multiple disc graft. Bone 46:1457–63
36. Masquelet AC, Begue T (2010) The concept of induced membrane for reconstruction of long bone defects. The Orthop Clinic North America 41:27–37. Table of contents
37. Klaue K, Knothe U, Anton C et al (2009) Bone regeneration in long-bone defects: tissue compartmentalisation? In vivo study on bone defects in sheep. Injury 40(Suppl 4):S95–102

38. Jager M, Westhoff B, Wild A, Krauspe R (2005) Knochenspanentnahme am Becken. Techniken und Probleme. Der Orthopade 34:976–982, 84, 86–90, 92–94
39. Caplan AI, Bruder SP (2001) Mesenchymal stem cells: building blocks for molecular medicine in the 21st century. Trends MolMed 7:259–64
40. Jager M, Hernigou P, Zilkens C et al (2010) Cell therapy in bone healing disorders. Orthop Rev 2:e20
41. Hernigou P, Poignard A, Beaujean F, Rouard H (2005) Percutaneous autologous bone-marrow grafting for nonunions. Influence of the number and concentration of progenitor cells. J Bone Joint Surg Am 87:1430–1437
42. Hernigou P, Poignard A, Manicom O, Mathieu G, Rouard H (2005) The use of percutaneous autologous bone marrow transplantation in nonunion and avascular necrosis of bone. J Bone Joint Surg Br 87:896–902
43. Gobbi A (2012) One-step full-thickness knee chondral lesions repair with BMAC and collagen matrix. In: Annual meeting of the AAOS, San Francisco
44. Lacitignola L, Crovace A, Rossi G, Francioso E (2008) Cell therapy for tendinitis, experimental and clinical report. Vet Res Commun 32(Suppl 1):S33–8
45. Cooper K, Viswanathan C (2011) Establishment of a mesenchymal stem cell bank. Stem Cells Int 2011:905621
46. Colnot C (2011) Cell sources for bone tissue engineering: insights from basic science. Tissue Eng Part B Rev 17:449–57
47. Kon E, Muraglia A, Corsi A et al (2000) Autologous bone marrow stromal cells loaded onto porous hydroxyapatite ceramic accelerate bone repair in critical-size defects of sheep long bones. J Biomed Mater Res 49:328–37
48. Rucker M, Laschke MW, Junker D et al (2008) Vascularization and biocompatibility of scaffolds consisting of different calcium phosphate compounds. J Biomed Mater Res A 86:1002–11
49. Holtorf HL, Jansen JA, Mikos AG (2005) Ectopic bone formation in rat marrow stromal cell/ titanium fiber mesh scaffold constructs: effect of initial cell phenotype. Biomaterials 26:6208–16
50. Kandziora F, Schmidmaier G, Schollmeier G, et al. (2002) IGF-I and TGF-beta1 application by a poly-(D,L-lactide)-coated cage promotes intervertebral bone matrix formation in the sheep cervical spine. Spine (Phila Pa 1976) 27:1710–1723
51. Jagodzinski M, Breitbart A, Haasper C, Hankemeier S, Hurschler C, Zeichen J (2006) Effects of cyclic compression and perfusion on bone marrow stromal cells in a three dimensional matrix-bioreactor system. In: 2006 2006/10/03/; 70th annual meeting of the german trauma society, Berlin
52. Sena K, Leven RM, Mazhar K, Sumner DR, Virdi AS (2005) Early gene response to low-intensity pulsed ultrasound in rat osteoblastic cells. Ultrasound Med Biol 31:703–8
53. Kim IS, Cho TH, Kim K, Weber FE, Hwang SJ (2010) High power-pulsed Nd:YAG laser as a new stimulus to induce BMP-2 expression in MC3T3-E1 osteoblasts. Lasers Surg Med 42:510–8
54. Jagodzinski M, Breitbart A, Wehmeier M et al (2008) Influence of perfusion and cyclic compression on proliferation and differentiation of bone marrow stromal cells in 3-dimensional culture. J Biomech 41:1885–91
55. Altman GH, Horan RL, Martin I et al (2002) Cell differentiation by mechanical stress. FASEB J 16:270–2
56. Jagodzinski M, Berding G, Baillot G, Liodakis E, Knobloch K, Krettek C (2012) Therapie von Segmentdefekten der Röhrenknochen mit Stammzellkonzentraten: Eine klinische und nuklearmedizinische Analyse. In: Deutscher Kongress für Orthopädie und Unfallchirurgie. Berlin
57. Schieker M, Heiss C, Mutschler W (2008) Knochenersatzmaterialien. Unfallchirurg 111:613–619. quiz 20
58. Caplan AI (1991) Mesenchymal Stem Cells. J Orthop Res 9:641–50
59. Yoo JU, Barthel TS, Nishimura K et al (1998) The chondrogenic potential of human bone-marrow-derived mesenchymal progenitor cells. JBone Joint SurgAm 80:1745–57

60. Malgieri A, Kantzari E, Patrizi MP, Gambardella S (2010) Bone marrow and umbilical cord blood human mesenchymal stem cells: state of the art. Int J Clin Exp Med, In, pp 248–69
61. Johnstone B, Hering TM, Caplan AI, Goldberg VM, Yoo JU (1998) In vitro chondrogenesis of bone marrow-derived mesenchymal progenitor cells. ExpCell Res 238:265–72
62. Nöth U, Steinert AF, Tuan RS (2008) Technology insight: adult mesenchymal stem cells for osteoarthritis therapy. Nat Clin Pract Rheumatol, In, pp 371–80
63. Haasper C, Zeichen J, Meister R, Krettek C, Jagodzinski M (2008) Tissue engineering of osteochondral constructs in vitro using bioreactors. Injury 39(Suppl 1):S66–76
64. Pittenger MF, Mackay AM, Beck SC et al (1999) Multilineage potential of adult human mesenchymal stem cells. Science 284:143–7
65. Indrawattana N, Chen G, Tadokoro M et al (2004) Growth factor combination for chondrogenic induction from human mesenchymal stem cell. Biochem Biophys Res Commun 320:914–9
66. Park Y, Sugimoto M, Watrin A, Chiquet M, Hunziker EB (2005) BMP-2 induces the expression of chondrocyte-specific genes in bovine synovium-derived progenitor cells cultured in three-dimensional alginate hydrogel. OsteoarthritisCartilage 13:527–36
67. Stevens MM, Marini RP, Martin I, Langer R, Prasad SV (2004) FGF-2 enhances TGF-beta1-induced periosteal chondrogenesis. J Orthop Res 22:1114–9
68. Anders S (2011) AMIC versis ACT: Gibt es einen Unterschied im funkitonellen Ergebnis? In: Lunchsymposium im Rahmen des 28 Kongresses der Deutschsprachigen Arbeitsgemeinschaft für Arthroskopie (AGA), Thieme, Regensburg
69. Redman SN, Oldfield SF, Archer CW (2005) Current strategies for articular cartilage repair. Eur Cell Mater 9:23–32
70. Knutsen G, Drogset JO, Engebretsen L et al (2007) A randomized trial comparing autologous chondrocyte implantation with microfracture. Findings at five years. J Bone Joint Surg Am 89:2105–12
71. Knutsen G, Engebretsen L, Ludvigsen TC et al. (2004) Autologous chondrocyte implantation compared with microfracture in the knee. A randomized trial. J Bone Joint Surg Am 86A:455–464
72. Kon E, Gobbi A, Filardo G, Delcogliano M, Zaffagnini S, Marcacci M (2009) Arthroscopic second-generation autologous chondrocyte implantation compared with microfracture for chondral lesions of the knee: prospective nonrandomized study at 5 years. Am J Sports Med 37:33–41
73. Mithoefer K, McAdams T, Williams RJ, Kreuz PC, Mandelbaum BR (2009) Clinical efficacy of the microfracture technique for articular cartilage repair in the knee: an evidence-based systematic analysis. Am J Sports Med 37:2053–63
74. Pascual-Garrido C, Slabaugh MA, L'Heureux DR, Friel NA, Cole BJ (2009) Recommendations and treatment outcomes for patellofemoral articular cartilage defects with autologous chondrocyte implantation: prospective evaluation at average 4-year follow-up. Am J Sports Med 37(Suppl 1):33S–41S
75. Schneider U, Rackwitz L, Andereya S et al (2011) A prospective multicenter study on the outcome of type I collagen hydrogel-based autologous chondrocyte implantation (CaReS) for the repair of articular cartilage defects in the knee. Am J Sports Med 39:2558–65
76. Steadman JR, Briggs KK, Rodrigo JJ, Kocher MS, Gill TJ, Rodkey WG (2003) Outcomes of microfracture for traumatic chondral defects of the knee: average 11-year follow-up. Arthroscopy 19:477–84
77. de Girolamo L, Bertolini G, Cervellin M, Sozzi G, Volpi P (2010) Treatment of chondral defects of the knee with one step matrix-assisted technique enhanced by autologous concentrated bone marrow: in vitro characterisation of mesenchymal stem cells from iliac crest and subchondral bone. In: Injury, pp 1172–1177
78. Lennon DP, Edmison JM, Caplan AI (2001) Cultivation of rat marrow-derived mesenchymal stem cells in reduced oxygen tension: effects on in vitro and in vivo osteochondrogenesis. J Cell Physiol 187:345–55

79. Angele P, Yoo JU, Smith C et al (2003) Cyclic hydrostatic pressure enhances the chondrogenic phenotype of human mesenchymal progenitor cells differentiated in vitro. J Orthop Res 21:451–7
80. Madry H, Cucchiarini M (2011) Clinical potential and challenges of using genetically modified cells for articular cartilage repair. Croat Med J 52(3):245–261
81. Evans CH, Robbins PD, Ghivizzani SC, Wasko MC, Tomaino MM, Kang R, et al. (2005) Gene transfer to human joints: progress toward a gene therapy of arthritis. G Proc Natl Acad Sci U S A 102:8698–8703
82. Kon E, Filardo G, Delcogliano M et al (2009) Platelet-rich plasma: new clinical application: a pilot study for treatment of jumper's knee. Injury 40:598–603
83. Wakitani S, Mitsuoka T, Nakamura N, Toritsuka Y, Nakamura Y, Horibe S (2004) Autologous bone marrow stromal cell transplantation for repair of full-thickness articular cartilage defects in human patellae: two case reports. Cell transplantation 13:595–600
84. Wakitani S, Imoto K, Yamamoto T, Saito M, Murata N, Yoneda M (2002) Human autologous culture expanded bone marrow mesenchymal cell transplantation for repair of cartilage defects in osteoarthritic knees. Osteoarthr Cartil/OARS Osteoarthr Res Soc 10:199–206
85. Schnettler R, Dingeldein E, Herr G (1998) Defektaufbau mit demineralisierter Knochenmatrix. Experimentelle Untersuchungen beim Minischwein. Der Orthopade 27:80–8
86. Acharya NK, Kumar RJ, Varma HK, Menon VK (2008) Hydroxyapatite-bioactive glass ceramic composite as stand-alone graft substitute for posterolateral fusion of lumbar spine: a prospective, matched, and controlled study. J Spinal Disord Tech 21:106–11
87. Kasten P, Luginbuhl R, Vogel J et al (2004) Induktion von Knochengewebe auf unterschiedlichen Matrizes: Eine In vitro- und In vivo-Pilotstudie in der SCID Maus. Zeitschrift fur Orthopadie und ihre Grenzgebiete 142:467–75
88. Giannoudis PV, Einhorn TA, Marsh D (2007) Fracture healing: a harmony of optimal biology and optimal fixation? Injury 38(Suppl 4):S1–2

Adv Biochem Eng Biotechnol (2013) 130: 89–103
DOI: 10.1007/10_2012_151

Published Online: 1 September 2012

Adult Mesenchymal Stem Cells Explored in the Dental Field

K. M. Fawzy El-Sayed, C. Dörfer, F. Fändrich, F. Gieseler, M. H. Moustafa and H. Ungefroren

Abstract During the last decade it was realized that stem cell-based therapies hold an enormous therapeutic potential, improving the life of patients with conditions ranging from neurodegenerative and traumatic diseases to regenerative medicine requiring replacement of complex structures such as bones and teeth. Based on their ability to regenerate and/or repair damaged tissue and eventually restore organ function, multiple types of stem/progenitor cells have been discovered. In the field of periodontal regeneration and tooth engineering, several types of adult multipotent mesenchymal stem cells from various sources are currently being investigated. These include the bone marrow stromal stem cells (BMSSCs), adipose-derived stromal cells (ADSCs), dental pulp stem cells (DPSCs), dental follicle stem cells (DFSCs), stem cells from human exfoliated deciduous teeth (SHEDs), stem cells from the apical papilla (SCAP), periodontal ligament stem cells (PDLSCs), alveolar bone proper-derived stem cells, and gingival stem cells.

An erratum to this chapter is available at 10.1007/10_2012_177.

K. M. Fawzy El-Sayed · C. Dörfer
Clinic for Conservative Dentistry and Periodontology, School of Dental Medicine, UKSH, Campus Kiel, Arnold-Heller Strasse 3 Hs. 26, 24105 Kiel, Germany
e-mail: christof.doerfer@uk-sh.de

F. Fändrich · H. Ungefroren (✉)
Clinic for Applied Cellular Medicine, UKSH, Campus Kiel, Arnold-Heller Strasse 3, Hs. 18, 24105 Kiel, Germany
e-mail: hendrik.ungefroren@uksh.de

F. Fändrich
e-mail: fred.faendrich@uksh-kiel.de

F. Gieseler · H. Ungefroren
First Department of Medicine, UKSH, Campus Lübeck, Ratzeburger Allee 160, 23538 Lübeck, Germany
e-mail: frank.gieseler@uksh.de

K. M. Fawzy El-Sayed · M. H. Moustafa
Department of Oral Medicine and Periodontology, Faculty of Oral and Dental Medicine, Cairo University, 1 Mathaf el Manial Street, Giza Egypt
e-mail: karim.fawzy@gmail.com

M. H. Moustafa
e-mail: mhosnymoustafa@gmail.com

The potential of these different MSCs as precursors for regenerative purposes in the dental field is discussed in this chapter.

Keywords Adipose-derived stromal cell · Alveolar bone proper-derived stem cell · Bone marrow stromal stem cell · Dental follicle stem cell · Dental pulp stem cell · Gingival stem cell · Mesenchymal stem cell · Periodontal ligament stem cell · Periodontium · Stem cells from human exfoliated deciduous tooth · Stem cells from the apical papilla

Contents

1 Introduction

Over the last decade, various medical disciplines have begun to explore the possible applications of stem cells and tissue engineering in the fields of repair and regeneration of damaged/injured tissues of the human body. The defined long-term goal is to make regenerative medicine take its place in clinical practice as an important future therapeutic modality.

Stem cells are capable of self-renewal through mitosis and they can give rise to cells that have the potential to differentiate into specialized cell types. Embryonic stem cells (ESCs) are pluripotent and can differentiate into almost every cell type of the human body. However, due to ethical and legal issues the use of ESCs is controversial, thus restricting their application for regenerative purposes in the clinic. Unlike ESCs, adult stem cells have the potential to be used for the treatment of various diseases. They have several advantages over ESCs: (i) their use is less problematic because they can be retrieved without destroying an embryo (ii) they reside in almost all tissues of the human body including dental tissues, and

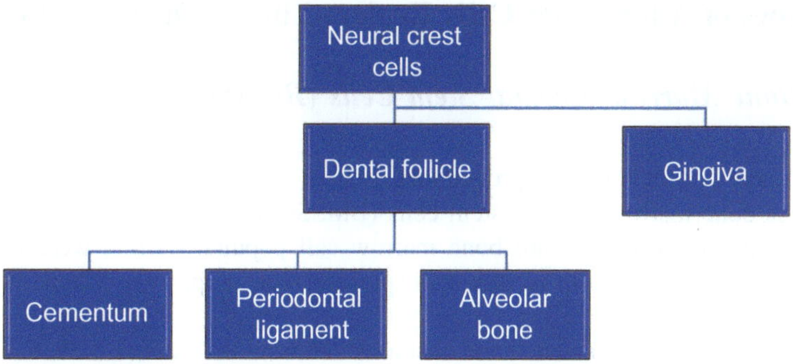

Fig. 1 Developmental origin of the dental tissues

(iii) their use in an autologous setting circumvents any problems with rejection by the host immune system.

Adult stem cells, also known as mesenchymal stromal cells, mesenchymal stem cells, or multipotent stromal cells (MSCs) are a heterogeneous subset of pluripotent stromal cells that can be isolated from many different adult tissues and demonstrate the potential to give rise to cells of various lineages [13]. These cell populations do not develop to sizable proportions under normal culture conditions but their isolation and expansion requires enriched specific culture media under special inductive culture conditions.

Morphologically, MSCs may be either large and flat or elongated and fibroblastlike. This is not a defining or distinguishing feature of these cells. Their identification is based on the positive expression of specific surface markers (CD44, CD73, CD90, CD105, CD106, STRO-1) and the absence of expression of hematopoietic cell surface markers (CD34, CD45, CD11a, CD19) and HLA-DR, as well as on their ability of self-renewal and multipotency. Human MSCs display plastic adherence under standard culture conditions and can form colonies (i.e., they are clonogenic). Their multipotent nature is evident from the ability to differentiate along various lineages including those for osteoblasts, adipocytes, myelosupportive stroma, chondrocytes, and neuronal cells in response to specific stimuli [13].

The neural crest cells, a transient, migratory, multipotent cell population in vertebrates, participated in the embryonic development of most dental tissues including the gingiva, the dental follicle, the periodontal ligament, and the alveolar bone (Fig. 1). Several adult cell populations with stem cell properties have recently been isolated and partially characterized from these tissues. The intention of this review is to give an overview of the stem cell types investigated in the dental field including their tissue sources, properties, differentiation potential, and comparative assessment of their advantages for tissue engineering.

2 Types of Adult Stem Cells Explored in the Dental Field

2.1 Bone Marrow Stromal Stem Cells (BMSSCs)

In addition to hematopoietic progenitors or stem cells (HSCs), the bone marrow contains bone marrow stromal stem cells (BMSSCs) that give rise to nonhematopoietic tissues. BMSSCs are bone marrow cell populations that were the first mesenchymal stem cells to be isolated exploiting their property to adhere to tissue culture plastics [18].

BMSSCs have been isolated and characterized from the extra [18, 76, 6, 42, 69] as well as the intra- oral [1, 27, 51] bone marrow. They are cable of forming colony-forming unit-fibroblasts (CFU-Fs) in vitro [18] and express Oct-4, Nanog, STRO-1, CD73, CD90, CD105, CD146 and are negative for CD14, CD34, CD45 and HLA-DR [19, 13, 74, 4, 24]. They are capable of differentiation into multiple mesenchymal lineages including osteoblasts, adipocytes, chondrocytes, muscle cells, tenocytes, or nerve cells [6, 69, 49, 39, 10, 42, 80, 18].

BMSSCs cultures usually encompass a mixture of fibroblasts, osteoblasts, adipocyte progenitors and reported range of up to 4-19% stem cells [66]. The majority of attempts to engineer teeth initially employed purified BMSSCs cell populations [65]. Indeed, bone as well as soft tissues could be formed from heterogeneous populations. Ohazama and colleagues [61], were able to generate tooth-like structures after transferring whole transplants from bone marrow-derived cells into the renal capsule. Moreover, they amalgamated embryonic oral epithelium with three types of stem cells, namely neural stem cells, ESCs, and adult bone marrow-derived cells. They transferred the combination into adult jaw and renal capsules and observed formation of tooth-like structures and bone. A study conducted by Li and coworkers [45] yielded similar results, demonstrating that the combination of oral epithelial cells from rat embryos with BMSSCs resulted in the expression of a variety odontogenic genes such as Pax-9, dentine sialophosphoprotein (DSPP), and dentine matrix protein 1 (DMP1) and histologically produced tooth-like structures.

In the field of periodontal regeneration BMSSCs have shown great promise. The auto-transplantation of BMSSCs in an animal study resulted in almost complete regeneration of periodontal defects in only four weeks. Histologically, the presence of cementum, periodontal ligament (PDL), and alveolar bone was confirmed [35]. Therefore, BMSSCs represent a competitive MSC source for the regenerative treatment of periodontal diseases, despite showing a major limitation in their application, having a strongly age-dependent differentiation capability which considerably decreases with increasing donor age [33].

2.2 Adipose-derived Stromal Cells (ADSCs)

Adipose tissues represent a readily available source of multipotent post-natal stem cells first described in 2001 [91]. Adipose-derived stromal cells (ADSCs) are characterized by stable proliferation doubling kinetics in vitro [65]. The good accessibility and tissue abundance is clearly an advantage of ADSCs. ADSCs can be obtained via minimally invasive methods, including the increasingly popular cosmetic liposuction procedure, and in larger quantities than BMSSCs, making their utilization as a stem cell source very attractive [91].

In accordance with the criteria for multipotent stromal cells defined by Dominici et al. [13], ADSCs exhibited a multilineage differentiation potential into osteogenic, chondrogenic, and adipogenic directions in vitro [46] and were able to form osteoid matrix [28] and bone [34] in vivo. ADSCs further strongly expressed multiple important bone marker proteins including alkaline phosphatase (ALP), type I collagen, osteopontin, and osteocalcin [83].

In 2008 Jing and co-workers found that ADSCs could be differentiated into the odontogenic lineage and might represent a promising alternative for seeding cells for tooth regeneration to replace lost teeth in elderly patients [33].

In the field of tooth tissue engineering, a recent study further demonstrated that incubating primary cultures of human ADSCs in a dental-inducing medium and subsequently culturing the aggregates in three-dimensional conditions can transdifferentiate the cells to produce a specific three dimensional organization and phenotype resembling a dental bud in vitro [16].

2.3 Dental Pulp Stem Cells (DPSCs)

It is well known that upon pulpal injury, reparative/tertiary dentine forms as a protective barrier for the pulpal chamber [59]. This natural regenerative aptitude of the dentin/pulp complex points to the possibility that dental pulp may contain stem cells or progenitors responsible for its regeneration/repair [65].

Dental pulp stem cells (DPSCs) were first identified by Gronthos et al. [23] who showed that DPSCs from CFU-F and could produce dentine-pulplike structures. DPSCs when compared to BMSSCs cultured under the same conditions showed a higher proliferation rate which could be attributed to their strong expression of cyclin-dependent kinase 6 [74].

The expression by these cells of a range of perivascular cell markers including STRO-1, CD146/MUC-18, VCAM-1, and α-smooth muscle actin pointed to the fact that DPSCs are located in the perivascular niches within the dentin/pulp complex and represent a heterogeneous population of MSCs [23, 74].

DPSCs possess a self-renewal capability and multilineage differentiation potential into chondrocytes, adipocytes, odontoblasts, and neural-like cells under appropriate induction conditions [21, 31, 29]. DPSCs loaded on a hydroxyapatite/

tricalcium phosphate (HA/TCP)-scaffold formed bone after transplantation in immunocompromised mice. In addition, it was revealed that even after two years of storage, DPSCs were still able to differentiate into pre-osteoblasts and form woven bone, while preserving their cellular integrity [64, 63]. A recent study showed that the Coculture of dental pulp stem cells with endothelial cells enhances osteo-/odontogenic and angiogenic potential in vitro with greater ALP activity, greater amount of calcification, higher expression of ALP, BSP, and DSPP genes and stabilized vessel-like structures formed by endothelial cells [12]. A further study demonstarted that DPSCs derived from maxillary premolar in combination with anorganic scaffolds could regenerate experimentally-created periodontal defects [54].

Yet, in a contrasting study by Zhang and colleagues, DPSCs seeded onto three-dimensional spongeous collagen, fibrous titanium mesh, and porous ceramic scaffolds, and implanted in nude mice for six or twelve weeks did not form the expected dentine-pulplike complex but differentiated into tissues that resembled connective tissue [89].

2.4 Dental Follicle Stem Cells (DFSCs)

The dental follicle is a mesenchymal component that surrounds the tooth germ during development in its socket prior to eruption [65] and from which cementum, PDL, and alveolar bone arises through complex interactions [87]. Dental follicle stem cells (DFSCs) were initially isolated from follicles of human impacted third molars scheduled for extraction. They were shown to express the stem cell markers STRO-1, Notch-1, and nestin [55, 56]. DFSC cell lines were found to be heterogeneous and to consist of three main lineages: a highly undifferentiated, periodontal ligament type lineage, a cementoblastic, and an osteoblastic lineage [48].

DFSCs, similar to other MSCs, demonstrated a multilineage differentiation ability into osteoblasts/cementoblasts [83, 36], adipocytes, and neurons [36, 57, 86, 9] as well as PDL-like tissue [87].

Compared to DPSCs, DFSCs showed a faster proliferation rate (as was evidenced by a higher number of population doublings), a greater percentage of cells expressing the surface marker STRO-1, and an increased capacity for in vivo dentine regeneration. However, DFSCs exhibit telomerase activity, a characteristic feature of ESCs [77, 78, 85]. Telomerase is an enzyme that adds DNA sequence TTAGGG to the 5’ end in the telomere regions of the chromosomes. Normally the telomere region in each chromosome is shortened with every replication cycle (mitosis). Due to the action of telomerase in some cells expressing it, including ESCs and cancer cells, this region is not significantly shortened during mitosis and aging of the chromosomes is hindered, which principally confers immortality to the cells. Whether this expression is an advantage or may pose a potential risk for malignant tumor formation similar to the situation in ESCs in tissue engineering still needs to be extensively investigated.

2.5 Stem Cells from Human Exfoliated Deciduous Teeth (SHEDs)

Stem cells from human exfoliated deciduous teeth (SHEDs) were identified in freshly exfoliated deciduous teeth containing living pulp remnants by Miura and colleagues. They linger alive inside the dental pulp for a very short time after tooth exfoliation during which they can be harvested, representing an interesting and easily accessible stem cell source.

SHEDs show major advantages over other types of MSCs including a higher proliferation rate compared to DPSCs and BMSSCs, (SHED > DPSCs > BMMSCs) [30], a similar multilineage differentiation capacity to other MSCs with the ability to differentiate into neurons, adipocytes, osteoblasts, and odontoblasts, in addition to easier accessibility with little or no morbidity [40, 53].

SHEDs express CD146/MUC18 and STRO-1 similar to other MSCs [74] and a variety of osteoblastic and odontoblastic markers including Runx2, ALP, matrix phosphoglycoprotein, bone sialoprotein (BSP), and DSPP. They further exhibit the embryonic stem cell markers Nanog, Oct4, stage-specific embryonic antigens (SSEA-3, SSEA-4), and tumor recognition antigens (TRA-1-60 and TRA-1-81) [37].

SHEDs show adipogenic, neurogenic, myogenic as well as chondrogenic differentiation potential similar to other stem cell populations [37, 53]. Regarding their osteogenic potential, Miura et al. [37] stated that SHEDs could not be differentiated directly into osteoblasts, but had distinctive osteoinductive abilities, inducing new bone formation by recruiting host osteogenic cells. In contrast, Cordeiro and co-workers showed that when SHEDs were seeded in poly-*L*-lactide acid (PLLA)-scaffolds and transplanted into the subcutaneous tissue of immunodeficient mice, they differentiated into odontoblast like cells and into blood vessels that anastomosed with the host vasculature forming a continuous vascular supply to the newly implanted construct. These studies show that SHEDs might be promising source of stem cells for tooth structure repair and bone regeneration [65].

2.6 Stem cells from the apical papilla (SCAP)

Stem cells from the apical papilla (SCAP) were first described in 2008 [78]. Compared to DPSCs and BMMSCs, SCAP showed similar osteo/dentinogenic with lower adipogenic differentiation potential. SCAP further expressed a higher proliferation rate and mineralization potential compared to DPSCs [2]. Similar to other stem cell populations, SCAP expressed STRO-1 and CD146, were positive for CD34 and negative for CD45 as well as showed multiple dentinogenic markers including ALP, bone sialophosphoprotein, osteocalcin [2], and the growth factors TGFbetaRI and FGFR1 [78]. Compared to DPSCs, SCAP express lower levels of DSP, matrix extracellular phosphoglycoprotein (MEPE), transforming growth factor β receptor II (TGFβRII), FGFR3, Flt-1 (VEGF receptor 1), Flg (FGFR1), and melanoma-associated glycoprotein (MUC18) [30]. Upon stimulation with a neurogenic medium, SCAP expressed neurogenic markers as nestin and neurofilament M [78].

2.7 Periodontal Ligament Stem Cells (PDLSCs)

The periodontium, one of the highly specialized and complex connective tissues of the human body, is derived from the dental follicle and the neural crest cells [65]. The PDL harbors a heterogeneous population of progenitor cells [44, 58], which are thought to be responsible for maintaining tissue homeostasis and to play a crucial role in periodontal regeneration [5]. A study by Seo and colleagues initially identified and characterized human PDL-derived stem cells from extracted teeth as periodontal ligament stem cells (PDLSCs).

PDLSCs exhibited an approximately 30 % higher number of population doublings compared to BMSSCs. It appeared that PDLSCs retain this high growth potential beyond 100 population doublings before they become senescent, compared to approximately 50 population doublings for BMSSCs [3]. In addition, PDLSCs showed a higher frequency of fibroblastic colony-forming units (aggregates of 50 cells or more) than that noted for BMSSCs (170 for PDLSCs and 14 for BMSSCs per 10^5 cells plated; [72].

PDLSCs express the stem cell markers STRO-1 and CD146/MUC18 [72, 84] entailing a perivascular origin similar to all MSCs. A proportion of PDLSCs also co-expressed α-smooth muscle actin (similar to DPSCs), the pericyte-associated antigen 3G5, and were negative for the hematopoietic markers CD14, CD45, and CD34 [3]. PDLSCs express mature mineralized tissue markers such as ALP, type I and III collagens, osteonectin, osteopontin, osteocalcin, and BSP [22, 32, 72, 75] and high levels of scleraxis, a tendon-specific transcription factor associated with tendon cells [60, 72]. PDLSCs are multipotent, possessing the ability to differentiate into adipocytes, cementoblast like cells, osteoblasts, and collagen-forming cells [72].

Multiple studies on PDLSCs confirmed their aptitude for tissue regeneration and periodontal repair [38, 46, 72]. In the study by Seo et al. [72], human PDLSCs were loaded onto a HA/TCP-scaffold and subcutaneously implanted into immunocompromised mice, resulting in a cementum and PDL-like structure being produced. Orciani and colleagues demonstrated that osteogenically differentiated cells were marked by an increase in Ca^{2+} and nitric oxide production and that the implantation of PDLSCs together with a nitric oxide donor could be a promising regimen for periodontal regeneration [62]. When PDLSCs were transplanted into surgically created periodontal defects, these cells were reported to integrate into the PDL, connect to the surrounding alveolar bone and cementum via Sharpey's fibers and regenerated the experimental defects [11, 47, 72].

This characteristic feature of PDLSCs to produce cementum and PDL-like tissue [72], in contrast to the dentine or pulplike structure and lamellar bone and marrowlike structure generated by DPSCs and BMSSCs, respectively [23, 25, 42], verified that PDLSCs embody a distinctive MSC population [9].

Recently, Park and co-workers successfully isolated and characterized human PDLSCs from healthy (hPDLSCs) and inflamed (ihPDLSCs) PDL tissues and evaluated their regenerative potential. Both ihPDLSCs and hPDLSCs were successfully differentiated under an osteogenic/cementogenic and adipogenic microenvironment. The proliferative potential did not differ between healthy hPDLSCs and ihPDLSCs.

2.8 Alveolar Bone proper-derived Stem Cells

The alveolar bone proper similar to the PDL is embryonically derived from the dental follicle. Recently, a scheme for the minimally invasive isolation of alveolar bone margin-derived stem cells was introduced [14]. The isolated cells showed plastic adherence and colony formation, and expressed the surface markers CD73, CD90, CD105, STRO-1, and CD146/MUC18, while lacking the expression of the hematopoietic markers CD14, CD34, and CD45. The cells could be differentiated into osteoblastic, adipocytic, and chondroblastic lineages and demonstrated a high expression of ALP, type I, III, and V collagens. The isolation scheme of alveolar bone margin-derived stem cells described in this study constituted a conservative alternative to many previously described isolation techniques for adult stem/progenitor cells from the dental pulp or periodontal ligament [20, 72, 77, 81] as well as the intra- [1, 27, 51] and extra-oral [41, 43] bone marrow. Further studies are needed to verify the regenerative potential of these cells as well as to compare them with other stem cell populations.

2.9 Gingival Stem Cells

Representing a key component of the periodontium, one of the gingiva's eminent characteristics is its remarkable regenerative and wound healing capacity with a rapid reconstitution of tissue architecture, with little evidence of scarring [7]. The multiple functions of gingival fibroblasts, their diversity in responsiveness to growth factors and in the ability to produce specific extracellular matrix proteins during healing, demonstrated that gingival connective tissue fibroblasts constitute a heterogeneous population of cells [26, 67, 68, 70, 71]. This also entails the existence of a population of stem/progenitor cells that give rise to these heterogeneous cell populations. Because the lamina propria of oral mucosa including alveolar mucosa, gingival, and palatal mucosa originates from the embryonic neural crest this may suggest that a primitive population is retained in the adult human gingiva. Recent studies reported on the isolation of progenitors from the oral soft tissue, such as the rugae and incisive papillae of the palate [82], the maxillary tuberosity [52], the oral mucosa [50], and the attached gingiva [17, 79].

The isolated gingival stem cells expressed CD73, CD90, and CD105 and lacked expression of CD14, CD34, and CD45. They demonstrated a multilineage differentiation capacity into adipogenic, osteogenic, and chondrogenic lineages [52]. The immunomodulatory properties of gingival margin-derived stem cells were exploited experimentally in the therapy of inflammatory destructive diseases including arthritis and colitis through inhibiting the proliferation of T-lymphocytes and promoting the proliferation of regulatory T cells [89]. A recent study further demonstrated a remarkable periodontal regenerative potential of these cells in conjunction with collagen and demineralized bovine cancellous bone matrices [15]. Human gingiva is one of the most convenient tissues for biopsy and is considered an ideal source for stem cell isolation. The major advantage of this stem cell source appears to be the ability to obtain a large quantity without the need to sacrifice a tooth irreversibly to obtain its pulp, periodontal ligament, or dental follicle.

3 Concluding Remarks

Dental stem cells offer several advantages and promising facets over other types of stem cells, for example, a high proliferation rate, easy accessibility, and a relative ease of differentiation induction into distinct cell lineages. There is still much to learn about the nature, basic biology, and developmental potency of dental stem/progenitor cells. However, the perspectives for their exploitation in dental tissue regeneration are far-reaching. It is to be hoped that a better understanding of their biology will result in significant benefits for the management of dental diseases in patients.

References

1. Akintoye SO, Lam T, Shi S, Brahim J, Collins MT, Robey PG (2006) Skeletal site-specific characterization of orofacial and iliac crest human bone marrow stromal cells in same individuals. Bone 38:758–768
2. Bakopoulou A, Leyhausen G, Volk J, Tsiftsoglou A, Garefis P, Koidis P, Geurtsen W (2011) Comparative analysis of in vitro osteo/odontogenic differentiation potential of human dental pulp stem cells (DPSCs) and stem cells from the apical papilla (SCAP). Arch Oral Biol 56:709–721 doi:S0003-9969(10)00383-3 [pii] 10.1016/j.archoralbio.2010.12.008.
3. Bartold PM, Shi S, Gronthos S (2006) Stem cells and periodontal regeneration. Periodontol 2000 40:164–172
4. Battula VL, Bareiss PM, Treml S, Conrad S, Albert I, Hojak S, Abele H, Schewe B, Just L, Skutella T, Buhring HJ (2007) Human placenta and bone marrow derived MSC cultured in serum-free, b-FGF-containing medium express cell surface frizzled-9 and SSEA-4 and give rise to multilineage differentiation. Differentiation 75:279–291. doi:S0301-4681(09)60122-5 [pii]
5. Beertsen W, McCulloch CA, Sodek J (1997) The periodontal ligament: a unique, multifunctional connective tissue. Periodontol 2000 13:20–40

6. Bruder SP, Jaiswal N, Haynesworth SE (1997) Growth kinetics, self-renewal, and the osteogenic potential of purified human mesenchymal stem cells during extensive subcultivation and following cryopreservation. J Cell Biochem 64:278–294
7. Cobb CM (2006) Lasers in periodontics: a review of the literature. J Periodontol 77:545–564
8. Cordeiro MM, Dong Z, Kaneko T, Zhang Z, Miyazawa M, Shi S, Smith AJ, Nör JE (2008) Dental pulp tissue engineering with stem cells from exfoliated deciduous teeth. J Endod 34:962–969
9. Coura GS, Garcez RC, de Aguiar CB, Alvarez-Silva M, Magini RS, Trentin AG (2008) Human periodontal ligament: a niche of neural crest stem cells. J Periodontal Res 43:531–536
10. Deans RJ, Moseley AB (2000) Mesenchymal stem cells: biology and potential clinical uses. Exp Hematol 28:875–884
11. Ding G, Liu Y, Wang W, Wei F, Liu D, Fan Z, An Y, Zhang C, Wang S (2010) Allogeneic periodontal ligament stem cell therapy for periodontitis in swine. Stem Cells 28:1829–1838
12. Dissanayaka WL, Zhan X, Zhang C, Hargreaves KM, Jin L, Tong EH (2012) Coculture of dental pulp stem cells with endothelial cells enhances osteo-/odontogenic and angiogenic potential in vitro. J Endod 38:454–463. doi:S0099-2399(11)01457-9 [pii]
13. Dominici M, Le Blanc K, Mueller I, Slaper-Cortenbach I, Marini F, Krause D, Deans R, Keating A, Prockop D, Horwitz E (2006) Minimal criteria for defining multipotent mesenchymal stromal cells. The international society for cellular therapy position statement. Cytotherapy 8:315–317
14. El-Sayed KM, Paris S, Becker S, Kassem N, Ungefroren H, Fändrich F, Wiltfang J, Dörfer C (2012) Isolation and characterization of multipotent postnatal stem/progenitor cells from human alveolar bone proper. J Craniomaxillofac Surg [Epub ahead of print]
15. Fawzy El-Sayed KM, Paris S, Becker ST, Neuschl M, De Buhr W, Salzer S, Wulff A, Elrefai M, Darhous MS, El-Masry M, Wiltfang J, Dorfer CE (2012) Periodontal regeneration employing gingival margin-derived stem/progenitor cells: an animal study. J Clin Periodontol. doi:10.1111/j.1600-051X.2012.01904.x
16. Ferro F, Spelat R, Falini G, Gallelli A, D'Aurizio F, Puppato E, Pandolfi M, Beltrami AP, Cesselli D, Beltrami CA, Ambesi-Impiombato FS, Curcio F (2011) Adipose tissue-derived stem cell in vitro differentiation in a three-dimensional dental bud structure. Am J Pathol 178:2299–2310 doi:S0002-9440(11)00171-4 [pii]
17. Fournier BP, Ferre FC, Couty L, Lataillade JJ, Gourven M, Naveau A, Coulomb B, Lafont A, Gogly B (2010) Multipotent progenitor cells in gingival connective tissue. Tissue Eng Part A 16:2891–2899
18. Friedenstein AJ, Deriglasova UF, Kulagina NN, Panasuk AF, Rudakowa SF, Luria EA, RuadkowIA (1974) Precursors for fibroblasts in different populations of hematopoietic cells as detected by the in vitro colony assay method. Exp Hematol 2:83–92
19. Greco SJ, Liu K, Rameshwar P (2007) Functional similarities among genes regulated by OCT4 in human mesenchymal and embryonic stem cells. Stem Cells 25:3143–3154. doi:2007-0351 [pii]
20. Gronthos S, Arthur A, Bartold PM, Shi S (2011) A method to isolate and culture expand human dental pulp stem cells. Methods Mol Biol 698:107–121
21. Gronthos S, Brahim J, Li W, Fisher LW, Cherman N, Boyde A, DenBesten P, Robey PG, Shi S (2002) Stem cell properties of human dental pulp stem cells. J Dent Res 81:531–535
22. Gronthos S, Chen S, Wang CY, Robey PG, Shi S (2003a) Telomerase accelerates osteogenesis of bone marrow stromal stem cells by upregulation of CBFA1, osterix, and osteocalcin. J Bone Miner Res 18:716–722
23. Gronthos S, Mankani M, Brahim J, Robey PG, Shi S (2000) Postnatal human dental pulp stem cells (DPSCs) in vitro and in vivo. Proc Natl Acad Sci USA 97:13625–13630
24. Gronthos S, Zannettino AC (2008) A method to isolate and purify human bone marrow stromal stem cells. Methods Mol Biol 449:45–57 doi:10.1007/978-1-60327-169-1_3
25. Gronthos S, Zannettino AC, Hay SJ, Shi S, Graves SE, Kortesidis A, Simmons PJ (2003b) Molecular and cellular characterisation of highly purified stromal stem cells derived from human bone marrow. J Cell Sci 116:1827–1835

26. Häkkinen L, Uitto VJ, Larjava H (2000) Cell biology of gingival wound healing. Periodontol 24:127–152
27. Han J, Okada H, Takai H, Nakayama Y, Maeda T, Ogata Y (2009) Collection and culture of alveolar bone marrow multipotent mesenchymal stromal cells from older individuals. J Cell Biochem 107:1198–1204
28. Hicok KC, Du Laney TV, Zhou YS, Halvorsen YD, Hitt DC, Cooper LF, Gimble JM (2004) Human adipose-derived adult stem cells produce osteoid in vivo. Tissue Eng 10:371–380
29. Hosoya A, Nakamura H, Ninomiya T, Hoshi K, Yoshiba K, Yoshiba N, Takahashi M, Okabe T, Sahara N, Yamada H, Kasahara E, Ozawa H (2007) Hard tissue formation in subcutaneously transplanted rat dental pulp. J Dent Res 86:469–474
30. Huang GT, Gronthos S, Shi S (2009) Mesenchymal stem cells derived from dental tissues vs. those from other sources: their biology and role in regenerative medicine. J Dent Res 88:792–806. doi:88/9/792 [pii]
31. Iohara K, Zheng L, Ito M, Tomokiyo A, Matsushita K, Nakashima M (2006) Side population cells isolated from porcine dental pulp tissue with self-renewal and multipotency for dentinogenesis, chondrogenesis, adipogenesis, and neurogenesis. Stem Cells 24:2493–2503
32. Ivanovski S, Haase HR, Bartold PM (2001) Expression of bone matrix protein mRNAs by primary and cloned cultures of the regenerative phenotype of human periodontal fibroblasts. J Dent Res 80:1665–1671
33. Jing W, Wu L, Lin Y, Liu L, Tang W, Tian W (2008) Odontogenic differentiation of adipose-derived stem cells for tooth regeneration: necessity, possibility, and strategy. Med Hypotheses 70:540–542
34. Kakudo N, Shimotsuma A, Miyake S, Kushida S, Kusumoto K (2008) Bone tissue engineering using human adipose-derived stem cells and honeycomb collagen scaffold. J Biomed Mater Res A 84:191–197
35. Kawaguchi H, Hirachi A, Hasegawa N, Iwata T, Hamaguchi H, Shiba H, Takata T, Kato Y, Kurihara H (2004) Enhancement of periodontal tissue regeneration by transplantation of bone marrow mesenchymal stem cells. J Periodontol 75:1281–1287
36. Kémoun P, Laurencin-Dalicieux S, Rue J, Farges JC, Gennero I, Conte-Auriol F, Briand-Mesange F, Gadelorge M, Arzate H, Narayanan AS, Brunel G, Salles JP (2007) Human dental follicle cells acquire cementoblast features under stimulation by BMP-2/-7 and enamel matrix derivatives (EMD) in vitro. Cell Tissue Res 329:283–294
37. Kerkis I, Kerkis A, Dozortsev D, Stukart-Parsons GC, Gomes Massironi SM, Pereira LV, Caplan AI, Cerruti HF (2006) Isolation and characterization of a population of immature dental pulp stem cells expressing OCT-4 and other embryonic stem cell markers. Cells Tissues Organs 184:105–116. doi:000099617 [pii]
38. Kim AC, Hammer GD (2007) Adrenocortical cells with stem/progenitor cell properties: recent advances. Mol Cell Endocrinol 265–266:10–16
39. Kopen GC, Prockop DJ, Phinney DG (1999) Marrow stromal cells migrate throughout forebrain and cerebellum, and they differentiate into astrocytes after injection into neonatal mouse brains. Proc Natl Acad Sci USA 96:10711–10716
40. Koyama N, Okubo Y, Nakao K, Bessho K (2009) Evaluation of pluripotency in human dental pulp cells. J Oral Maxillofac Surg 67:501–506
41. Kramer PR, Kramer SF, Puri J, Grogan D, Guan G (2009) Multipotent adult progenitor cells acquire periodontal ligament characteristics in vivo. Stem Cells Dev 18:67–75
42. Kuznetsov SA, Krebsbach PH, Satomura K, Kerr J, Riminucci M, Benayahu D, Robey PG (1997) Single-colony derived strains of human marrow stromal fibroblasts form bone after transplantation in vivo. J Bone Miner Res 12:1335–1347
43. Larsen KH, Frederiksen CM, Burns JS, Abdallah BM, Kassem M (2010) Identifying a molecular phenotype for bone marrow stromal cells with in vivo bone-forming capacity. J Bone Miner Res 25:796–808
44. Lekic P, Rojas J, Birek C, Tenenbaum H, McCulloch CA (2001) Phenotypic comparison of periodontal ligament cells in vivo and in vitro. J Periodontal Res 36:71–79

45. Li ZY, Chen L, Liu L, Lin YF, Li SW, Tian WD (2007) Odontogenic potential of bone marrow mesenchymal stem cells. J Oral Maxillofac Surg 65:494–500
46. Liu Y, Zheng Y, Ding G, Fang D, Zhang C, Bartold PM, Gronthos S, Shi S, Wang S (2008a) Periodontal ligament stem cell-mediated treatment for periodontitis in miniature swine. Stem Cells 26:1065–1073
47. Liu Y, Zheng Y, Ding G, Fang DJ, Zhang CM, Bartold PM, Gronthos S, Shi ST, Wang SL (2008b) Periodontal ligament stem cell-mediated treatment for periodontitis in miniature swine. Stem Cells 26:1065–1073 doi:DOI 10.1634/stemcells.2007-0734
48. Luan X, Ito Y, Dangaria S, Diekwisch TG (2006) Dental follicle progenitor cell heterogeneity in the developing mouse periodontium. Stem Cells Dev 15:595–608
49. Makino S, Fukuda K, Miyoshi S, Konishi F, Kodama H, Pan J, Sano M, Takahashi T, Hori S, Abe H, Hata J, Umezawa A, Ogawa S (1999) Cardiomyocytes can be generated from marrow stromal cells in vitro. J Clin Invest 103:697–705
50. Marynka-Kalmani K, Treves S, Yafee M, Rachima H, Gafni Y, Cohen MA, Pitaru S (2010) The lamina propria of adult human oral mucosa harbors a novel stem cell population. Stem Cells 28:984–995
51. Matsubara T, Suardita K, Ishii M, Sugiyama M, Igarashi A, Oda R, Nishimura M, Saito M, Nakagawa K, Yamanaka K, Miyazaki K, Shimizu M, Bhawal UK, Tsuji K, Nakamura K, Kato Y (2005) Alveolar bone marrow as a cell source for regenerative medicine: differences between alveolar and iliac bone marrow stromal cells. J Bone Miner Res 20:399–409
52. Mitrano TI, Grob MS, Carrión F, Nova-Lamperti E, Luz PA, Fierro FS, Quintero A, Chaparro A, Sanz A (2010) Culture and characterization of mesenchymal stem cells from human gingival tissue. J Periodontol 81:917–925
53. Miura M, Gronthos S, Zhao M, Lu B, Fisher LW, Robey PG, Shi S (2003) SHED: stem cells from human exfoliated deciduous teeth. Proc Natl Acad Sci USA 100:5807–5812
54. Mohamadreza BE, Khorsand A, Arabsolghar M, Paknejad M, Ghaedi B, Rokn AR, Moslemi N, Nazarian H, Jahangir S (2012) Autologous Dental Pulp Stem Cells in Regeneration of Defect Created in Canine Periodontal Tissue. J Oral Implantol. doi:10.1563/AAID-JOI-D-12-00027.1
55. Morsczeck C, Götz W, Schierholz J, Zeilhofer F, Kühn U, Möhl C, Sippel C, Hoffmann KH (2005) Isolation of precursor cells (PCs) from human dental follicle of wisdom teeth. Matrix Biol 24:155–165
56. Morsczeck C, Schmalz G, Reichert TE, Völlner F, Galler K, Driemel O (2008) Somatic stem cells for regenerative dentistry. Clin Oral Investig 12:113–118
57. Morsczeck C, Völlner F, Saugspier M, Brandl C, Reichert TE, Driemel O, Schmalz G (2010) Comparison of human dental follicle cells (DFCs) and stem cells from human exfoliated deciduous teeth (SHED) after neural differentiation in vitro. Clin Oral Investig 14:433–440
58. Murakami Y, Kojima T, Nagasawa T, Kobayashi H, Ishikawa I (2003) Novel isolation of alkaline phosphatase-positive subpopulation from periodontal ligament fibroblasts. J Periodontol 74:780–786
59. Murray PE, About I, Franquin JC, Remusat M, Smith AJ (2001) Restorative pulpal and repair responses. J Am Dent Assoc 132:482–491
60. Nagatomo K, Komaki M, Sekiya I, Sakaguchi Y, Noguchi K, Oda S, Muneta T, Ishikawa I (2006) Stem cell properties of human periodontal ligament cells. J Periodontal Res 41:303–310
61. Ohazama A, Modino SA, Miletich I, Sharpe PT (2004) Stem-cell-based tissue engineering of murine teeth. J Dent Res 83:518–522
62. Orciani M, Trubiani O, Vignini A, Mattioli-Belmonte M, Di Primio R, Salvolini E (2009) Nitric oxide production during the osteogenic differentiation of human periodontal ligament mesenchymal stem cells. Acta Histochem 111:15–24
63. Otaki S, Ueshima S, Shiraishi K, Sugiyama K, Hamada S, Yorimoto M, Matsuo O (2007) Mesenchymal progenitor cells in adult human dental pulp and their ability to form bone when transplanted into immunocompromised mice. Cell Biol Int 31:1191–1197

64. Papaccio G, Graziano A, d'Aquino R, Graziano MF, Pirozzi G, Menditti D, De Rosa A, Carinci F, Laino G (2006) Long-term cryopreservation of dental pulp stem cells (SBP-DPSCs) and their differentiated osteoblasts: a cell source for tissue repair. J Cell Physiol 208:319–325
65. Peng L, Ye L, Zhou XD (2009) Mesenchymal stem cells and tooth engineering. Int J Oral Sci 1:6–12 doi:10.4248/ijos.08032
66. Pereira RF, O'Hara MD, Laptev AV, Halford KW, Pollard MD, Class R, Simon D, Livezey K, Prockop DJ (1998) Marrow stromal cells as a source of progenitor cells for nonhematopoietic tissues in transgenic mice with a phenotype of osteogenesis imperfecta. Proc Natl Acad Sci USA 95:1142–1147
67. Phipps RP, Borrello MA, Blieden TM (1997) Fibroblast heterogeneity in the periodontium and other tissues. J Periodontal Res 32:159–165
68. Pitaru S, McCulloch CA, Narayanan SA (1994) Cellular origins and differentiation control mechanisms during periodontal development and wound healing. J Periodontal Res 29:81–94
69. Pittenger MF, Mackay AM, Beck SC, Jaiswal RK, Douglas R, Mosca JD, Moorman MA, Simonetti DW, Craig S, Marshak DR (1999) Multilineage potential of adult human mesenchymal stem cells. Science 284:143–147
70. Schor SL, Ellis I, Irwin CR, Banyard J, Seneviratne K, Dolman C, Gilbert AD, Chisholm DM (1996) Subpopulations of fetal-like gingival fibroblasts: characterisation and potential significance for wound healing and the progression of periodontal disease. Oral Dis 2:155–166
71. Sempowski GD, Borrello MA, Blieden TM, Barth RK, Phipps RP (1995) Fibroblast heterogeneity in the healing wound. Wound Repair Regen 3:120–131
72. Seo BM, Miura M, Gronthos S, Bartold PM, Batouli S, Brahim J, Young M, Robey PG, Wang CY, Shi S (2004) Investigation of multipotent postnatal stem cells from human periodontal ligament. Lancet 364:149–155
73. Seo BM, Miura M, Sonoyama W, Coppe C, Stanyon R, Shi S (2005) Recovery of stem cells from cryopreserved periodontal ligament. J Dent Res 84:907–912
74. Shi S, Gronthos S (2003) Perivascular niche of postnatal mesenchymal stem cells in human bone marrow and dental pulp. J Bone Miner Res 18:696–704
75. Shi S, Robey PG, Gronthos S (2001) Comparison of human dental pulp and bone marrow stromal stem cells by cDNA microarray analysis. Bone 29:532–539
76. Simmons PJ, Torok-Storb B (1991) Identification of stromal cell precursors in human bone marrow by a novel monoclonal antibody, STRO-1. Blood 78:55–62
77. Sonoyama W, Liu Y, Fang D, Yamaza T, Seo BM, Zhang C, Liu H, Gronthos S, Wang CY, Wang S, Shi S (2006) Mesenchymal stem cell-mediated functional tooth regeneration in swine. PLoS One 1:e79
78. Sonoyama W, Liu Y, Yamaza T, Tuan RS, Wang S, Shi S, Huang GT (2008) Characterization of the apical papilla and its residing stem cells from human immature permanent teeth: a pilot study. J Endod 34:166–171
79. Tomar GB, Srivastava RK, Gupta N, Barhanpurkar AP, Pote ST, Jhaveri HM, Mishra GC, Wani MR (2010) Human gingiva-derived mesenchymal stem cells are superior to bone marrow-derived mesenchymal stem cells for cell therapy in regenerative medicine. Biochem Biophys Res Commun 393:377–383
80. Tsutsumi S, Shimazu A, Miyazaki K, Pan H, Koike C, Yoshida E, Takagishi K, Kato Y (2001) Retention of multilineage differentiation potential of mesenchymal cells during proliferation in response to FGF. Biochem Biophys Res Commun 288:413–419
81. Wang L, Shen H, Zheng W, Tang L, Yang Z, Gao Y, Yang Q, Wang C, Duan Y, Jin Y (2011) Characterization of stem cells from alveolar periodontal ligament. Tissue Eng Part A 17:1015–1026
82. Widera D, Zander C, Heidbreder M, Kasperek Y, Noll T, Seitz O, Saldamli B, Sudhoff H, Sader R, Kaltschmidt C, Kaltschmidt B (2009) Adult palatum as a novel source of neural crest-related stem cells. Stem Cells 27:1899–1910

83. Wu L, Zhu F, Wu Y, Lin Y, Nie X, Jing W, Qiao J, Liu L, Tang W, Zheng X, Tian W (2008) Dentin sialophosphoprotein-promoted mineralization and expression of odontogenic genes in adipose-derived stromal cells. Cells Tissues Organs 187:103–112
84. Xu J, Wang W, Kapila Y, Lotz J, Kapila S (2009) Multiple differentiation capacity of STRO-1 +/CD146 + PDL mesenchymal progenitor cells. Stem Cells Dev 18:487–496
85. Yang ZH, Zhang XJ, Dang NN, Ma ZF, Xu L, Wu JJ, Sun YJ, Duan YZ, Lin Z, Jin Y (2009) Apical tooth germ cell-conditioned medium enhances the differentiation of periodontal ligament stem cells into cementum/periodontal ligament-like tissues. J Periodontal Res 44:199–210
86. Yao S, Pan F, Prpic V, Wise GE (2008) Differentiation of stem cells in the dental follicle. J Dent Res 87:767–771
87. Yokoi T, Saito M, Kiyono T, Iseki S, Kosaka K, Nishida E, Tsubakimoto T, Harada H, Eto K, Noguchi T, Teranaka T (2007) Establishment of immortalized dental follicle cells for generating periodontal ligament in vivo. Cell Tissue Res 327:301–311
88. Zhang Q, Shi S, Liu Y, Uyanne J, Shi Y, Shi S, Le AD (2009) Mesenchymal stem cells derived from human gingiva are capable of immunomodulatory functions and ameliorate inflammation-related tissue destruction in experimental colitis. J Immunol 183:7787–7798
89. Zhang W, Walboomers XF, van Kuppevelt TH, Daamen WF, Bian Z, Jansen JA (2006) The performance of human dental pulp stem cells on different three-dimensional scaffold materials. Biomaterials 27:5658–5668
90. Zuk PA, Zhu M, Ashjian P, De Ugarte DA, Huang JI, Mizuno H, Alfonso ZC, Fraser JK, Benhaim P, Hedrick MH (2002) Human adipose tissue is a source of multipotent stem cells. Molecular Biology of the Cell 13:4279–4295. doi:DOI 10.1091/mbc.E02-02-0105
91. Zuk PA, Zhu M, Mizuno H, Huang J, Futrell JW, Katz AJ, Benhaim P, Lorenz HP, Hedrick MH (2001) Multilineage cells from human adipose tissue: implications for cell-based therapies. Tissue Eng 7:211–228 doi:10.1089/107632701300062859

Adv Biochem Eng Biotechnol (2013) 130: 105–129
DOI: 10.1007/10_2012_140

Published Online: 8 July 2012

Mesenchymal Stem Cell Therapy and Lung Diseases

Khondoker M. Akram, Sohel Samad, Monica Spiteri and Nicholas R. Forsyth

Abstract Mesenchymal stem cells (MSCs), a distinct population of adult stem cells, have amassed significant interest from both medical and scientific communities. An inherent multipotent differentiation potential offers a cell therapy option for various diseases, including those of the musculoskeletal, neuronal, cardiovascular and pulmonary systems. MSCs also secrete an array of paracrine factors implicated in the mitigation of pathological conditions through anti-inflammatory, anti-apoptotic and immunomodulatory mechanisms. The safety and efficacy of MSCs in human application have been confirmed through small- and large-scale clinical trials. However, achieving the optimal clinical benefit from MSC-mediated regenerative therapy approaches is entirely dependent upon adequate understanding of their healing/regeneration mechanisms and selection of appropriate clinical conditions.

Keywords Mesenchymal stem cells · Multipotent · Lung · Acute lung injury · Chronic lung disease · Chronic obstructive pulmonary disease · Cystic fibrosis · Idiopathic pulmonary fibrosis

Abbreviations

MSC	Mesenchymal stem cell
CFU-F	Colony-forming unit fibroblast
AEC	Alveolar epithelial cell
AECI	Type I alveolar epithelial cells
AECII	Type II alveolar epithelial cells
VEGF	Vascular endothelial growth factor
eNOS	Endothelial nitric oxide synthase
ALI	Acute lung injury
ARDS	Acute respiratory distress syndrome
LPS	Lipopolysaccharide
KGF	Keratinocyte growth factor

K. M. Akram · S. Samad · M. Spiteri · N. R. Forsyth (✉)
Guy Hilton Research Centre, Institute of Science and Technology in Medicine, Keele University, Stoke-on-Trent ST4 7QB, UK
e-mail: n.r.forsyth@keele.ac.uk

COPD	Chronic obstructive pulmonary disease
CF	Cystic fibrosis
CFTR	Cystic fibrosis transmembrane conductance regulator
IPF	Idiopathic pulmonary fibrosis
SP-C	Surfactant Protein-C
HSCs	Haematopoietic stem cells
PH	Pulmonary hypertension

Contents

1 Introduction

Mesenchymal stem cells (MSCs) are a population of adult stem cells that have amassed significant interest from the medical and scientific community since their initial discovery [44]. The interest in MSCs arises from their potential applications in regenerative medicine, consequent to their proposed ability to aid in the regeneration and repair of otherwise incurable diseases and physiological damage, including articular cartilage damage, neurological disorders, immunological diseases, and the development of irreversible lung fibrosis (a hallmark of idiopathic pulmonary fibrosis). Through continuing research, many new insights have been gained in our understanding of MSCs; however, there are still many unanswered questions regarding the functionality of MSCs and how best to use their clinical potential. Due to the scope of this chapter, we limit our discussion to the general properties of MSCs and their potential applications in the treatment of selected pulmonary diseases.

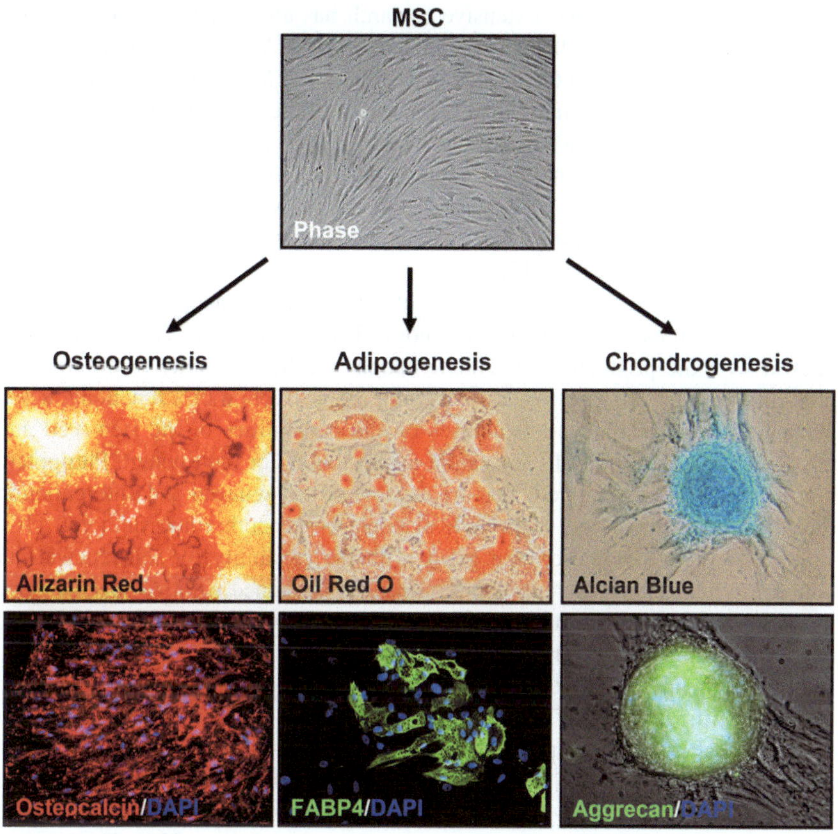

Fig. 1 Morphology of human MSCs and their classical tri-lineage differentiation. Phase images show the typical spindle-shaped morphology of adherent human MSCs. Osteogenesis: deposited calcium by differentiated osteoblasts was stained with Alizarin Red and osteocalcin was labelled by anti-osteocalcin antibody. Adipogenesis: differentiated adipocytes produce triglyceride which was stained with Oil Red O and adipocytes were stained with anti-FABP4 antibody. Chondrogenesis: chondrogenic nodules were stained with Alcian Blue and anti-aggrecan antibody. N.B. Histological images and immunofluorescence images are taken from representative and not identical fields of view

2 History, Definition and Properties of MSC

The first descriptions of MSCs took shape with the work of Friedenstein and colleagues with the discovery of multi-potential precursor cells that were spindle-shaped in nature within bone marrow samples. Further in vitro experiments demonstrated a colony-forming capacity associated with these cells, defined as colony-forming unit fibroblast (CFU-F) [44]. The cells within the CFU-F had the potential to differentiate into chondrocytes, adipocytes, osteoblasts (Fig. 1) and were also postulated to form a stromal layer that is essential in maintaining haematopoiesis. However, it was the capacity for differentiation that accrued most interest [95].

Since their initial discovery, extensive research has attempted to understand and harness the enormous medicinal potential of MSCs. Although well studied and documented, no agreement on the true definition of the MSC has yet been reached.

To define the MSC, the individual components of the mesenchyme and the stem cell should first be considered. Mesenchyme or stroma describes the tissue that provides structural and functional support for the growth and development of numerous organ systems. The bone marrow mesenchyme is a layer of cells that delivers the essential support that haematopoietic stem cells require for self-renewal and differentiation [115].

The definition of a stem cell has evolved largely through increased understanding of haematopoietic stem cell biology. The cell line must demonstrate self-renewal with the production of a clone daughter cell, the ability to differentiate into multi-lineage cell lines, and also in vivo reconstruction of a functional tissue [129]. Functional classification focuses on the capabilities of the stem cell and begins with a description of their nature as pluripotent or multipotent. Pluripotent describes a group of stem cells that are capable of self-renewal and differentiation into all three germ layers; a classical example is the embryonic stem cell [16]. Multipotent stem cells describe a group with the capability for self-renewal but their ability to differentiate is limited to lineages contained within a specific germ layer; an example is, the haematopoietic stem cell, which can differentiate into cells of the immune and haematological cell lines [66].

A difficulty in defining the MSC is the variation in nomenclature that is used within scientific literature. Terminology used to refer to MSCs includes multipotent stromal cells, multipotent progenitor cells, non-haematopoietic stem cells, and stromal progenitor cells [19]. All of these terms are essentially synonymous with the term MSC. This variation within the literature may stem from the lack of evidence for in vivo self-renewal and reconstruction of functional tissue [28]. The International Society for Cellular Therapy has categorized MSCs with a broad definition: "firstly MSCs must be adherent to plastic when maintained in culture, secondly they must express surface antigens CD105, CD73 and CD90 and lack the surface markers for monocytes, macrophages, and B cells in addition to lacking markers of the haematopoietic antigens CD45 and CD34. Finally the MSC must have the potential to differentiate into osteoblast, chondrocytes and adipocytes" [56].

By definition, MSCs under the influence of appropriate growth factors can differentiate into multiple cell lines, in particular to osteoblasts, chondrocytes and adipocytes. Therefore, through demonstration of the presence of these three cell lineages after directed differentiation of a colony of cells in vitro, one can retrospectively deduce that the original cells are MSCs [103]. However, this technique in itself contains numerous pitfalls as it is often very difficult to isolate and culture MSCs without altering and manipulating their original phenotype. A further difficulty in the definition of the MSC is that no single marker has been described that is specific to the MSC, thus making them extremely difficult to identify in vitro and in vivo [136]. Current practice is to define MSCs based on a combination of their differentiation potential, phenotype features, and morphological features—often in a retrospective manner (Figs. 1, 2).

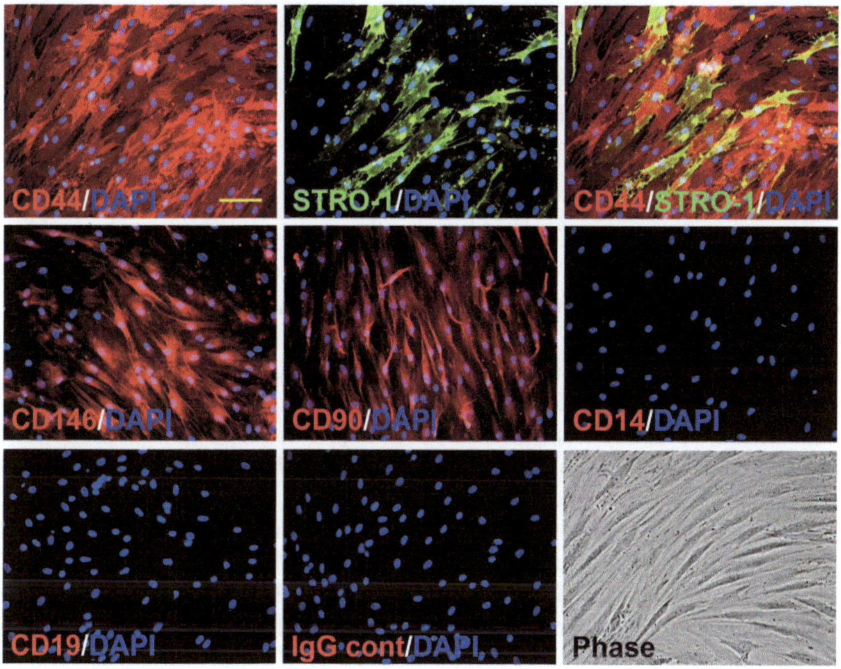

Fig. 2 Phenotypic antigenic markers of MSCs. Human MSCs demonstrate positive expression of CD44, STRO-1, CD90, CD146 and negative expression to haematopoietic markers CD14 and CD19. Nuclei are stained with DAPI. Phase images show typical morphology of MSC. Scale bar = 100 μm

A number of recent reports suggest an additional differentiation capacity of MSCs into a wide range of mesodermal and non-mesodermal adult phenotypes, including cardiomyocytes [34, 121], neurons [37, 135], hepatocytes [105] and lung epithelial cells [65, 72]. The benchmark definitions of MSCs may evolve over the coming years to reflect these descriptions.

3 Sources of MSCs

Friedenstein and others used cells that were isolated from collected bone marrow. Further locations and sites for MSC isolation have emerged, but bone marrow-derived MSCs are still the most frequently used MSCs in experimental research and are considered to be the criterion standard against which the newer sources of MSCs are compared [96].

There are numerous drawbacks and disadvantages associated with MSC isolation from bone marrow. Firstly, the procedure required for collecting bone marrow is through needle aspiration; this is accompanied by a mild discomfort that can be painful and can cause patient distress. Furthermore, there is a risk of infection as a result of the procedure, with osteomyelitis posing a particular threat [24]. Because of the risk of the aforementioned difficulties and other potential complications, there has been extensive research into searching for other potential sources of MSCs.

Adipose tissue, peripheral blood, the lung, deciduous teeth, and the myocardium are all documented as potential sources of adult MSCs, while the placenta, amnion, umbilical cord and cord blood have been studied as potential birth-associated sources of MSCs. However, there does appear to be differences within the phenotypes, quality, and quantity of the MSCs collected at the various sites [15].

Adipose tissue is a potential source of adult MSCs. One of the main advantages of using adipose tissue is the relative ease with which it can be collected and the quantity of adipose tissue available [68]. Adipose tissue is collected through liposuction, which is a commonly performed and safe procedure with minimal patient distress or risk [106]. Furthermore, the frequency of MSCs in the adipose tissue is 1–10 in 100 stromal vascular fractions [50, 86], whereas, in bone marrow it is 1–10 in 1,00,000 mononuclear cells [13, 74]. In addition, adipose tissue-derived MSCs have a greater proliferative potential than bone marrow-derived MSCs, particularly in long-term cultures [62]. However, there are variations within the markers of the bone marrow and adipose tissue MSCs [12, 74].

Peripheral blood is another potential site for MSC collection. Collection of peripheral blood is performed through venipuncture and is thus a minimally invasive procedure with a low risk of complications [107]. However, studies have determined that although MSCs can be isolated from peripheral blood with subsequent differentiation, the frequency of peripheral blood MSC is much lower than that of adipose tissue and bone marrow, thus meaning a much larger sample of peripheral blood is required to evaluate MSC quality [127].

To negate the requirement for bone marrow-derived MSCs, numerous studies have investigated the effectiveness of using MSCs derived from birth-associated tissue with some promising results. Much interest has developed in isolating MSCs in this manner as it negates the use of invasive procedures such as bone marrow aspiration and is also more readily available. Furthermore, the cells collected from birth-associated tissues have been documented to demonstrate an improved capacity for self-renewal, differentiation, and an increased rate of proliferation when compared to their adult bone marrow-derived counterparts [15].

Human placental tissue is a potential source of birth-associated MSCs. Placental tissue has been characterised from four different locations: amniotic epithelial, amniotic mesenchymal stromal cells, chorionic mesenchymal stromal cells, and chorionic trophoblastic tissue [94]. There are four potential sources of placental tissue MSCs, but only the chorionic and amniotic mesenchymal stromal cells have been shown to demonstrate MSC properties [119]. Placental MSCs are reported as having a limited proliferative lifespan and as lacking adipogenic differentiation potential; further research is required to achieve a comprehensive conclusion [94].

Umbilical cord blood can be subdivided into whole umbilical cord, umbilical cord blood, and Wharton's jelly [40]. Umbilical cord MSCs demonstrate distinct features in comparison to bone marrow-derived MSCs. Umbilical MSCs and cord blood MSCs display an initially higher proliferative capacity when compared to bone marrow MSCs, but similar to placental MSCs they appear to lack an adipogenic differentiation capacity [13, 29].

In conclusion, although bone marrow was the original site for isolation of MSCs, recent advances in our understanding of MSC biology have determined that there are other locations that may also yield MSCs. However, despite these recent advances, bone marrow remains the standard location for MSC isolation; further research will evaluate alternative locations and determine their value in practical and functional applications.

4 Bio-markers of MSC

Although no specific marker for MSCs has yet been identified, there are an abundance of non-specific surface antigens on MSCs. To provide clarification, the International Society for Cellular Therapy has provided guidance on MSC markers; MSCs must express CD73, CD90, CD105 and lack the expression of CD34, CD45, CD14, CD11b, CD19 or MHC class II antigens [39]. However, there are MSC marker variations readily located within the literature; STRO-1 provides a good example [25, 56, 57, 112] (Fig. 2).

5 Reparative Mechanistic Properties of MSC

Preclinical studies and clinical trials demonstrate that the application of MSCs stimulates wound repair and regeneration with efficient amelioration of a number of clinical conditions [18, 65, 80, 100], (www.clinicaltrials.gov). However, the precise mechanism of MSC-mediated wound repair and regeneration is not clear. One of the unique properties of MSCs is their site-specific migration and engraftment to injured tissues and differentiation into specific cell types. A variety of experimental animal models suggest active participation in wound repair and tissue regeneration [65, 80, 100]. On the other hand, some studies postulate that MSC-secreted paracrine factors play a vital role for wound repair, most likely through their anti-inflammatory, anti-apoptotic, angiogenic and immunomodulatory properties [9, 24, 83, 90, 137]. Additional reports suggest that MSC secretory products are capable of stimulating tissue-specific regional progenitor cells propagating tissue regeneration [47, 118].

5.1 Functional Contribution of MSCs in Tissue Repair

In 2002, Toma and colleagues injected human bone marrow MSCs isolated from healthy donors into the myocardium of healthy mice. They observed that MSC had differentiated into cardiomyocyte-like cells after a week [121]. Berry and colleagues injected MSCs into the infarct region of the cardiac wall of myocardial infarction rat models and demonstrated that MSC treatment improved cardiac function; it reduced cardiomyocyte apoptosis and fibrosis scars in comparison to non-MSC treated control groups [18]. They also showed that transplanted MSCs expressed the cardiomyocyte-specific protein 'troponin T' while lacking a cardiomyocyte morphology, suggestive of a putative paracrine role that underpinned the reparative process.

MSC differentiation into type I and type II alveolar epithelial cells (AECI and AECII respectively) in vivo has been reported [65, 100]. Studies on bleomycin-induced animal lung fibrosis models demonstrated that following intratracheal and intravenous administration of MSCs, a small proportion of transplanted cells were engrafted to the affected lung and differentiated into AECI and AECII cells with an accompanying amelioration of pulmonary fibrosis [65, 100]. Human MSCs are capable of in vitro differentiation into Surfactant Protein-C (SP-C; a bio-marker of AECII)—expressing AECII-like cells when co-cultured with fetal lung mesenchymal cells [72]. In addition, the systemic application of murine MSCs in a cisplatin-induced acute renal failure mouse model resulted in migration and engraftment to the affected kidneys. This migration and engraftment was associated with differentiation into renal tubular epithelial cells and amelioration of renal dysfunction with augmentation of renal tubular regeneration. This is suggestive of the MSC as a potential candidate cell for a regenerative medicine-based therapy for the treatment of acute renal failure [80].

The differentiation of MSCs into hepatocytes was demonstrated when Sato and colleagues injected human MSCs directly into an alcohol-induced injury in the rat liver and assessed for expression of hepatocyte-specific bio-markers over an ensuing time-course [105]. From 7 days post-transplant, MSCs displayed expression of hepatocyte-specific and linked proteins including human-specific alpha-fetoprotein (AFP), albumin (Alb), cytokeratin-19 (CK-19), cytokeratin-18 (CK-18), and asialoglycoprotein receptor (AGPR) [105]. In addition, MSCs have been shown to differentiate into functional neuronal phenotypes [37, 135], retinal pigment epithelial cells [7] and skin epithelial cells [81].

Increased reports describe differentiation of MSCs into a variety of adult cell phenotypes. In many of these instances, differentiation into the desired cell-type was confirmed based on their cell-type specific biomarkers. Although some markers are specific for certain cells, this is not the case in every instance. Empirical analysis on both human and rodent MSCs demonstrated that the MSC is, by nature, primed for osteogenic, chondrogenic, adipogenic, and vascular smooth muscle differentiation and can undergo active differentiation under appropriate culture condition via activation of either transforming growth factor-beta, hedgehog, peroxisome proliferation-activated receptor-mediated interaction, and mitogen-activated protein kinase pathways, respectively [36]. Thereby, precaution must be taken in the application of MSCs in vivo to avoid any unwanted ectopic differentiation as a consequence of their relatively non-specific responsiveness to external cues.

5.2 Tissue Repair by MSC-Mediated Paracrine Mechanism

A growing body of evidence supports the hypothesis that paracrine mechanisms may underpin the role that the MSCs play in tissue repair and the regenerative process. MSCs possess an immunomodulatory function that has been demonstrated through their therapeutic efficacy in alleviation of graft-versus-host disease and animal models of bronchial asthma through putative roles in modulating Type-1 (Th1) and Type-2 (Th2) immune responses [84]. MSC-secreted factors are cyto-protective as demonstrated in the cardiac injury animal model driven by anti-apoptotic and inotropic effects [47]. The MSC-mediated anti-apoptotic effect can be driven by up-regulation of the anti-apoptotic gene Bcl-2, which was demonstrated in an animal model of emphysema [137]. Animal models of myocardial infarction and pulmonary hypertension have demonstrated that transplanted MSCs improve cardiac function and pulmonary vasculature by stimulating neovascularisation possibly via their secretory VEGF (vascular endothelial growth factor) and eNOS (endothelial nitric oxide synthase) [9, 24, 61]. The anti-inflammatory function of MSCs has been documented in many animal model studies, in which the mechanism is paracrine in nature and occurs via blocking of anti-inflammatory cytokines such as TNF-α and IL-1 [52, 90].

6 MSC Therapy in Pulmonary Disease

6.1 Acute Lung Injury

Acute lung injury (ALI) represents a continuum of clinical and radiological changes that affect the lungs. ALI can occur at any age and is characterised by a rapid onset of severe hypoxemia that is not secondary to left atrial hypertension [17]. Acute respiratory distress syndrome represents the most severe form of ALI. The definition of ALI has evolved through time as our understanding of the condition has improved. ALI was first described by Ashbaugh in 1967 with the description of a group of 12 patients who had refractory hypoxemia with abnormal changes on radiographic and pulmonary function tests [8].

6.1.1 Acute Respiratory Distress Syndrome

Acute respiratory distress syndrome (ARDS) is a common and devastating clinical syndrome of ALI caused by various direct and indirect insults including infection, trauma, and major surgery. It can result in respiratory failure and ultimately death [128]. The pathological hallmarks of ARDS include diffuse alveolar damage with presence of neutrophils, macrophages, erythrocytes, formation of hyaline membranes, accumulation of protein rich oedema fluid in the alveolar spaces, capillary injury and disruption of the alveolar epithelium [4, 10, 11]. ARDS is a leading cause of death and disability in critically ill adults and children [101]. In the United States, there are 2,00,000 new cases of ARDS diagnosed each year, with a high mortality rate of 40 % (comparable to that seen in breast cancer; [102]). To date, there is no curative treatment for this devastating disease and the management is widely supportive [55].

A growing number of animal model studies demonstrate compelling data on the beneficial effects of MSCs in resolving acute lung injuries induced by endotoxin [32, 52, 70, 78], hyperoxia [26], pneumonia [67] and systemic sepsis [83]. In a recent description, endotoxin-induced lung injury in explanted human lungs was ameliorated with the infusion of MSCs [70]. The accumulation of this pre-clinical data offers considerable hope that MSCs could be a potential candidate for the effective therapy of ARDS. However, MSCs have not yet been evaluated for the therapeutic efficacy for ARDS in clinical trials.

In the ALI model, injury is induced by administration of bacterial endotoxin lipopolysaccharide (LPS) either via the intraperitoneal or intratracheal route, which drives the development of acute pulmonary inflammation within 24–48 h of LPS challenge in mice [100]. Evaluation of the LPS-induced mouse ALI model demonstrated that intravenous or intratracheal administration of MSCs within 1–4 h of LPS challenge significantly attenuated pulmonary inflammation, alveolar injuries, improved alveolar fluid clearance, and reduced mortality [52]. This improvement of the pulmonary condition was observed in the absence of significant engraftment of MSCs in the lung, suggesting a paracrine role of MSCs in the alleviation of ALI. This alleviation could be through down-regulation of pro-inflammatory responses via repression of TNF-α and increased anti-inflammatory cytokine IL-10 [52]. In support of MSC-paracrine mediated anti-inflammatory effects, Ortiz and colleagues demonstrated that MSCs and/or acellular conditioned media collected from cultured MSCs attenuated acute pulmonary inflammation. This attenuation was via suppression of both IL-1α-dependent T-lymphocyte proliferation and inhibition of TNF-α secretion by activated macrophages via MSC-secreted IL-1 receptor antagonist in vitro and in the bleomycin-induced murine lung injury model [90].

Nemeth and colleagues demonstrated that MSCs were stimulated by pro-inflammatory cytokines and endotoxins such as TNF-α and LPS. MSC endotoxin-based activation occurred via toll-like receptor-4, resulting in increased production of cyclooxygenase-2 and increased prostaglandin-E2 release. MSC-secreted prostaglandin-E2 drove increased macrophage IL-10 secretion and attenuated sepsis and sepsis-associated lung injury [83]. The explanted human lung model provided the demonstration that MSC enhanced LPS-induced ALI repair had likely occurred in a keratinocyte growth factor (KGF)-dependent manner [70]. Preclinical data are promising; however, clinical trials will decide the ultimate fate of MSCs as a therapeutic modality for ARDS in the near future.

6.2 Chronic Lung Disease

Chronic lung disease refers to any condition resulting in the long-term impairment of the lung that affects an individual's daily functioning [134]. The conditions that result in chronic lung disease are varied in their etiology, progression, clinical features and management [117]. For example, cystic fibrosis occurs due to a genetic defect, chronic obstructive pulmonary disease may occur as a result of an environmental irritant such as cigarette smoke [92], and finally some chronic diseases, such as idiopathic pulmonary fibrosis, may occur due an unknown cause. Although some chronic lung diseases such as asthma can be controlled and treated, many eventually result in respiratory failure.

6.2.1 Chronic Obstructive Pulmonary Disease

Chronic obstructive pulmonary disease (COPD) is the fourth leading cause of death worldwide and has been projected to be the third leading cause in 2020 [27]. No curative therapy is available for COPD at this time. COPD is characterised by an ongoing cycle of repeated destruction and repair of bronchilo-alveolar regions with subsequent tissue remodelling and sustained irreversible airway obstruction [2]. Approximately 20 % of patients with COPD present with emphysema, which is characterised by destruction of terminal bronchioles and alveolar walls resulting in an irreversible enlargement of alveolar spaces. The pathogenesis of COPD is not well understood. However, a significant reduction of circulating CD34+ progenitor cells has been observed in patients with end-stage COPD [92] and an elastase-induced experimental lung emphysema model [1]. Circulating bone marrow-derived CD34+ cells are haemopoietic progenitors thought to play a role in tissue repair [92]. The causes of progenitor cell destruction in COPD are not clear; however, it is assumed that the products of smoking create oxidative stress that may cause or contribute to progenitor cell destruction and apoptosis [63].

Systemic administration of bone marrow-derived MSCs was reported to ameliorate the emphysematous changes in the irradiation and papain-induced experimental mouse models [137]. Here Zhen and colleagues demonstrated that transplanted MSCs were localised to the emphysematous lung parenchyma and had differentiated into AECIIs. This was accompanied by reduced alveolar epithelial cell apoptosis, via Bcl-2 expression, and reduced enlargement of alveolar spaces [137]. Autologous intratracheal transplantation of bone marrow stem cells significantly mitigated elastase-induced pulmonary emphysema in the rabbit model [133]. The transplantation of bone marrow stem cells was associated with improved lung function, an attenuation of inflammation, an inhibition of epithelial apoptosis, a decrease in matrix metalloproteinase-2 expression, and the stimulation of alveolar and bronchiolar cell proliferation where engraftment and differentiation of the transplanted stem was negligible [133].

A Phase II, multicenter, randomized, double-blind, placebo-controlled clinical trial for the evaluation of safety and efficacy of MSCs for the treatment of moderate to severe COPD has recently been completed (www.clinicaltrial.gov). The trial enrolled 62 patients with COPD in six different centers in the United States. MSCs were administrated through an intravenous route. The complete report has yet to be published; however, preliminary reports are indicative of an improvement of quality of life with reduction of serum C-reactive protein, suggestive of a mitigation of inflammation (http://copsonlinenews.blogspot.com/2011/04/osiris-therapeutics-reports-interim.html).

6.2.2 Cystic Fibrosis

Cystic fibrosis (CF) of the lung is an autosomal recessive disorder caused by a mutation in the gene encoding the CF transmembrane conductance regulator (CFTR). CFTR is expressed in airway epithelial cells and the protein located on the luminal side of the plasma membrane, where it serves as a regulator of the Cl^- channel to maintain fluid and ions transport [75, 111, 116]. Activation of CFTR negatively regulates the epithelial Na^+ channel, which is why mutation of CFTR causes dysfunction of both Na^+ and Cl^- channels [75, 116]. According to current hypotheses on CF lung disease, the loss of Cl^- ion secretion and increased Na^+ ion absorption by airway epithelia reduce the thickness of the airway surface liquid layer overlying the epithelia, resulting in impaired mucociliary clearance [77]. Loss of CFTR function also suppresses mucous and antimicrobial factors secretion by airway submucosal glands [130]. Therefore, dysfunction of CFTR causes formation of thick and dehydrated mucous membranes that provides an ideal environment for persistent bacterial infection, triggering chronic inflammation and ultimately resulting in organ failure. At present, there is no curative treatment for CF. Because a genetic mutation underpins the pathogenesis of this disease, gene therapy is thought to be a valid option for the cure of CF [30]. Stem cell therapy has also been proposed to restore CFTR defective airway epithelia and to alleviate the concomitant inflammation [113, 117].

The main hurdle for stem cell therapy in the restoration of CFTR-defective epithelial cells is their low engraftment efficiency in the lung. Animal models demonstrate that transplantation of wild-type CFTR-expressing engineered bone marrow-derived MSCs in the CFTR knock-out transgenic mice results in a lung engraftment rate of about 0.025 % [71]. Moreover, the CFTR-expressing airway epithelial cells represented less than 0.01 % of the total airway epithelial cells, which was insufficient to replenish the lung with CFTR-expressing epithelial cells [71]. A low engraftment efficacy of CFTR-expressing MSCs in the intestinal epithelia of CFTR knock-out mice was reported by Bruscia and colleagues, where the engraftment was less than 0.01 % [22, 23]. These two studies indicate that complete restoration of CFTR-defective lung epithelial cells by transplanted CFTR-expressing MSCs, at least in the current animal models of CF, is virtually impossible.

An in vitro study has suggested that to restore epithelial ion and fluid clearance, it is not necessary to replace 100 % of CFTR-defective cells. The restoration of 6–20 % of CFTR-expressing epithelial cells was sufficient for effective Cl^- secretion by airway epithelial cells [42, 60]. Conversely, earlier reports suggested that all cells must express CFTR to re-establish the negative regulatory effects on airway Na^+ channel for effective Na^+ ion absorption homeostasis [48, 59].

Patients with CF frequently suffer from severe repeated pulmonary infections and chronic inflammation. This is often the main cause of sickness, disability, and mortality due to failure of lung function. It has been suggested that the anti-inflammatory and immunomodulatory functions of MSCs (discussed above) could serve a role in the ablation of the inflammatory conditions of CF lungs with potential therapeutic benefits.

6.2.3 Idiopathic Pulmonary Fibrosis

Idiopathic pulmonary fibrosis (IPF) is a specific form of chronic, progressive fibrosing interstitial pneumonia of unknown etiology, occurring primarily in older adults, limited to the lungs, and associated with the histopathological and/or radiological pattern of usual interstitial pneumonias [98]. This disease was also referred as cryptogenic fibrosing alveolitis before being displaced by the term IPF [110]. IPF is characterised by repeated microinjuries to the alveolar epithelium and consequent abnormal wound repair. This is accompanied by the accumulation of fibroblasts and myofibroblasts, with the deposition of excessive extracellular matrix resulting in the replacement of normal lung tissue with fibrotic scars. Accompanying the alteration of normal lung architecture is the clinical manifestation of progressive dyspnea worsening and reduced lung function resulting in respiratory failure [46, 98]. Unlike other inflammatory and fibrotic lung diseases, IPF does not respond to steroids and other potent immunosuppressive agents largely fail to reduce death rates in patients with IPF; the only potential curative treatment option at the moment is lung transplantation [108].

Patients with IPF are generally more than 50 years of age and two-thirds are older than 60 years at disease presentation. The median survival of patients with IPF is 2.8 years [20]. In the United States, the incidence and prevalence of IPF are 16.3 and 42.7 per 1,00,000 people, respectively [99]. Higher incidences are noted for the 75 years or older age group, in which it is 76.4 per 1,00,000 people, as compared to age group 18–34 years, with 1.2 per 1,00,000 people. An estimated 48,000 new IPF cases are diagnosed annually in the United States alone [98, 99]. In the United Kingdom, the overall incidence rate of IPF is 4.6 per 1,00,000 cases per annum [49]. More than 4,000 new IPF cases are currently diagnosed each year in the United Kingdom. The mortality rate from IPF has also increased over the last two decades [87]; death rates from IPF have reported to be higher than death rates from some cancers [58].

Unlike CF, to date, no specific genetic or acquired cause has been identified for IPF; however, mutation in the genes encoding for hTERT [6, 38, 123] and SP-C [120] have been reported in the familial form of IPF. The pathophysiological process of IPF is widely unknown. Previously, IPF was thought to be a consequence of chronic pulmonary inflammation. However, non-responsiveness to anti-inflammatory or anti-fibrotic drugs and lack of histopathological evidence of inflammation in IPF lungs suggest that inflammation may not be an initiating trigger in the pathogenesis of this disease [109]. An evolving hypothesis describes IPF as a consequence of aberrant alveolar wound repair and regeneration, most likely due to a combination of repeated AEC injury [31, 109], increased AEC apoptosis [14, 69, 125], dysregulated epithelial-mesenchymal cross-talk [110], polarised immune response [114, 124] and altered coagulation cascade [27, 64].

Stem cell-mediated regenerative therapeutic approaches have been proposed for the treatment of IPF. To assist in these studies, several animal models of pulmonary fibrosis have been developed [79], including the bleomycin-induced pulmonary fibrosis model [53, 79], radiation-induced fibrosis [54], silica-induced fibrosis [33] and asbestos-induced lung fibrosis models [21].

The bleomycin-induced pulmonary fibrosis mouse model provided a demonstration of migration and engraftment of endotracheal or systematically transplanted MSCs towards the site of injuries of the lung and subsequent attenuation of pulmonary fibrosis [91, 100]. Systemic administration of bone marrow-derived MSCs after 4 h of bleomycin administration attenuated pulmonary inflammation, reduced fibrosis, and decreased mortality after 14 days of injury. Transplanted MSCs had engrafted into the injured alveoli with accompanying differentiation into type II AEC-like phenotype [91]. However, when MSCs were administered after 7 days of injury, the MSC-mediated protective function was abrogated [91]. Complementary results were noted in independent studies [100]. In 2007, Ortiz and colleagues showed that MSCs protected against bleomycin-induced lung injury and reduced fibrosis by blocking pro-inflammatory cytokines such as TNF-α and IL-1 by MSC-associated IL-1 receptor antagonist [90].

The administration of KGF-expressing MSCs or HSCs (haematopoietic stem cells) in the bleomycin-induced mouse lung fibrosis model was associated with reduced fibrosis via suppression of collagen accumulation [3]. KGF has an established role in the repair of alveolar epithelium through stimulation of type II AEC proliferation, migration and spreading [51, 93, 132]. This proof-of-concept experiment demonstrated that genetically modified MSCs or HSCs with suitable cytokine/growth factor have potential as a therapeutic strategy for pulmonary fibrosis [3].

The pre-clinical studies described previously suggest a role for MSCs as a potential candidate for regenerative therapy for IPF. There are remaining concerns that MSC have pro-fibrotic effects and could deteriorate the pathological condition if they are applied in chronic lung fibrosis [131]. Yan and colleagues demonstrated that after systemic application of MSCs at 4 h of irradiation-induced lung injury, transplanted cells engrafted in the alveolar and bronchiolar epithelium and differentiated into epithelial phenotype; however, MSCs administered at 60 and 120 days post-injury localised in interstitial spaces and differentiated into myofibroblasts, a fibrotic cell that plays major role in fibrogenesis [131]. These authors concluded that fate of MSC differentiation is controlled by the microenvironment milieu and warned that MSC therapy might be ideal for ALI but may augment fibrosis in chronic lung fibrosis, such as IPF.

Supporting the putative profibrotic nature of MSCs, an in vitro study demonstrated that human and mouse MSCs secrete TGF-β1 and Wnt proteins that stimulate both human/mouse lung fibroblast proliferation and collagen production—two major hallmarks of lung fibrosis [104]. Prostaglandin E2 treatment significantly inhibited resident MSC proliferation and collagen secretion and abrogated fibrotic differentiation into myofibroblasts [126]. If this is true for MSCs from common sources such as bone marrow and cord blood, prostaglandin E2 could be administered concomitantly with MSCs to reduce putative fibrotic effects.

Conversely, no TGF-β1 expression was detected in MSCs isolated from the bone marrow of normal healthy individuals or patients with IPF; the expression of fibroblast growth factor and VEGF was not significantly different in either case [5]. However, CXCR4, a potent chemokine receptor, was significantly over-expressed in patients with IPF. The increased CXCR4 expression by IPF MSCs suggests that the bone marrow is probably implicated in the pathophysiology of IPF by mobilising resident MSCs in response to or preceding lung injury [5]. Further study will confirm that whether this MSC mobilisation is a mere attempt to repair lung injury or solely aggravates fibrosis in IPF.

6.2.4 Bronchial Asthma

Bronchial asthma, one of the most common chronic inflammatory lung diseases, affects over 300 million people world-wide [76]. Asthma is characterised by reversible airway obstruction, hyper-responsiveness of airway smooth muscle, and airway inflammation. There is no permanent curative treatment for asthma; most of the patients remain symptomatically controlled by combined mediation of bronchodilator and steroids. However, approximately 5 % of patients with asthma are resistant to conventional therapy and suffer from substantial morbidity and mortality [117]. The ability of MSCs to modulate the immune system encouraged researchers to explore the potential of MSCs as an anti-asthmatic therapy.

The ragweed-induced mouse asthma model was used to demonstrate that administration of bone marrow-derived MSCs ameliorated allergic and inflammatory responses in the airway [84]. After transplantation, animals were protected from the majority of asthma-specific pathological changes, including inhibition of eosinophil infiltration and excess mucus production in the lung, decreased levels of Th2 cytokines (IL-4, IL-5, and IL-13) in bronchial lavage, and lowered serum levels of Th2 immunoglobulins (IgG1 and IgE) [84].

6.3 Pulmonary Vascular Disease

Pulmonary vascular disease is an umbrella term used to describe a group of conditions associated with damage or alterations to the lung vasculature [35]. Diseases within this realm include pulmonary hypertension, pulmonary embolism, pulmonary veno-occlusive diseases, arterio-venous malformation and pulmonary edema [41]. Pulmonary hypertension is frequently associated with lung parenchymal damage and can present as a secondary complication of chronic lung disease [35, 41]. The remaining conditions within the group are frequently associated with vascular and cardiac pathologies.

6.3.1 Pulmonary Hypertension

Pulmonary hypertension (PH) is rapidly progressive and often fatal disease characterised by increased pulmonary arterial pressure, right heart dysfunction, and lung vasculature remodelling leading to loss of alveolar vasculature [97]. MSC-based therapy has been explored for application in the regeneration of pulmonary vasculature because they secrete VEGF a potent stimulator of neovascularisation. Intratracheal administration of bone marrow-derived MSCs in the monocrotaline-induced rat PH model attenuated PH [9]. Transplantation of MSCs reduced monocrotaline-induced pulmonary arterial pressure and improved pulmonary vasculature through paracrine mediator(s). Immunohistochemistry showed no evidence of endothelial differentiation of MSCs [9].

Intravenous administration of MSCs and eNOS-overexpressing MSCs in the monocrotaline-induced rat PH model also resulted in attenuation of PH and improved right ventricular hypertrophy in comparison to un-treated control groups [61]. Interestingly, the reduction of right ventricular hypertrophy was significantly higher in the eNOS-overexpressing MSC treated group in comparison to the MSC groups, suggesting that MSC-mediated improvement of pulmonary vasculature in PH could be driven by modulation of nitric oxide secretion by the vascular endothelium [61].

7 Tissue Engineered Lung Tissue

Current research in the tissue engineering field is focused on exploration of 3-dimensional tissue culture systems for use in development of functional lung tissue. The ultimate ambition of these studies is to reduce donor-dependent lung transplantation [85, 122]. Because of the unique architecture of the lung and its anatomical and physiological complexity, this presents a major challenge. Tissue-engineered tracheas (wind pipe) have been developed using MSCs isolated from various sources before being cultured on biodegradable and biosynthetic scaffolds to generate tracheal cartilage for the repair of congenital tracheal defects in both animal and human clinical trial models [88, 89]. Very recently, a group of tissue engineers, stem cell researchers, and medical professionals developed a functional human airway by culturing MSC-derived chondrocytes on an acellular tracheal scaffold, which was subsequently transplanted in a female patient who had suffered airway damage from tuberculosis [73]. Macchiarini and colleagues first decellularised a 7-cm long segment of human trachea taken from a 51-year-old white female donor who had died of cerebral hemorrhage. The recipient's bone marrow-derived MSCs were differentiated into chondrocytes and airway epithelial cells cultured using *in vitro* tissue culture system. The MSC-derived chondrocytes were seeded on the external surface of the acellular trachea and epithelial cells seeded on the luminal surface, in an equal ratio, and cultured in an air–liquid interface rotating bioreactor for 96 h. After *in vitro* preparation, the tracheal construct was transplanted to the left bronchus of the recipient, which improved breathing difficulties without graft rejection [73]. This achievement should encourage the development of more complicated parts of the lung, such the alveoli and pulmonary vasculature, in the near future.

8 Challenges for MSC Therapy in Pulmonary Disease

Although pre-clinical data provide evidence of promising therapeutic benefits of MSCs in various pulmonary diseases, many hurdles remain. Some important parameters such as MSC choice, dose, timing, route of administration, and selection of suitable clinical conditions for cell therapy need to be established before clinical application [2]. As a route of administration, intravenous, intra-arterial, and intra-tracheal routes have all been implemented in animal models for MSC delivery. MSC engraftment was higher when administered into injured lungs through the intravenous route [43], whereas administration through the intra-arterial route was accompanied by complications associated with microvasculature occlusion [45]. The intratracheal route was also demonstrated to be suitable for efficient engraftment [52, 70]. Clinical trials of MSCs in pulmonary diseases, such as COPD, although safe, have not yet evidenced an appropriate efficacy of repair. The prospects of MSC-based regenerative cell therapy for the treatment of pulmonary diseases will be determined by the outcome of future large-scale clinical trials.

References

1. Abe S, Boyer C, Liu X, Wen FQ, Kobayashi T, Fang Q, Wang X, Hashimoto M, Sharp JG, Rennard SI (2004) Cells derived from the circulation contribute to the repair of lung injury. Am J Respir Crit Care Med 170(11):1158–1163
2. Abreu SC, Antunes MA, Pelosi P, Morales MM, Rocco PR (2011) Mechanisms of cellular therapy in respiratory diseases. Intensive Care Med 37(9):1421–1431
3. Aguilar S, Scotton CJ, McNulty K, Nye E, Stamp G, Laurent G, Bonnet D, Janes SM (2009) Bone marrow stem cells expressing keratinocyte growth factor via an inducible lentivirus protects against bleomycin-induced pulmonary fibrosis. PLoS ONE 4(11):e8013

4. Anderson WR, Thielen K (1992) Correlative study of adult respiratory distress syndrome by light, scanning, and transmission electron microscopy. Ultrastruct Pathol 16(6):615–628
5. Antoniou KM, Papadaki HA, Soufla G, Kastrinaki MC, Damianaki A, Koutala H, Spandidos DA, Siafakas NM (2010) Investigation of bone marrow mesenchymal stem cells (BM MSCs) involvement in idiopathic pulmonary fibrosis (IPF). Respir Med 104(10): 1535–1542
6. Armanios MY, Chen JJ, Cogan JD, Alder JK, Ingersoll RG, Markin C, Lawson WE, Xie M, Vulto I, Phillips JA 3rd, Lansdorp PM, Greider CW, Loyd JE (2007) Telomerase mutations in families with idiopathic pulmonary fibrosis. N Engl J Med 356(13):1317–1326
7. Arnhold S, Absenger Y, Klein H, Addicks K, Schraermeyer U (2007) Transplantation of bone marrow-derived mesenchymal stem cells rescue photoreceptor cells in the dystrophic retina of the rhodopsin knockout mouse. Graefes Arch Clin Exp Ophthalmol 245(3): 414–422
8. Ashbaugh DG, Bigelow DB, Petty TL, Levine BE (1967) Acute respiratory distress in adults. The Lancet Crit Care Resusc 7(1):60–61
9. Baber SR, Deng W, Master RG, Bunnell BA, Taylor BK, Murthy SN, Hyman AL, Kadowitz PJ (2007) Intratracheal mesenchymal stem cell administration attenuates monocrotaline-induced pulmonary hypertension and endothelial dysfunction. Am J Physiol Heart Circ Physiol 292(2):120–128
10. Bachofen M, Bachofen H (1979) Follow-up of severe adult respiratory distress syndrome. Schweiz Med Wochenschr 109(50):1982–1989
11. Bachofen M, Weibel ER (1977) Alterations of the gas exchange apparatus in adult respiratory insufficiency associated with septicemia. Am Rev Respir Dis 116(4):589–615
12. Baksh D, Song L, Tuan RS (2004) Adult mesenchymal stem cells: characterization, differentiation, and application in cell and gene therapy. J Cell Mol Med 8(3):301–316
13. Baksh D, Yao R, Tuan RS (2007) Comparison of proliferative and multilineage differentiation potential of human mesenchymal stem cells derived from umbilical cord and bone marrow. Stem Cells 25(6):1384–1392
14. Barbas-Filho JV, Ferreira MA, Sesso A, Kairalla RA, Carvalho CR, Capelozzi VL (2001) Evidence of type II pneumocyte apoptosis in the pathogenesis of idiopathic pulmonary fibrosis (IFP)/usual interstitial pneumonia (UIP). J Clin Pathol 54(2):132–138
15. Barlow S, Brooke G, Chatterjee K, Price G, Pelekanos R, Rossetti T, Doody M, Venter D, Pain S, Gilshenan K, Atkinson K (2008) Comparison of human placenta- and bone marrow-derived multipotent mesenchymal stem cells. Stem Cells Dev 17(6):1095–1107
16. Beltrami AP, Barlucchi L, Torella D, Baker M, Limana F, Chimenti S, Kasahara H, Rota M, Musso E, Urbanek K, Leri A, Kajstura J, Nadal-Ginard B, Anversa P (2003) Adult cardiac stem cells are multipotent and support myocardial regeneration. Cell 114(6):763–776
17. Bernard GR, Artigas A, Brigham KL, Carlet J, Falke K, Hudson L, Lamy M, Legall JR, Morris A, Spragg R (1994) The American-European consensus conference on ARDS. Definitions, mechanisms, relevant outcomes, and clinical trial coordination. Am J Respir Crit Care Med 149(3):818–824
18. Berry MF, Engler AJ, Woo YJ, Pirolli TJ, Bish LT, Jayasankar V, Morine KJ, Gardner TJ, Discher DE, Sweeney HL (2006) Mesenchymal stem cell injection after myocardial infarction improves myocardial compliance. Am J Physiol Heart Circ Physiol 290(6): 196–203
19. Bianco P, Robey PG, Simmons PJ (2008) Mesenchymal stem cells: revisiting history, concepts, and assays. Cell Stem Cell 2(4):313–319
20. Bjoraker JA, Ryu JH, Edwin MK, Myers JL, Tazelaar HD, Schroeder DR, Offord KP (1998) Prognostic significance of histopathologic subsets in idiopathic pulmonary fibrosis. Am J Respir Crit Care Med 157(1):199–203
21. Bozelka BE, Sestini P, Gaumer HR, Hammad Y, Heather CJ, Salvaggio JE (1983) A murine model of asbestosis. Am J Pathol 112(3):326–337
22. Bruscia EM, Price JE, Cheng EC, Weiner S, Caputo C, Ferreira EC, Egan ME, Krause DS (2006) Assessment of cystic fibrosis transmembrane conductance regulator (CFTR) activity

in CFTR-null mice after bone marrow transplantation. Proc Natl Acad Sci U S A 103(8): 2965–2970
23. Bruscia EM, Ziegler EC, Price JE, Weiner S, Egan ME, Krause DS (2006) Engraftment of donor-derived epithelial cells in multiple organs following bone marrow transplantation into newborn mice. Stem Cells 24(10):2299–2308
24. Burdon TJ, Paul A, Noiseux N, Prakash S, Shum-Tim D (2011) Bone marrow stem cell derived paracrine factors for regenerative medicine: current perspectives and therapeutic potential. Bone Marrow Res. doi:10.1155/2011/207326
25. Chamberlain G, Fox J, Ashton B, Middleton J (2007) Concise review: mesenchymal stem cells: their phenotype, differentiation capacity, immunological features, and potential for homing. Stem Cells 25(11):2739–2749
26. Chang YS, Oh W, Choi SJ, Sung DK, Kim SY, Choi EY, Kang S, Jin HJ, Yang YS, Park WS (2009) Human umbilical cord blood-derived mesenchymal stem cells attenuate hyperoxia-induced lung injury in neonatal rats. Cell Transplant 18(8): 869-886
27. Chapman HA, Allen CL, Stone OL (1986) Abnormalities in pathways of alveolar fibrin turnover among patients with interstitial lung disease. Am Rev Respir Dis 133(3):437–443
28. Clark BR, Jamieson C, Keating A (1997) Human long-term bone marrow culture. Methods Mol Biol 75:249–256
29. Conconi MT, Burra P, Di Liddo R, Calore C, Turetta M, Bellini S, Bo P, Nussdorfer GG, Parnigotto PP (2006) CD105(+) cells from Wharton's jelly show in vitro and in vivo myogenic differentiative potential. Int J Mol Med 18(6):1089–1096
30. Conese M, Ascenzioni F, Boyd AC, Coutelle C, De Fino I, De Smedt S, Rejman J, Rosenecker J, Schindelhauer D, Scholte BJ (2011) Gene and cell therapy for cystic fibrosis: from bench to bedside. J Cyst Fibros 10(2):114–128
31. Corrin B, Dewar A, Rodriguez-Roisin R, Turner-Warwick M (1985) Fine structural changes in cryptogenic fibrosing alveolitis and asbestosis. J Pathol 147(2):107–119
32. Danchuk S, Joni H, Hossain FF, Sorge RF, Ramsey A, Ryan W, Sullivan DE (2010) Human multipotent stromal cells attenuate lipopolysaccharide-induced acute lung injury in mice via secretion of tumor necrosis factor-alpha-induced protein 6. Stem Cell Res Ther 2(3):27–42
33. Davis GS, Leslie KO, Hemenway DR (1998) Silicosis in mice: effects of dose, time, and genetic strain. J Environ Pathol Toxicol Oncol 17(2):81–97
34. Deb A, Wang S, Skelding KA, Miller D, Simper D, Caplice NM (2003) Bone marrow-derived cardiomyocytes are present in adult human heart: a study of gender-mismatched bone marrow transplantation patients. Circulation 107(9):1247–1249
35. Deboeck G, Scoditti C, Huez S, Luc Vachiery J, Lamotte M, Sharples L, Melot C, Naeije R (2012) Exercise to predict outcome in idiopathic vs associated pulmonary arterial hypertension. Eur Respir J. doi:10.1183/09031936.00217911
36. Delorme B, Ringe J, Pontikoglou C, Gaillard J, Langonne A, Sensebe L, Noel D, Jorgensen C, Haupl T, Charbord P (2009) Specific lineage-priming of bone marrow mesenchymal stem cells provides the molecular framework for their plasticity. Stem Cells 27(5): 1142–1151
37. Dezawa M, Kanno H, Hoshino M, Cho H, Matsumoto N, Itokazu Y, Tajima N, Yamada H, Sawada H, Ishikawa H, Mimura T, Kitada M, Suzuki Y, Ide C (2004) Specific induction of neuronal cells from bone marrow stromal cells and application for autologous transplantation. J Clin Invest 113(12):1701–1710
38. Diaz de Leon A, Cronkhite JT, Katzenstein AL, Godwin JD, Raghu G, Glazer CS, Rosenblatt RL, Girod CE, Garrity ER, Xing C, Garcia CK (2010) Telomere lengths, pulmonary fibrosis and telomerase (TERT) mutations. PLoS ONE 5(5):e10680
39. Dominici M, Le Blanc K, Mueller I, Slaper-Cortenbach I, Marini F, Krause D, Deans R, Keating A, Prockop D, Horwitz E (2006) Minimal criteria for defining multipotent mesenchymal stromal cells. The international society for cellular therapy position statement. Cytotherapy 8(4):315–317

40. Ennis J, Sarugaser R, Gomez A, Baksh D, Davies JE (2008) Isolation, characterization, and differentiation of human umbilical cord perivascular cells (HUCPVCs). Methods Cell Biol 86:121–136
41. Fares WH, Trow TK (2012) Targeted approaches to the treatment of pulmonary hypertension. Ther Adv Respir Dis. doi:10.1177/1753465812441849
42. Farmen SL, Karp PH, Ng P, Palmer DJ, Koehler DR, Hu J, Beaudet AL, Zabner J, Welsh MJ (2005) Gene transfer of CFTR to airway epithelia: low levels of expression are sufficient to correct Cl^- transport and overexpression can generate basolateral CFTR. Am J Physiol Lung Cell Mol Physiol 289(6):1123–1130
43. Fischer UM, Harting MT, Jimenez F, Monzon-Posadas WO, Xue H, Savitz SI, Laine GA, Cox CS Jr (2009) Pulmonary passage is a major obstacle for intravenous stem cell delivery: the pulmonary first-pass effect. Stem Cells Dev 18(5):683–692
44. Friedenstein AJ, Chailakhjan RK, Lalykina KS (1970) The development of fibroblast colonies in monolayer cultures of guinea-pig bone marrow and spleen cells. Cell Tissue Kinet 3(4):393–403
45. Furlani D, Ugurlucan M, Ong L, Bieback K, Pittermann E, Westien I, Wang W, Yerebakan C, Li W, Gaebel R, Li RK, Vollmar B, Steinhoff G, Ma N (2009) Is the intravascular administration of mesenchymal stem cells safe? Mesenchymal stem cells and intravital microscopy. Microvasc Res 77(3):370–376
46. Gharaee-Kermani M, Gyetko MR, Hu B, Phan SH (2007) New insights into the pathogenesis and treatment of idiopathic pulmonary fibrosis: a potential role for stem cells in the lung parenchyma and implications for therapy. Pharm Res 24(5):819–841
47. Gnecchi M, Zhang Z, Ni A, Dzau VJ (2008) Paracrine mechanisms in adult stem cell signalling and therapy. Circ Res 103(11):1204–1219
48. Goldman MJ, Yang Y, Wilson JM (1995) Gene therapy in a xenograft model of cystic fibrosis lung corrects chloride transport more effectively than the sodium defect. Nat Genet 9(2):126–131
49. Gribbin J, Hubbard RB, Le Jeune I, Smith CJ, West J, Tata LJ (2006) Incidence and mortality of idiopathic pulmonary fibrosis and sarcoidosis in the UK. Thorax 61(11): 980–985
50. Guilak F, Lott KE, Awad HA, Cao Q, Hicok KC, Fermor B, Gimble JM (2006) Clonal analysis of the differentiation potential of human adipose-derived adult stem cells. J Cell Physiol 206(1):229–237
51. Guo J, Yi ES, Havill AM, Sarosi I, Whitcomb L, Yin S, Middleton SC, Piguet P, Ulich TR (1998) Intravenous keratinocyte growth factor protects against experimental pulmonary injury. Am J Physiol 275(4):800–805
52. Gupta N, Su X, Popov B, Lee JW, Serikov V, Matthay MA (2007) Intrapulmonary delivery of bone marrow-derived mesenchymal stem cells improves survival and attenuates endotoxin-induced acute lung injury in mice. J Immunol 179(3):1855–1863
53. Harrison JH Jr, Lazo JS (1987) High dose continuous infusion of bleomycin in mice: a new model for drug-induced pulmonary fibrosis. J Pharmacol Exp Ther 243(3):1185–1194
54. Haston CK, Travis EL (1997) Murine susceptibility to radiation-induced pulmonary fibrosis is influenced by a genetic factor implicated in susceptibility to bleomycin-induced pulmonary fibrosis. Cancer Res 57(23):5286–5291
55. Hayes M, Curley G, Laffey JG (2012) Mesenchymal stem cells—a promising therapy for acute respiratory distress syndrome. Med Rep 4:2
56. Horwitz EM, Le Blanc K, Dominici M, Mueller I, Slaper-Cortenbach I, Marini FC, Deans RJ, Krause DS, Keating A, International Society for Cellular Therapy (2005) Clarification of the nomenclature for MSC: the international society for cellular therapy position statement. Cytotherapy 7(5):393–395
57. Javazon EH, Beggs KJ, Flake AW (2004) Mesenchymal stem cells: paradoxes of passaging. Exp Hematol 32(5):414–425
58. Jemal A, Siegel R, Ward E, Murray T, Xu J, Thun MJ (2007) Cancer statistics, 2007. CA Cancer J Clin 57(1):43–66

59. Johnson LG, Boyles SE, Wilson J, Boucher RC (1995) Normalization of raised sodium absorption and raised calcium-mediated chloride secretion by adenovirus-mediated expression of cystic fibrosis transmembrane conductance regulator in primary human cystic fibrosis airway epithelial cells. J Clin Invest 95(3):1377–1382
60. Johnson LG, Olsen JC, Sarkadi B, Moore KL, Swanstrom R, Boucher RC (1992) Efficiency of gene transfer for restoration of normal airway epithelial function in cystic fibrosis. Nat Genet 2(1):21–25
61. Kanki-Horimoto S, Horimoto H, Mieno S, Kishida K, Watanabe F, Furuya E, Katsumata T (2006) Implantation of mesenchymal stem cells overexpressing endothelial nitric oxide synthase improves right ventricular impairments caused by pulmonary hypertension. Circulation 114(1):181–185
62. Kern SF, Eichler HF, Stoeve JF, Kluter HF, Bieback K (2006) Comparative analysis of mesenchymal stem cells from bone marrow, umbilical cord blood, or adipose tissue. Stem cells 24(5): 1294-1301
63. Kondo T, Hayashi M, Takeshita K, Numaguchi Y, Kobayashi K, Iino S, Inden Y, Murohara T (2004) Smoking cessation rapidly increases circulating progenitor cells in peripheral blood in chronic smokers. Arterioscler Thromb Vasc Biol 24(8):1442–1447
64. Kotani I, Sato A, Hayakawa H, Urano T, Takada Y, Takada A (1995) Increased procoagulant and antifibrinolytic activities in the lungs with idiopathic pulmonary fibrosis. Thromb Res 77(6):493–504
65. Kotton DN, Ma BY, Cardoso WV, Sanderson EA, Summer RS, Williams MC, Fine A (2001) Bone marrow-derived cells as progenitors of lung alveolar epithelium. Development 128(24):5181–5188
66. Kovacic JC, Muller DW, Harvey R, Graham RM (2005) Update on the use of stem cells for cardiac disease. Intern Med J 35(6):348–356
67. Krasnodembskaya A, Song Y, Fang X, Gupta N, Serikov V, Lee JW, Matthay MA (2010) Antibacterial effect of human mesenchymal stem cells is mediated in part from secretion of the antimicrobial peptide LL-37. Stem Cells 28(12):2229–2238
68. Kuhbier JW, Weyand B, Radtke C, Vogt PM, Kasper C, Reimers K (2010) Isolation, characterization, differentiation, and application of adipose-derived stem cells. Adv Biochem Eng Biotechnol 123:55–105
69. Kuwano K, Miyazaki H, Hagimoto N, Kawasaki M, Fujita M, Kunitake R, Kaneko Y, Hara N (1999) The involvement of Fas-Fas ligand pathway in fibrosing lung diseases. Am J Respir Cell Mol Biol 20(1):53–60
70. Lee JW, Fang X, Gupta N, Serikov V, Matthay MA (2009) Allogeneic human mesenchymal stem cells for treatment of *E. coli* endotoxin-induced acute lung injury in the ex vivo perfused human lung. Proc Natl Acad Sci U S A 106(38):16357–16362
71. Loi R, Beckett T, Goncz KK, Suratt BT, Weiss DJ (2006) Limited restoration of cystic fibrosis lung epithelium in vivo with adult bone marrow-derived cells. Am J Respir Crit Care Med 173(2):171–179
72. Ma N, Gai H, Mei J, Ding FB, Bao CR, Nguyen DM, Zhong H (2011) Bone marrow mesenchymal stem cells can differentiate into type II alveolar epithelial cells in vitro. Cell Biol Int 35(12):1261–1266
73. Macchiarini P, Jungebluth P, Go T, Asnaghi MA, Rees LE, Cogan TA, Dodson A, Martorell J, Bellini S, Parnigotto PP, Dickinson SC, Hollander AP, Mantero S, Conconi MT, Birchall MA (2008) Clinical transplantation of a tissue-engineered airway. Lancet 372(9655): 2023–2030
74. Madonna R, Geng YJ, De Caterina R (2009) Adipose tissue-derived stem cells: characterization and potential for cardiovascular repair. Arterioscler Thromb Vasc Biol 29(11):1723–1729
75. Mall MF, Bleich MF, Greger RF, Schreiber RF, Kunzelmann K (1998) The amiloride-inhibitable Na^+conductance is reduced by the cystic fibrosis transmembrane conductance regulator in normal but not in cystic fibrosis airways. J Clin Investig 102(1):15–21

76. Masoli M, Fabian D, Holt S, Beasley R, Global Initiative for Asthma (GINA) Program (2004) The global burden of asthma: executive summary of the GINA Dissemination Committee report. Allergy 59(5):469–478
77. Matsui H, Grubb BR, Tarran R, Randell SH, Gatzy JT, Davis CW, Boucher RC (1998) Evidence for periciliary liquid layer depletion, not abnormal ion composition, in the pathogenesis of cystic fibrosis airways disease. Cell 95(7):1005–1015
78. Mei SHJ, McCarter SD, Deng Y, Parker CH, Liles WC, Stewart DJ (2007) Prevention of LPS-induced acute lung injury in mice by mesenchymal stem cells overexpressing angiopoietin 1. PLoS Med 4(9):e269–e269
79. Moore BB, Hogaboam CM (2008) Murine models of pulmonary fibrosis. Am J Physiol Lung Cell Mol Physiol 294(2):L152–L160
80. Morigi M, Imberti B, Zoja C, Corna D, Tomasoni S, Abbate M, Rottoli D, Angioletti S, Benigni A, Perico N, Alison M, Remuzzi G (2004) Mesenchymal stem cells are renotropic, helping to repair the kidney and improve function in acute renal failure. J Am Soc Nephrol 15(7):1794–1804
81. Nakagawa H, Akita S, Fukui M, Fujii T, Akino K (2005) Human mesenchymal stem cells successfully improve skin-substitute wound healing. Br J Dermatol 153(1):29–36
82. Nash JR, McLaughlin PJ, Butcher D, Corrin B (1993) Expression of tumour necrosis factor-alpha in cryptogenic fibrosing alveolitis. Histopathology 22(4):343–347
83. Nemeth K, Leelahavanichkul A, Yuen PS, Mayer B, Parmelee A, Doi K, Robey PG, Leelahavanichkul K, Koller BH, Brown JM, Hu X, Jelinek I, Star RA, Mezey E (2009) Bone marrow stromal cells attenuate sepsis via prostaglandin E(2)-dependent reprogramming of host macrophages to increase their interleukin-10 production. Nat Med 15(1):42–49
84. Nemeth K, Keane-Myers A, Brown JM, Metcalfe DD, Gorham JD, Bundoc VG, Hodges MG, Jelinek I, Madala S, Karpati S, Mezey E (2010) Bone marrow stromal cells use TGF-beta to suppress allergic responses in a mouse model of ragweed-induced asthma. Proc Natl Acad Sci U SA 107(12):5652–5657
85. Nichols JE, Cortiella J (2008) Engineering of a complex organ: progress toward development of a tissue-engineered lung. Proc Am Thorac Soc 5(6):723–730
86. Oedayrajsingh-Varma MJ, van Ham SM, Knippenberg M, Helder MN, Klein-Nulend J, Schouten TE, Ritt MJ, van Milligen FJ (2006) Adipose tissue-derived mesenchymal stem cell yield and growth characteristics are affected by the tissue-harvesting procedure. Cytotherapy 8(2):166–177
87. Olson AL, Swigris JJ, Lezotte DC, Norris JM, Wilson CG, Brown KK (2007) Mortality from pulmonary fibrosis increased in the United States from 1992 to 2003. Am J Respir Crit Care Med 176(3):277–284
88. Omori K, Tada Y, Suzuki T, Nomoto Y, Matsuzuka T, Kobayashi K, Nakamura T, Kanemaru S, Yamashita M, Asato R (2008) Clinical application of in situ tissue engineering using a scaffolding technique for reconstruction of the larynx and trachea. Ann Otol Rhinol Laryngol 117(9):673–678
89. Omori K, Nakamura T, Kanemaru S, Magrufov A, Yamashita M, Shimizu Y (2008) In situ tissue engineering of the cricoid and trachea in a canine model. Ann Otol Rhinol Laryngol 117(8):609–613
90. Ortiz LA, Dutreil M, Fattman C, Pandey AC, Torres G, Go K, Phinney DG (2007) Interleukin 1 receptor antagonist mediates the anti-inflammatory and anti-fibrotic effect of mesenchymal stem cells during lung injury. Proc Natl Acad Sci U S A 104(26): 11002–11007
91. Ortiz LA, Gambelli F, McBride C, Gaupp D, Baddoo M, Kaminski N, Phinney DG (2003) Mesenchymal stem cell engraftment in lung is enhanced in response to bleomycin exposure and ameliorates its fibrotic effects. Proc Natl Acad Sci U S A 100(14):8407–8411
92. Palange P, Testa U, Huertas A, Calabro L, Antonucci R, Petrucci E, Pelosi E, Pasquini L, Satta A, Morici G, Vignola MA, Bonsignore MR (2006) Circulating haemopoietic and endothelial progenitor cells are decreased in COPD. Eur Respir J 27(3):529–541

93. Panos RJ, Rubin JS, Csaky KG, Aaronson SA, Mason RJ (1993) Keratinocyte growth factor and hepatocyte growth factor/scatter factor are heparin-binding growth factors for alveolar type II cells in fibroblast-conditioned medium. J Clin Invest 92(2):969–977
94. Parolini O, Alviano F, Bagnara GP, Bilic G, Buhring HJ, Evangelista M, Hennerbichler S, Liu B, Magatti M, Mao N, Miki T, Marongiu F, Nakajima H, Nikaido T, Portmann-Lanz CB, Sankar V, Soncini M, Stadler G, Surbek D, Takahashi TA, Redl H, Sakuragawa N, Wolbank S, Zeisberger S, Zisch A, Strom SC (2008) Concise review: isolation and characterization of cells from human term placenta: outcome of the first international workshop on placenta derived stem cells. Stem Cells 26(2):300–311
95. Patt HM, Maloney MA (1975) Bone marrow regeneration after local injury: a review. Exp Hematol 3(2):135–148
96. Pittenger MF, Mackay AM, Beck SC, Jaiswal RK, Douglas R, Mosca JD, Moorman MA, Simonetti DW, Craig S, Marshak DR (1999) Multilineage potential of adult human mesenchymal stem cells. Science 284(5411):143–147
97. Rabinovitch M (2008) Molecular pathogenesis of pulmonary arterial hypertension. J Clin Invest 118(7):2372–2379
98. Raghu G, Collard HR, Egan JJ, Martinez FJ, Behr J, Brown KK, Colby TV, Cordier JF, Flaherty KR, Lasky JA, Lynch DA, Ryu JH, Swigris JJ, Wells AU, Ancochea J, Bouros D, Carvalho C, Costabel U, Ebina M, Hansell DM, Johkoh T, Kim DS, King TE Jr, Kondoh Y, Myers J, Muller NL, Nicholson AG, Richeldi L, Selman M, Dudden RF, Griss BS, Protzko SL, Schunemann HJ, ATS/ERS/JRS/ALAT Committee on Idiopathic Pulmonary Fibrosis (2011) An official ATS/ERS/JRS/ALAT statement: idiopathic pulmonary fibrosis: evidence-based guidelines for diagnosis and management. Am J Respir Crit Care Med 183(6):788–824
99. Raghu G, Weycker D, Edelsberg J, Bradford WZ, Oster G (2006) Incidence and prevalence of idiopathic pulmonary fibrosis. Am J Respir Crit Care Med 174(7):810–816
100. Rojas M, Xu J, Woods CR, Mora AL, Spears W, Roman J, Brigham KL (2005) Bone marrow-derived mesenchymal stem cells in repair of the injured lung. Am J Respir Cell Mol Biol 33(2):145–152
101. Rubenfeld GD (2003) Epidemiology of acute lung injury. Crit Care Med 31(4):276–284
102. Rubenfeld GD, Caldwell E, Peabody E, Weaver J, Martin DP, Neff M, Stern EJ, Hudson LD (2005) Incidence and outcomes of acute lung injury. N Engl J Med 353(16):1685–1693
103. Sacchetti B, Funari A, Michienzi S, Di Cesare S, Piersanti S, Saggio I, Tagliafico E, Ferrari S, Robey PG, Riminucci M, Bianco P (2007) Self-renewing osteoprogenitors in bone marrow sinusoids can organize a hematopoietic microenvironment. Cell 131(2):324–336
104. Salazar KD, Lankford SM, Brody AR (2009) Mesenchymal stem cells produce Wnt isoforms and TGF-beta1 that mediate proliferation and procollagen expression by lung fibroblasts. Am J Physiol Lung Cell Mol Physiol 297(5):1002–1011
105. Sato Y, Araki H, Kato J, Nakamura K, Kawano Y, Kobune M, Sato T, Miyanishi K, Takayama T, Takahashi M, Takimoto R, Iyama S, Matsunaga T, Ohtani S, Matsuura A, Hamada H, Niitsu Y (2005) Human mesenchymal stem cells xenografted directly to rat liver are differentiated into human hepatocytes without fusion. Blood 106(2):756–763
106. Schaffler A, Buchler C (2007) Concise review: adipose tissue-derived stromal cells—basic and clinical implications for novel cell-based therapies. Stem Cells 25(4):818–827
107. Secco M, Zucconi E, Vieira NM, Fogaca LL, Cerqueira A, Carvalho MD, Jazedje T, Okamoto OK, Muotri AR, Zatz M (2008) Multipotent stem cells from umbilical cord: cord is richer than blood. Stem Cells 26(1):146–150
108. Selman M, Carrillo G, Salas J, Padilla RP, Perez-Chavira R, Sansores R, Chapela R (1998) Colchicine, D-penicillamine, and prednisone in the treatment of idiopathic pulmonary fibrosis: a controlled clinical trial. Chest 114(2):507–512
109. Selman M, King TE, Pardo A, American Thoracic Society, European Respiratory Society and American College of Chest Physicians (2001) Idiopathic pulmonary fibrosis: prevailing and evolving hypotheses about its pathogenesis and implications for therapy. Ann Intern Med 134(2):136–151

110. Selman M, Pardo A (2002) Idiopathic pulmonary fibrosis: an epithelial/fibroblastic cross-talk disorder. Respir Res 3:31. doi:10.1186/rr175
111. Sheppard DN, Welsh MJ (1999) Structure and function of the CFTR chloride channel. Physiol Rev 79(1):23–45
112. Sordi V, Malosio ML, Marchesi F, Mercalli A, Melzi R, Giordano T, Belmonte N, Ferrari G, Leone BE, Bertuzzi F, Zerbini G, Allavena P, Bonifacio E, Piemonti L (2005) Bone marrow mesenchymal stem cells express a restricted set of functionally active chemokine receptors capable of promoting migration to pancreatic islets. Blood 106(2):419–427
113. Spencer HF, Jaffe A (2004) The potential for stem cell therapy in cystic fibrosis. J R Soc Med 97(44):52–56
114. Strieter RM (2005) Pathogenesis and natural history of usual interstitial pneumonia: the whole story or the last chapter of a long novel. Chest 128(5):526–532
115. Strobel ES, Gay RE, Greenberg PL (1986) Characterization of the in vitro stromal microenvironment of human bone marrow. Int J Cell Cloning 4(5):341–356
116 Stutts MJ, Canessa CM, Olsen JC, Hamrick M, Cohn JA, Rossier BC, Boucher RC (1995) CFTR as a cAMP-dependent regulator of sodium channels. Science 269(5225):847–850
117. Sueblinvong V, Weiss DJ (2010) Stem cells and cell therapy approaches in lung biology and diseases. Transl Res 156(3):188–205
118. Suzuki G, Iyer V, Lee TC, Canty JM Jr (2011) Autologous mesenchymal stem cells mobilize cKit+ and CD133+ bone marrow progenitor cells and improve regional function in hibernating myocardium. Circ Res 109(9):1044–1054
119. Tamagawa T, Oi S, Ishiwata I, Ishikawa H, Nakamura Y (2007) Differentiation of mesenchymal cells derived from human amniotic membranes into hepatocyte-like cells in vitro. Hum Cell 20(3):77–84
120. Thomas AQ, Lane K, Phillips J, Prince M, Markin C, Speer M, Schwartz DA, Gaddipati R, Marney A, Johnson J, Roberts R, Haines J, Stahlman M, Loyd JE (2002) Heterozygosity for a surfactant protein C gene mutation associated with usual interstitial pneumonitis and cellular nonspecific interstitial pneumonitis in one kindred. Am J Respir Crit Care Med 165(9):1322–1328
121. Toma C, Pittenger MF, Cahill KS, Byrne BJ, Kessler PD (2002) Human mesenchymal stem cells differentiate to a cardiomyocyte phenotype in the adult murine heart. Circulation 105(1):93–98
122. Tomei AA, Boschetti F, Gervaso F, Swartz MA (2009) 3D collagen cultures under well-defined dynamic strain: a novel strain device with a porous elastomeric support. Biotechnol Bioeng 103(1):217–225
123. Tsakiri KD, Cronkhite JT, Kuan PJ, Xing C, Raghu G, Weissler JC, Rosenblatt RL, Shay JW, Garcia CK (2007) Adult-onset pulmonary fibrosis caused by mutations in telomerase. Proc Natl Acad Sci U S A 104(18):7552–7557
124. Tzouvelekis A, Koliakos G, Ntolios P, Baira I, Bouros E, Oikonomou A, Zissimopoulos A, Kolios G, Kakagia D, Paspaliaris V, Kotsianidis I, Froudarakis M, Bouros D (2011) Stem cell therapy for idiopathic pulmonary fibrosis: a protocol proposal. J Transl Med 9:182–191
125. Uhal BD, Joshi I, Hughes WF, Ramos C, Pardo A, Selman M (1998) Alveolar epithelial cell death adjacent to underlying myofibroblasts in advanced fibrotic human lung. Am J Physiol 275(6):1192–1199
126. Walker NM, Badri LN, Wadhwa A, Wettlaufer S, Peters-Golden M, Lama VN (2012) Prostaglandin E2 as an inhibitory modulator of fibrogenesis in human lung allografts. Am J Respir Crit Care Med 185(1):77–84
127. Wan C, He Q, Li G (2006) Allogenic peripheral blood derived mesenchymal stem cells (MSCs) enhance bone regeneration in rabbit ulna critical-sized bone defect model. J Orthop Res 24(4):610–618
128. Ware LB, Matthay MA (2000) The acute respiratory distress syndrome. N Engl J Med 342(18):1334–1349
129. Weissman IL (2000) Translating stem and progenitor cell biology to the clinic: barriers and opportunities. Science 287(5457):1442–1446

130. Wine JJ, Joo NS (2004) Submucosal glands and airway defense. Proc Am Thorac Soc 1(1):47–53
131. Yan X, Liu Y, Han Q, Jia M, Liao L, Qi M, Zhao RC (2007) Injured microenvironment directly guides the differentiation of engrafted Flk-1(+) mesenchymal stem cell in lung. Exp Hematol 35(9):1466–1475
132. Yano T, Mason RJ, Pan T, Deterding RR, Nielsen LD, Shannon JM (2000) KGF regulates pulmonary epithelial proliferation and surfactant protein gene expression in adult rat lung. Am J Physiol Lung Cell Mol Physiol 279(6):1146–1158
133. Yuhgetsu H, Ohno Y, Funaguchi N, Asai T, Sawada M, Takemura G, Minatoguchi S, Fujiwara H, Fujiwara T (2006) Beneficial effects of autologous bone marrow mononuclear cell transplantation against elastase-induced emphysema in rabbits. Exp Lung Res 32(9): 413–426
134. Zeki AA, Schivo M, Chan AL, Hardin KA, Kenyon NJ, Albertson TE, Rosenquist GL, Louie S (2010) Geoepidemiology of COPD and idiopathic pulmonary fibrosis. J Autoimmun 34(3):327–338
135. Zeng R, Wang LW, Hu ZB, Guo WT, Wei JS, Lin H, Sun X, Chen LX, Yang LJ (2011) Differentiation of human bone marrow mesenchymal stem cells into neuron-like cells in vitro. Spine 36(13):997–1005
136. Zhang Q, Shi S, Liu Y, Uyanne J, Shi Y, Shi S, Le AD (2009) Mesenchymal stem cells derived from human gingiva are capable of immunomodulatory functions and ameliorate inflammation-related tissue destruction in experimental colitis. J Immunol 183(12): 7787–7798
137. Zhen G, Liu H, Gu N, Zhang H, Xu Y, Zhang Z (2008) Mesenchymal stem cells transplantation protects against rat pulmonary emphysema. Front Biosci 13:3415–3422

Adv Biochem Eng Biotechnol (2013) 130: 131–162
DOI: 10.1007/10_2012_158

Published Online: 1 March 2013

Mesenchymal Stem Cells as Cellular Immunotherapeutics in Allogeneic Hematopoietic Stem Cell Transplantation

Claudia Papewalis, Daniela Topolar, Barbara Götz, Stefan Schönberger and Dagmar Dilloo

Abstract Allogeneic hematopoietic stem cell transplantation (allo-HSCT) is a curative treatment option in hematopoietic disorders, immunodeficiencies and leukemia. To date graft-versus-host disease (GvHD) represents a life-threatening complication even if associated with beneficial antileukemic reactivity. GvHD is the clinical manifestation of donor cells reacting against host tissue. Because of their ability to facilitate endogenous repair and to attenuate inflammation, MSC have evolved as a highly attractive cellular therapeutic in allo-HSCT. Here we report on the clinical experience in the use of MSC to enhance engraftment and prevent and treat acute and chronic GvHD. In early clinical trials, MSC have shown considerable benefit in the setting of manifest GvHD. These encouraging results warrant further exploration.

Keywords Graft-versus-host disease · Clinical grade MSC · Immunosuppresion

Abbreviations

APC	Antigen presenting cells
BM	Bone marrow
CR	Complete response
DC	Dendritic cell
EBMT	European Group for Blood and Marrow Transplantation
GvHD	Graft-versus-host disease
GvL	Graft-versus-leukemia
(allo-)HSCT	(allogeneic) Hematopoietic stem cell transplantation
HLA(-G)	Human leukocyte antigen(-G)
HSC	Hematopoietic stem cell
IDO	Indoleamine-2,3,-dioxygenase

C. Papewalis · D. Topolar · B. Götz · S. Schönberger · D. Dilloo (✉)
Department of Pediatric Hematology and Oncology, Center for Pediatrics, University Hospital, Friedrich-Wilhelm University, Bonn Germany
e-mail: Dagmar.Dilloo@ukb.uni-bonn.de

ISCT	International Society for Cellular Therapy
IFN	Interferon
IL	Interleukin
MSC	Mesenchymal stem cell
OR	Overall response
PR	Partial response
PBSC	Peripheral blood stem cell
PD-L	Ligand of the programmed death receptor
PGE_2	Prostaglandin E_2
PL	Platelet lysate
TGF	Transforming growth factor
Th	T helper cells
TNF	Tumor necrosis factor
UCB	Umbilical cord blood

Contents

1 Introduction

Allogeneic hematopoietic stem cell transplantation (allo-HSCT) has evolved as a potentially curative treatment option for patients with malignant and nonmalignant hematological and immunological disorders. In bone marrow failure syndromes and immunodeficiencies, hematopoietic stem cells (HSC) from a healthy donor are transplanted with the intent to reconstitute the patient with a functional hematopoietic and immunological system. In leukemia and other hematological malignancies, the aim is to eliminate residual neoplastic disease in a twofold manner. Thus, treatment with cytotoxic radio-/chemotherapy pre-transplant is consolidated by antineoplastic immunological attack mediated by donor-derived immune system cells and myeloablative conditioning regimens. The allogeneic HSC-graft also serves to compensate for treatment-related lethal hematopoietic failure. Although the graft-versus-leukemia (GvL) reaction is a critical therapeutic component of allo-HSCT, it is associated with the potentially detrimental effects of graft-versus-host disease (GvHD) [65, 132]. GvHD results from cytotoxic allo-reactivity of grafted immune cells against normal host tissue. Severe donor versus host reactions lead to massive tissue injury and ultimately to impaired immunological recovery. As greater HLA disparity between recipient and donor is associated with an enhanced risk for GvHD, related and unrelated donors are generally chosen by a close degree of human-leukocyte-antigen (HLA) match [95].

Currently diverse sources of allogeneic stem cells, namely bone marrow (BM), cytokine-mobilized peripheral blood stem cells (PBSC) [113, 137], as well as umbilical cord blood (UCB) [9, 115, 127] are in use. In UCB-transplantation, the potential to cross significant HLA-barriers safely due to the relative immaturity of donor T cells in the graft has extended the access to suitable HSC products even in populations with rare tissue phenotypes. Also the possibility of mobilizing HSC to the periphery by growth factor stimulation has opened the avenue to harvest large quantities of HSC that lend themselves to further selection with the aim of enriching the stem cell population and/or depleting potentially allo-reactive T lymphocytes [21, 64, 92]. This has cleared the way for transplantation of HSC from donors with a full HLA-haplo type mismatch such as patients' parents [27, 105] and has further expanded the use of allogeneic HSCT over the last several years. To date already more than 25,000 patients per year worldwide have been transplanted with allogeneic HSC [39, 108]. Given current trends, the number of transplants from unrelated donors is expected to double within the next five years which will also significantly increase the population of patients at risk for GvHD [28].

With continuous improvement in anti-infectious, particularly antiviral and antimycotic therapy [108, 114, 123] and concepts of reduced intensity conditioning [41, 131, 136], the treatment-related mortality (TRM) of allo-HSCT has decreased considerably compared to its early beginnings [13, 106]. Yet, even with enhanced accuracy in HLA-typing and improved donor selection [96], the various possibilities of graft manipulations, and optimized immunosuppressive prophylaxis and therapy, GvHD remains a therapeutic challenge.

In addition to HSC, bone marrow and umbilical cord blood also harbor a mesenchymal stem cell (MSC) population with self-renewal and multilineage-differentiation ability [16, 109]. MSC further possess immunomodulatory potential that is not constitutive but specifically triggered in an inflammatory milieu. As MSC are able to migrate to sites of cellular injury and inflammation [135] and to exert their immunosuppressive activity in an environment of tissue damage [59, 90], MSC have gained considerable interest as cellular immunotherapeutics in allo-HSCT, particularly in the setting of GvHD.

2 Clinical GvHD

Graft-versus-host disease describes the clinical manifestations of recipient cells under attack by grafted donor immune cells. To date, it is still a life-threatening complication.

Acute GvHD (aGvHD) is defined to occur within the first 100 days after HSCT, and chronic GvHD (cGvHD) thereafter. In principle, the acute and chronic forms of GvHD may have overlapping symptoms and merge into each other. Acute GvHD can also resolve completely and still be followed later by cGvHD [28, 47]. In aGvHD, skin is most commonly affected and is usually the first organ involved. Acute GvHD of the skin often coincides with engraftment of donor cells. The characteristic of skin disease is a pruritic rash that can spread all over the body. In severe cases, the skin may blister and ulcerate. Gastrointestinal tract involvement usually presents as diarrhea combined with vomiting, anorexia, and abdominal pain. Depending on the severity, bloody diarrhea as a result of mucosal ulceration carries a particularly poor prognosis [37]. Liver disease caused by aGvHD may be difficult to distinguish from other causes of liver dysfunction following allo-HSCT such as veno-occlusive disease (VOD), drug toxicity, viral infection, or sepsis [34].

A grading system for aGvHD was introduced in the 1970s by Glucksberg et al. [36]. Today, most institutions use sets of criteria previously established at the Keystone Consensus Conference of 1994 [103] or the consensus criteria issued by the Center for International Blood and Marrow Transplant Research [76]. Scoring aGvHD severity is carried out by first staging the affection of skin, liver, and gastrointestinal tract as a basis of an overall grade that acknowledges both the stage of organ pathology as well as the number of organs involved. These overall grades are classified as I (mild), II (moderate), III (severe), and IV (very severe), or A–D, respectively. Severe aGvHD carries a poor prognosis, with 25 % long-term survival for grade III and 5 % for grade IV [18]. The incidence of aGvHD is related to the degree of mismatch between HLA-proteins and the degree of ex and in vivo graft manipulation [18]. Acute GvHD ranges from 35 to 45 % in BM or PBSC recipients of fully matched siblings to 60–80 % in T-replete $\geq$1 HLA-mismatched unrelated transplant recipients [31, 47, 70, 113, 137]. The same degree of mismatch causes less GvHD using UCB grafts. Thus, the incidence of aGvHD is lower following the transplant of partially matched UCB units and ranges from 25 to 65 % depending on the overall transplant setting such as intensity of conditioning, and in haploidentical PBSC-transplantation also on the extent of graft manipulation [9, 10, 27, 105].

Treatment of primary aGvHD largely comprises the same agents used for prophylaxis such as calcineurin-inhibitors and mycophenolate mofetil [130] with glucocorticoids representing the backbone of aGvHD treatment [76, 110]. Overall less than 50 % of patients respond to glucocorticoids with slightly higher response rates in children [47].

Chronic GvHD remains the major cause of late nonrelapse death following HSCT [63]. The syndrome has features resembling autoimmune and other immunological disorders such as scleroderma, Sjögren syndrome, primary biliary cirrhosis, wasting syndrome, bronchiolitis obliterans, immune cytopenias, and chronic immunodeficiency. Manifestations of cGvHD may be restricted to a single organ which is classified as limited or mild cGvHD. Chronic GvHD can also be widespread affecting many organ sites and is then termed extended or severe. It can lead to debilitating consequences, for example joint contractures, loss of sight, end-stage lung disease, or mortality due to profound chronic immune suppression with recurrent and ultimately life-threatening infections [29]. aGvHD consensus criteria for grading the severity of cGvHD have been published but are as yet not employed consistently [5].

Treatment of cGvHD follows along the same lines as in aGvHD. Yet the response rate is even lower, with a third of patients [4] not responding to first-line therapy often consisting of corticosteroid and calcineurin inhibitor therapy either alone or in combination [57]. Although for refractory cGvHD a variety of therapies have been evaluated [20, 46, 50, 71], efficacy has been limited. Long-term survival is poor due to toxicity related to profound and prolonged immunosuppression. Thus, treatment of GvHD remains a therapeutic challenge warranting the evaluation of novel treatment options [82, 134]. To date, glucocorticoid-resistant GvHD is among the most challenging complications in allo-HSCT.

3 Pathophysiology of GvHD

The paradigm of GvHD development has been conceptualized as a three-step process [28]. The initiation phase is characterized by tissue damage caused by intensive conditioning therapy pre-transplant. As a result, host antigen-presenting cells (APC) such as dendritic cells (DC) and monocytes become activated: HLA-antigens as well as co-stimulatory and adhesion molecules are up-regulated on their cell surface. In addition pro-inflammatory cytokines such as interleukin (IL)-1β, and tumor necrosis factor (TNF)-α and chemokines are released. Treatment-related mucositis with destruction of the gastrointestinal mucosal barrier results in systemic translocation of inflammatory stimuli derived from microbial products. These pathogen-associated molecular patterns serve as "danger-signals" and further enhance the activation and maturation of host APC [25, 43]. Recipient APC seem to be sufficient to induce GvHD, however, murine models suggest that donor APC may also contribute by indirect antigen presentation [3].

The second phase of GvHD-development is characterized by activation of mature donor T cells recognizing cognate antigens presented by host APC. In response, T cells proliferate and differentiate into activated effector cells within a danger-signal-rich milieu. They contribute to this by release of additional cytokines. Indeed, polymorphisms for critical cytokines such as TNF-α and interferon (IFN)-γ have been implicated as risk factors for GvHD [67]. Most of this process takes place within secondary lymphoid organs as early as three days after transplant, long before de novo regeneration of donor T cells has ensued [117].

The third phase is the effector phase of GvHD which leads to target organ destruction. Chemokines over-expressed by macrophages direct the migration of donor cells from lymphoid organs to the target tissues. Here cytotoxic cellular mediators, namely donor T and NK cells, and soluble inflammatory factors such as TNF-α, IFN-γ, IL-1β, and nitric oxide synergize and amplify local tissue damage and promote inflammation. In intestinal GvHD, integrins further facilitate homing of donor T cells to Peyer's patches [128]. Ultimately, end organ damage is predominately due to T cell-mediated tissue toxicity, which involves soluble mediators, including TNF-α, perforin, granzymes, Fas, and Fas ligand [7, 17, 40, 49, 77]. As hepatocytes express large amounts of Fas, in liver GvHD cytotoxic T cells preferentially use the Fas/FasL pathway for target cell lysis. In contrast, the perforin/granzyme pathway plays a dominant role in GvHD affecting skin and the gastrointestinal tract [125].

Thus in allo-HSCT, severe donor-versus-host immune reactions can result in massive end-organ injury. Based on the multitude of immunomodulatory activities and their capacity to support the healing process at sites of tissue injury, MSC are deemed highly attractive candidates for mitigation of both acute and chronic GvHD following allo-HSCT.

4 Immunomodulation in GvHD Mediated by MSC

MSC are pluripotent cells characterized by self-renewal and the multilineage differentiation capacity for a variety of cell types such as chondrocytes, adipocytes, and osteoblasts. MSC were originally isolated and characterized as nonhematopoietic multipotent progenitors of adult bone marrow [15, 16] and termed "multipotent stromal cells" [44]. They have been implicated in hematopoietic support [23].

Meanwhile, it is known that MSC can be effectively detected in almost every tissue such as umbilical cord blood, Wharton's jelly, amniotic fluid, adipose tissue, skeletal muscle, liver, brain, hair follicle, and dental pulp [42, 45, 104, 139]. Based on their ability to home to sites of organ injury, to facilitate tissue repair, and to critically modulate immune responses, MSC have generated considerable interest as cellular therapeutics. In an effort to harmonize MSC characterization, the International Society for Cellular Therapy (ISCT) has issued a consensus set of three minimal criteria to define MSC regardless of their tissue of origin: (I) plastic adherence, (II) maintenance of tri-lineage osteogenic, adipocytic, and chondroblastic differentiation potential after in vitro propagation, and (III) lack of the hematopoietic markers CD45, CD34, CD14, CD11b, CD79-α, CD19, and HLA-DR, and simultaneous expression of the surface molecules CD73, CD90, and CD105 on $\geq$95 % of the population [24]. The surface molecule CD73, an ecto-5′-nucleotidase, is involved in cellular crosstalk, migration, and modulation of adoptive immunity. The interaction between CD73 and adenosine A2A receptor results in the blockade of the adenosine pathway in activated T cells with a subsequent proliferation stop [16, 112]. CD90 (Thy-1) is viewed as a marker of "stemness". Its function on MSC is not entirely resolved but as a GPI-anchor it is known to mediate cell-to-cell interactions as well as monocyte and lymphocyte adhesion. CD105 (endogline) belongs to the TGF receptor family [16].

Of note, homing and immunosuppressive activity of MSC is not a constitutive phenomenon but requires a pro-inflammatory milieu. Expression and release of critical immunosuppressive factors such as prostaglandin E_2 (PGE_2), hepatocyte growth factor (HGF), IL-10, transforming growth factor-β (TGF-β), and leukemia inhibitory factor (LIF), human leukocyte antigen-G (HLA-G), and galectin-1 are dependent on MSC priming by cytokines such as by IFN-γ, TNF-α, and IL-1β [35, 118]. Also the enzyme Indoleamine 2,3-dioxygenase (IDO) is regulated by IFN-γ, IDO catabolizes tryptophan to kynurenine resulting in depletion of the cellular milieu from tryptophan and accumulation of cell-toxic kynurenine metabolites. We and others have previously shown that tryptophan starvation of the microenvironment down-tunes effector cell function such as proliferation, cytotoxicity, and cytokine production in activated T and NK cells [68, 84, 119, 120]. In addition, IFN-γ-dependent up-regulation of STRO-1 and ligand of the programmed death receptor-1 (PD-L1) are among the surface molecules involved in MSC-mediated T cell inhibition in a cell-contact-dependent manner [88, 111]. MSC further modulate the complement activation pathways by constitutive expression of factor H which again may be up-regulated by TNF-α and IFN-γ, key mediators of aGvHD [124].

Thanks to the plethora of immunosuppressive effects exerted on APC as well as effector cells, MSC are potentially capable of intercepting each of the individual stages in GvHD development (Fig. 1). In the first phase of GvHD, damage of the host leads to the accumulation of an array of chemokines such as CCL2, CCL5, CCL22, and CXCL12. The respective chemokine and growth factor receptors are expressed on MSC. They become up-regulated on TNF-α primed cells, thereby further enhancing their homing efficiency. All together, the migratory capacity of MSC is under the control of a large range of receptor tyrosine kinases, growth factors, and CC and CXC chemokines [100].

LPS, TNF-α, IL-1, and IL-6 are released at sites of injury. These cytokines stimulate maturation of host antigen presenting cells (APC) such as dendritic cells (DC) critical for subsequent activation of allo-reactive T lymphocytes. Here, MSC provide counter-regulatory signals, namely PGE_2, IL-6, and M-CSF that depress DC surface expression of HLA-DR and CD1a as well as of the co-stimulatory molecules CD80 and CD86 [90]. Also DC-expression of TNF-α, IL-12 is decreased whereas IL-10 release is up-regulated shifting the dendritic surface marker and cytokine profile towards a tolerogenic state. Here, the soluble factors IL-6 and M-CSF have been implicated not only in induction but also maintenance of the immature DC phenotype [89, 135]. In addition, MSC intervene with the effector phase of GvHD by inhibiting expansion of the effector cell pool and down-modulating cytokine production in T and NK cells. Suppression of NK cell

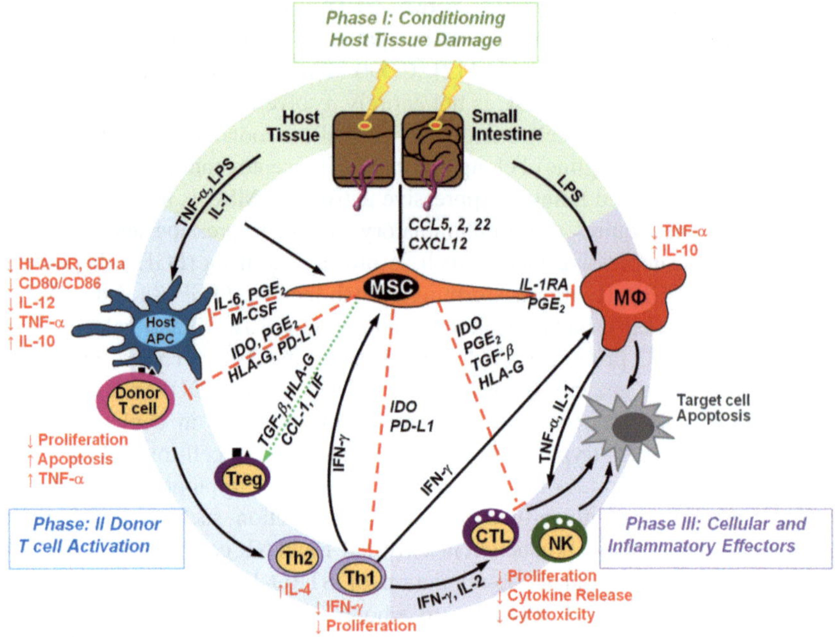

Fig. 1 Scheme of MSC attenuating all three phases of GvHD development. The tri-phasic model of acute GvHD evolution and maintenance is depicted as a self-perpetuating cycle of inflammation resulting in target organ damage mediated by allo-reactive effector cell responses (adapted from [43]). MSC are attracted by the pro-inflammatory milieu to sites of tissue damage. Once licensed by inflammatory cytokines such as IFN-γ and TNF-α, MSC actively modulate each phase of the immune response. *Black arrows* indicate the mode of interaction between different cellular players of the "GvHD-cycle." *Red lines* refer to MSC-mediated attenuating effects and the *green line* implies a supporting role of MSC. The resulting changes in effector functions are printed in *red* or *green*, respectively (Color figure online)

cytotoxicity is also due to MSC-mediated down-regulation of the activating NK receptors NKp 30, NKp 44, and NKD2D [90].

It is important to note that MSC dampen the self-perpetuating inflammatory mechanisms in GvHD by blocking the release of the critical cytokines TNF-α and IFN-γ. Thus, secretion of TNF-α by monocytes is suppressed by MSC-secretion of the IL-1a receptor antagonist (IL-1RA) [90]. Also, the T cell-dependent feedback loop of TNF-α production is intercepted by MSC. Here, TNF-α-induced PGE_2 expression in MSC not only down-modulates T cell proliferation but also T cell cytokine release including TNF-α [135]. Similarly, MSC deflect the IFN-dependent feedback mechanism, as IFN-γ-induced expression of IDO and PD-L1 in MSC [135] in turn reduces IFN production in Th1 cells and up-regulates IL-4 production in Th2 lymphocytes. This creates a tolerogenic milieu not only locally but also systemically tipping the balance towards an anti-inflammatory Th2 response [2]. Regulatory T (Treg) cells also contribute to this MSC-induced local and systemic network. MSC facilitate Treg-induction and expansion by release of

HLA-G, LIF, and CCL1 [90]. In addition, interaction between the surface molecules CD58 and CD52 expressed on MSC with CD2 and CD11a on T cells generates a FOXp3-negative CD4/CD8 double positive Treg population that has been found to be one hundredfold more T cell suppressive than FOXp3-positive CD4/CD25 double positive Treg [102].

Thus, MSC potentially interact with almost every immune cell population involved in GvHD initiation and perpetuation in an attenuating manner. At the same time, the need for so-called "licensing" by pro-inflammatory signals to trigger immunosuppressive activity renders MSC particularly attractive for cell therapy. In the absence of inflammation, MSC stay immunologically inert and thus do not contribute to generalized immune suppression as many pharmacological agents such as steroids do [82]. Moreover the MSC-mediated T cell inhibitory function is differentially directed against allo-specific T cell activity and does not attenuate antiviral recall responses [52]. MSC themselves exhibit profound antiviral and antimicrobial activity. Indeed IDO, one of the key IFN-γ dependent T cell inhibitory mechanisms formerly identified by us, also dampens the amplification of cytomegalovirus and toxoplasmosis, two highly critical infectious agents in allo-HSCT [22, 73, 86]. One of the major issues when introducing novel immunomodulatory cell therapeutics in clinical allo-HSCT is the increased risk of infection. The above-described pre-clinical insights partially address these concerns.

5 MSC for Clinical Application in Allo-HSCT

To date, GvHD remains a significant cause of nonrelapse morbidity and mortality following allo-HSCT. During the last decade, BM-derived MSC have been employed in a series of studies for prevention and treatment of GvHD in the allo-HSCT setting.

Initially, MSC were predominantly isolated from siblings or related haplo-identical donors. Separation of MSC from the bone marrow was performed by density gradient centrifugation of the mononuclear cell fraction and subsequent in vitro propagation of the plastic adherent cell fraction over 4–6 passages. Later, particularly in those studies employing MSC products provided by Osiris Therapeutics, Inc. (Prochymal®), MSC were obtained from unrelated healthy third-party volunteer donors with variable degrees of HLA-matching depending on the recipient's phenotype. With a frequency of 0.01–0.001 % mesenchymal progenitors in the BM, a 10-ml aspirate is generally sufficient to yield 50–300 $\times$ 10^6 MSC without loss of multidifferentiation potential [97]. In the earlier investigator-initiated trials (IIT), MSC were often used directly from the culture. Industrially prepared MSC are generally cryopreserved, off-the-shelf products that need to be defrosted prior to use. Also in some preliminary studies, other sources of MSC have been explored such as adipose tissue [26] and cord blood [62, 129].

Overall, there are 27 reports describing MSC application in allo-HSCT. In addition to a few case reports or case series [60, 87, 107, 129], the studies published are predominately pilot/phase I trial. There are few prospective phase II studies [53, 61, 69]; two of these are randomized open label studies [53, 69] and only one randomized phase III trial [81] which unfortunately has thus far only been published in an abstract format. About a third of the studies evaluate safety and feasibility of MSC transfusion in conjunction with transplantation of allogeneic HSC with the secondary aim to enhance engraftment and potentially prevent GvHD (Table 1). Another third of the studies assess MSC treatment for steroid-resistant aGvHD (Table 2) and only two studies focus entirely on MSC application for treatment of refractory and severe cGvHD [133, 140] (Table 3). Of note, there is one prospective phase II study employing MSC for first-line treatment of aGvHD [53] (Table 2b).

Two landmark reports introduced the medical community to the potential of MSC administration in allo-HSCT. A large multicenter feasibility study [56] documented that MSC expansion from BM to clinically relevant quantities was feasible within one month. A year before, Le Blanc's group from the Karolinska Institute, Stockholm, had reported on the first successful treatment of steroid-refractory severe acute GvHD in a nine-year-old boy with BM-derived MSC from his mother [60].

6 MSC for Enhanced Engraftment and Prevention of GvHD

Feasibility and safety of MSC/HSC co-transplantation was evaluated in two phase I studies [56, 91, 138] and one randomized phase II study following myeloablative conditioning in the context of HLA-matched sibling transplants. Secondary study endpoints assessed the kinetics of HSC engraftment and GvHD incidence. In all three studies, HSCT was performed for adult patients with high risk or relapsed hematological malignancies. GvHD prophylaxis comprised cyclosporine and MTX in all three studies. In the largest of these studies [56], patients were recruited in a multicenter study effort across the United States. For these 56 patients, BM-derived MSC were sampled from the respective HSC donors and prepared by Osiris Therapeutics Inc., Baltimore. Adequate expansion proved feasible in 91 % of sibling donors (51/56 donors) up to a dose of 2.5×10^6/kg within 30 days, even though only 46 patients were eventually transplanted with a combination of HSC and MSC.

Likewise in one of the two Chinese feasibility studies [138], MSC preparation up to a dose of 2×10^6/kg recipient body weight were obtained in 86 % of cases (12/14 donors). Yet, in the only randomized study [91], this target dose was not achieved, with only three patients transplanted with $\geq 1.0 \times 10^6$/kg. The MSC doses infused in this randomized open-label trial were considerably lower than in the other studies with 0.03–1.53×10^6/kg (median 0.34×10^6/kg). Moreover 5/15 patients of the "intend-to-treat" cohort had to be excluded from further

Table 1 MSC for enhanced engraftment and prevention of GvHD

First author, year Study type	pt/co (*n*)	Child/adult (*n*)	HSCT conditioning (*n*)	HSCT source (*n*)	MSC source MSC donor sib/haplo/MM (*n*)	MSC dose ($n \times 10^6$/kg)	Engraftment (%)	aGvHD [grade] (*n*/%)	cGvHD (*n*/%)	Relapse (*n*/%)	Infection (*n*)	OS/FU (%/months)
Lazarus, 2005 Phase I study, IIT	46/–	–/46	MAC	BM (19) PBSC (27)	Same as HSC donor 46/–/–	37pt: 1–2.5 5pt: 5	100	[I] 23/50 [II–IV] 13/28	ltd. 14/30 ext. 8/17	12/26	n.i.	53/24
Ning, 2008 Rand. phase II study, IIT	10/15	–/10	MAC	BM (4) PBSC (4) BM & PBSC (2)	Same as HSC donor 10/–/–	0.03–1.5	pt: 90 co: 100	pt: [II–IV] 1/10 co: [II–IV] 8/53	pt: 1/10 co: 4/26	pt: 6/60 co: 3/20	pt: CMV (2) Bacterial and/or fungal pneumonia (2) co: CMV (2) Pneumonia (3)	pt: 40/36 co: 67/36
Zhang, 2009 Pilot/phase I study, IIT	12/–	–/12	MAC	PBSC	Same as HSC donor 12/–/–	1.3–2.2	100	[I] 7/58 [II–III] 2/17	ltd. 2/17 ext. 2/17	4/33	CMV-viremia (4) Hepatitis B (1) Lung infection (1)	58/58
Gonzalo-Daganzo, 2009 Phase I/II study, IIT	9/46	–/9	MAC	UCB & PBSC	Same as PBSC donor –/7/2	1–2.2	pt: 100 co: 93	pt: [I] 1/11 [II] 4/44 co: [I] 18/39 [II] 5/11 [III–IV] 3/6.5	pt: ltd. 1/11 co: ltd. 8/17 ext. 3/7	pt: 1/11 co: 6/13	pt: CMV	pt: 89/22 co: 53/60
MacMillan, 2009 Phase I/II study	8/23	8/–	NMAC	UCB	–/8/– Plasme Lyte®	1pt: 0.9–5 3pt: 0.06–10 (2x)	pt: 100 co: 100	pt: [III] 3/38 co: [II–IV] 5/22	pt: – co: 4/17	pt: 1/13	pt: Aspergillosis (2) Clost. diff. colitis (3) Bacteremia (2) CMV (1) Sinusitis (1) Shingles (1)	pt: 75/12 co: 62/76–94

(continued)

Table 1 (continued)

First author, year Study type	pt/co (*n*)	Child/adult (*n*)	HSCT conditioning (*n*)	HSCT source (*n*)	MSC source MSC donor sib/haplo/MM (*n*)	MSC dose ($n \times 10^6$/kg)	Engraftment (%)	aGvHD [grade] (*n*/%)	cGvHD (*n*/%)	Relapse (*n*/%)	Infection (*n*)	OS/FU (%/months)
Lee, 2011(oral) Pilot study, IIT	7/22	7/–	n.i.	UCB	UCB pt: –/–/7	4pt: 1 (1x) 3pt: 1 (5x)	pt: 100 co: 86	pt: [III–IV] 1/14 co: [III–IV] 1/5	pt: ext. 1/14 co: ext. 4/19	n.i.	n.i.	pt: 75/24
Ball, 2007 Phase I/II study, IIT	14/47	14/–	MAC	sPBSC Haplo	As HSC donor 1/13/–	1–5	pt: 100 co: 85	pt: [I–II] 2/14 co: [I–II] 12/26 [III–IV] 2/4	pt: ltd. 1/7 co: ltd. 4/9 ext. 2/4	pt: 3/18 co: 12/26	pt: viral (7) Sepsis (1) ADV hepatitis (1) co: viral (5) ADV hepatitis (2)	pt: 72/3–28 co: n.i.
Baron, 2010 Pilot study, IIT	20/16	–/20	NMAC	PBSC MM	–/–/20	1	pt: 95 co: 100	pt: [II–III] 7/35 [IV] 2/10 co: [II–III] 6/32 [IV] 3/19	pt: n.i. co: 23/65	pt: 6/30 co: 4/25	pt: cerebral toxoplasmosis (1) Encephalopathy (1)	pt: 80/12 co: 44/12
Liu, 2011 Rand. phase II study, IIT	27/28	n.i./n.i.	MAC	BM & PBSC haplo	MM from rel. or unrel. donors –/4/23	0.3–0.5	pt: 100 co: 96	pt: [I–II] 16/70 co: [I–II] 15/53 [III] 1/4	pt: ltd. 9/33 ext. 4/15 co: ltd. 11/39 ext. 4/14	pt: 2/7 co: 1/4	pt: CMV-viremia (18) Pneumonia (3) EBV-PTLD (2) co: CMV-viremia (18) Pneumonia (3)	pt: 70/24 co: 64/24

(continued)

Table 1 (continued)

First author, year Study type	pt/co (n)	Child/adult (n)	HSCT conditioning (n)	HSCT source (n)	MSC source MSC donor sib/haplo/MM (n)	MSC dose (n × 10^6/kg)	Engraftment (%)	aGvHD [grade] (n/%)	cGvHD (n/%)	Relapse (n/%)	Infection (n)	OS/FU (%/months)
Wang, 2012 Case series in SAA, IIT	6/–	6/–	MAC	BM & PBSC (haplo) (3)/PBSC (MUD)(3)	–/1/5 1x BM 5x UCB	0.85–2.5	100	[I–II] 2/33	–	n.i	–	100/6–29
LeBlanc, 2007 Pilot study, IIT	7/–	4/3	MAC (3) NMAC (4)	BM (2) PBSC (4) UCB (1)	3/4/–	1	100	pt:[I–II] 6/85	1/14	n.i.	Bacterial (2) Aspergillosis (1)	86/≥1 12
Meuleman, 2009 Pilot/phase I study, IIT	6/–	–/6	MAC	PBSC	As HSC donor 3/3/–	1	33	n.i	n.i.	1/17	Pulmonary Aspergillosis (2) Retinitis Encephalitis CMV (1)	33/12 17/158
MSC transfusion at time of leukocyte recovery for prevention of GvHD												
Kuzmina, 2012 Pilot/phase I study, IIT	19/18	–/19	MAC (14) NMAC (4)	n.i.	As HSC donor Related donors **PL**	0.9–1.3	n.i.	pt: [I–II] 6/32 co: [I–II] 9/50 [III–IV] 1/5,5	pt: 5/26 co: 6/33	pt: 4/21 co: 5/28	n.i.	pt: 95/3 co: 78/3

Note With few exceptions, patients were transplanted for malignant hematological diseases. Controls were matched to patients co-transplanted with MSC for age, conditioning, HSC source, donors, and dose. MSC were BM-derived, unless otherwise indicated. Except for randomized studies, historical controls were used

Abbreviations: *ADV* adenovirus, *aGvHD* acute graft-versus-host-disease, *AS* autologous serum, *BM* bone marrow, *CB* cord blood, *cGvHD* chronic graft-versus-host-disease, *Clost. diff. colitis Clostridium difficile* colitis, *co* controls, *com* commercial, *CR* complete response, *EBV-PLTD* Epstein–Barr virus-related post-transplant lymphoproliferative disorder, *ext* extensive, *FCS* fetal calf serum, *FU* follow up, *GI* gastro-intestinal tract, *Haplo* HLA haploidentical related, *HD* high dose, *HM* hematological malignancies, *hist* historical, *HSCT* hematopoietic stem cell transplantation, *IIT* investigator initiated trial, *Inf* infection, *LD* low dose, *ltd* limited, *MAC* myeloablative chemotherapy, *Med* median, *MM* mismatched unrelated, *MOF* multiple organ failure, *MR* mixed response, *MSC* mesenchymal stem cells, *MUD* matched unrelated donor, *n* number, *n.i.* not indicated, *NMAC* nonmyeloablative chemotherapy, *OR* overall response: *sum of CR + PR*, *oral* oral presentation, *OS* overall survival, *PBSC* peripheral blood stem cells, *PL* platelet lysate, *PFS* progression free survival, *PR* partial response, *pt* patient, *rand* randomized, *rel.* related, *SAA* severe aplastic anemia, *SCT* stem cell transplantation, *Sib* HLA-identical sibling donor, *TBI* total body irradiation, *UCBT* umbilical cord blood transplantation, *VOD* veno-occlusive disease

Table 2 MSC for treatment of (a) steroid-resistant and (b) steroid-resistant or de novo* acute GvHD

First author, year Study type	pt/co (*n*)	aGvHD/ cGvHD (*n*)	Child/adult (*n*)	MSC donor Sib/Haplo/MM (*n*) Supplement	MSC application: Infusions (*n*)	MSC application: Dose (× 10^6/kg)	aGvHD pre MSC I/II/III/IV (*n*) I/II/III/IV (%)	Response to MSC CR/PR/MR/OR (*n*) CR/PR/MR/OR (%)	Infection (*n*)	OS/FU (%/months)
(a) *Bone marrow-derived MSC expanded with FCS*										
LeBlanc, 2004 Case report	1/–	1/–	1/–	–/1/–	1pt: 2	2–1	–/–/–/1 –/–/–/100	1/–/–/**1** 100/–/–/**100**	Repeated bacterial, viral, and invasive fungal infections	100/12
Ringden, 2006 Pilot/phase I study, IIT (pt incl. in LeBlanc, 2008)	9/16	8/1	2/7	4/3/1	5pt: 1 3pt: 2 1pt: 1	0.7–2.0 0.7–2.0 9.0	–/2/5/1 –/25/62/13	6/1/1/**7** 75/13/12/**88**	CMV (2) EBV-PTLD (1) ADV (1) Aspergillosis (1)	56/48
Müller, 2008 Pilot study, IIT	7/–	2/5	7/–	–/–/2	4pt: 2 2pt: 1 1pt: 3	0.4–3	–/1/1/– –/50/50/–	1/–/–/**1** 50/–/–/**50**	–	50/60
Fang, 2007 Phase I study, IIT	6/–	6/–	2/6	MSC source: adipose tissue –/2/4	6pt: 1–2	1–2	–/–/2/4 –/–/33/66 (children: n.i.)	5/–/–/**5** 83/–/–/**83**	n.i.	In phase I study 83/12
LeBlanc, 2008 Prospective multicenter phase II study, IIT	55/–	55/–	25/30	5/18/69	27pt: 1 22pt: 2 6pt: 3–5	0.6–9	–/5/25/25 –/9/45,5/45,5	30/9/–/**39** 55/16/–/**71**	CMV (4) EBV (3)-LPD (1) Aspergillosis (5) Septicemia (6) Inf.(viral + bact.) (9)	38/3–60
Prasad, 2011 Compassionate use multicenter study	12/–	12/–	12/–	–/–/12 Prochymal ® Formulated in Plasme Lyte ®	1pt: 3 8pt: 7–12 1pt: 2 2pt: 21	2 2 8 2–8	–/–/8/4 –/–/67/33	7/2/3/**9** 58/17/25/**75**	EBV-LPD (1) Resp. failure (6) Fusarium infection (2)	42/20
Kurtzberg, 2010[b] FDA expanded access program	59/–	59/–	59/–	–/–/59 Prochymal ® Formulated in Plasme Lyte ®	All pt: 8 (+4)	2	–/6/20/33 –/10/34/56	n.i./n.i./n.i./**38** n.i./n.i./n.i./**64**	n.i.	62/3

(continued)

Table 2 (continued)

First author, year Study type	pt/co (*n*)	aGvHD/ cGvHD (*n*)	Child/adult (*n*)	MSC donor Sib/Haplo/MM (*n*) Supplement	MSC application		aGvHD pre MSC I/II/III/IV (*n*) I/II/III/IV (%)	Response to MSC CR/PR/MR/OR (*n*) CR/PR/MR/OR (%)	Infection (*n*)	OS/FU (%/ months)
					Infusions (*n*)	Dose ($\times 10^6$/kg)				
Martin et al., 2010[b] Rand. placebo-controlled multicenter phase III study	163/81	163/–	–/163	–/–/163 Prochymal ® Formulated in Plasme Lyte ®	all pt: 8 (+4)	2	–/36/83/44 –/22/51/27	65/n.i./n.i./**103** 40/n.i./n.i./**63**	n.i.	n.i.
(b) *MSC expanded with platelet lysate (PL) or human autologous serum (AS)*										
Lucchini, 2010 Pilot study, IIT	11/–	8/3	11/–	–/–/7 **PL**	4pt: 1 4pt: 2	0.7–3.7 1–1.6	3/1/–/4 37/13/–/36	3/2/–/**5** 38/25/–/**63**	Pneumonia (1) Lung aspergillosis (1) Sepsis (1)	63/11
von Bonin, 2009 Pilot/phase I study, IIT	13/–	13/–	–/13	4/–/9 **PL**	13pt: 2	0.5–1.1	–/–/2/11 –/–/15/85	1/1/–/**2** 8/8/–/**16**	Infection (3)	69/12
Perez-Simon, 2011 Phase I/II study, IIT	18/–	10/8	–/18	–/8/2 **AS**	18 pt: 1–4	0.2–2.9	–/3/3/4 –/30/30/40	1/5/–/**6** 10/50/–/**60**	Sepsis (2)[a] Infection (3)[a]	10/11
Kebriaei, 2009 Rand. multicenter phase II study, IIT*	31/–	31/–	–/31	Prochymal ® **PL**	HD: 15pt: 1 LD: 16pt: 1	8 2	All: –/21/7/3 –/67/22/10 HD: –/10/4/1 –/66/26/6 LD: –/11/3/2 –/69/19/12	All: 24/7/–/**31** 77/22/–/**99** HD: 10/5/–/**15** 66/33/–/**99** LD: 14/2/–/**16** 87/13/–/**100**	BK(2) CMV(5) ADV (1) Bacteremia (4) Meningitis (1) Aspergillosis (1) Pneumonia (1)	71/3

Note With few exceptions, patients were transplanted for malignant hematological diseases except indicated patients who received myeloablative conditioning (MAC) and were transplanted with BM-derived MSC. MSC were cultured in FCS-supplemented medium unless stated otherwise. For abbreviations refer footnote in Table 1

[a] Mortality

[b] Short publication

Table 3 MSC for treatment of chronic GvHD

First author, year Study type	pt cGvHD/total (*n*)	Child/adult (*n*)	MSC Donor sib/haplo/MM (*n*) Supplement	MSC application Infusions (*n*)	 Dose ($\times 10^6$/kg)	Effect CR/PR/MR/OR (*n*) CR/PR/MR/OR (%)	Infection (*n*)	OS/FU (%/months)
Ringden, 2006 Pilot/phase I study, IIT	1/9	–/1	–/1/–	All pt: 1	1	–/–/1/– –/–/11/–	EBV-PLTD	–
Müller, 2008 Pilot study, IIT	3/7	3/–	–/2/1	All pt: 1	1.4–3	–/–/1/– –/–/33/–	EBV-PLTD (1)	33/48
Lucchini, 2010 Pilot study, IIT	5/11 (2/5 progr. aGvHD)	11/–	–/–/5	1pt: 4 4pt: 1–2	0.7–1.4 1–1.2	1/3/–/**4** 20/60/–/**80**	–	100/10
Perez-Simon, 2011 Phase I/II study, IIT	8/18	–/18	–/5/3 **AS**	7pt: 1–2 1pt: 4	0.2–1.2 0.8–1	1/3/–/**4** 5/16/–/**21**	Infection (1)	63/≥5
Zhou, 2010 Pilot/phase I study, IIT	4/4 Sclerodermatous cGvHD	–/4	–/–/4	All pts: 4–8	0.1–0.3	gradual response	n.i.	100/5–23
Weng, 2010 Pilot/phase I Study, IIT	19/19	–/19	–/–/19	All pts: 1–5	0.2–1.4	4/10/3/**14** 21/53/16/**74**	Infection (2)	100/3 77/≥24

Note With few exceptions, patients were transplanted for malignant hematological diseases, except indicated patients who received myeloablative conditioning (MAC) and were transplanted with BM-derived MSC. MSC were cultured in FCS-supplemented medium unless stated otherwise. For abbreviations refer footnote in Table 1 comparison because MSC preparations failed. In all three studies, no immediate side effects from MSC infusion and no ectopic tissue formation were observed.

Time to platelet and neutrophil engraftment was as expected for this type of transplant and did not differ significantly from the control group in the study by Ningh et al. Acute GvHD was low in all three studies with aGvHD II–IV in 24 % (16/68) of patients to whom MSC were administered. Overall 40 % (27/68) of patients were affected by cGvHD. Of these, about half suffered from the extensive form of the disease. In the three studies, relapse occurred in 35 % (24/68) of MSC/HSC co-transplanted patients. In the open-label randomized trial by Ning et al., however, there was a particularly high relapse rate with 60 % in the MSC group which was significantly different from the controls with only 20 %. Consequently, the three-year overall survival (OS) also differed significantly with 40 % in patients co-transplanted with MSC and 67 % for controls.

The study was closed early based on a potentially increased relapse risk associated with MSC. However, a generalized conclusion correlating relapse and MSC co-transplantation cannot be drawn, due to small patient numbers, the exclusion of five patients from the "intent-to-treat" population, and the use of historic controls. Accordingly, this study has caused considerable controversy [12]. It is valid to ask whether beyond the feasibility issue of timely large-scale MSC-preparation in patients, post-sibling donor HSCT with a low risk of graft failure and GvHD, MSC co-infusion as a prophylactic measure can be expected to provide any clinical benefit.

Over the last decade, transplant procedures have evolved that predominately rely on the GvL effect for elimination of malignant disease [131]. Intensity of pre-transplant radio-/chemotherapy has been significantly reduced to minimize conditioning-related toxicity. Following such nonmyeloablative conditioning, the risk of graft rejection is overcome by transplantation of large numbers of donor HSC. Still, in mismatched or haploidentical allo-HSCT the risk of graft failure has been higher than in HLA-matched transplants following myeloablative conditioning. Nonengraftment is also a concern in UCB transplantation, particularly in adults in whom adequate cell doses are not always readily available.

The notion that MSC might be employed to support hematopoietic engraftment in allo-HSCT is based on the longstanding concept that bone marrow stromal cells represent the key structural and regulatory components of the hematopoietic niche [16, 78]. This model has meanwhile been extended to include osteoblasts lining the bone surface, marrow endothelial cells, and primitive mesenchymal cells including CXCL12-abundant reticular and Nestin-expressing cells as HSC-niche forming cell populations [59, 122]. Yet, transplantation efficiency of stromal bone marrow cells has been a matter of longstanding debate [66, 93, 116]. The difficulty of detecting donor stromal cells may well be a result of different transplanted cellular doses and sensitivity of detection techniques. One recent study formally reported on 36 % donor stromal cell chimerism following HSCT from sibling donors. Donor stromal cell engraftment occurred in 3/8 BMT patients and in 5/18 patients transplanted with growth factor-mobilized PBSC [99]. This is in line with previous observations that MSC are also contained in peripheral blood [32]. Following

MSC transplantation the group from the Karolinska University Hospital, Stockholm, describes the autopsy results obtained from 18 patients. This includes 108 tissue samples analyzed by PCR for detection of donor DNA. Donor MSC engraftment was inversely correlated with the time from MSC infusion with 50 days seemingly a cut-off for donor MSC persistence. MSC distribution was limited to lung, lymph nodes, and intestine. In the BM, donor MSC were detected only in one patient in keeping with the results of Gonzalo-Daganzo et al. who after HSC/MSC co-transplantation submitted patients to serial bone marrow biopsies for chimerism analysis and found no MSC engraftment [38].

In spite of these incongruent results, MSC are deemed useful in the setting of UCB transplantation (UCBT) based on their graft-promoting effects. This hypothesis has also found support in a murine study [54]. Three small trials with 7–9 patients each were conducted to evaluate efficacy of MSC administration for improved engraftment and GvHD prophylaxis in UCBT (Table 1) [38, 62, 74]. In one study, patients received a transplant consisting of three cellular components, namely UCB, PBSC, and MSC [38]. In all studies, transplants were performed for high risk or relapsed hematological malignancies. Matched historic controls were provided for comparison of the outcome parameters in all three studies. Yet, no statistically significant difference in engraftment and acute and chronic GvHD was observed between UCB/MSC co-transplanted patients (pts.) and controls. Still it is noteworthy that in these three studies, only a single patient (1/24 pts.; 4 %) developed severe aGvHD III–IV in the MSC co-transplanted groups compared to the controls (9/91 pts.; 9 %). Likewise, only one patient suffered from limited and one from extensive cGvHD (2/24 pts.; 8 %) in the MSC cohorts. In the controls, the incidence of cGvHD was slightly higher (17/91 pts.; 18 %). In view of the favorable results in both the MSC and control groups, Gonzales-Daganzo et al. closed their study early based on the lack of evidence that MSC transplants are of benefit in UCBT in which hematopoietic engraftment is already bridged by co-transplantation of PBSC.

The MSC co-transplantation approach was further evaluated in mismatched/haploidentical HSCT [8, 11, 69, 129] with enhanced engraftment and GvHD prophylaxis as primary endpoints. In the three studies assessing this approach in the haploidentical setting, patients received myeloablative therapy prior to transplant. In the fourth study patients were transplanted with nonselected PBSC from $\geq$1 antigen-mismatched donor following reduced intensity conditioning. In none of these studies accelerated formal neutrophil or platelet engraftment was noted. Liu et al. observed, however, that in the MSC group platelets reached the 50×10^9/l threshold faster (22 days; range 12–58 days) than in the controls (28 days; range 10–99 days). In one trial, due to the typical NK cell surge, leukocyte counts rose to 10×10^6/l three days faster in the MSC co-transplantation group [8]. Acute GvHD in the two haploidentical transplant trials [8, 69] was generally low grade, both in the MSC co-transplant cohorts with no aGvHD III–IV and 43 % of aGvHD I–II (20/47 pts.) compared to the controls with only three patients with aGvHD III and aGvHD I–II in 36 % (27/75 pts.). Also in the study by Baron et al., severe aGvHD was comparable following mismatched unrelated

HSCT with aGvHD II–III in 35 % (7/20 pts.) and aGvHD IV in 10 % (2/20 pts.) following MSC co-transplantation versus aGvHD II–III in 32 % (6/16 pts.) and aGvHD IV in 19 % (3/16 pts.) of the controls. Yet, with 31 % the one-year probability of "dying from GvHD or infection while on GvHD therapy" was significantly higher in the controls compared to 10 % of patients co-transplanted with MSC. This translates into 37 % nonrelapse mortality in the controls compared to only 10 % in the MSC-transplanted patients at one-year post HSCT.

In summary, generation of clinical-scale quantities of MSC was feasible even when the HSC and MSC were harvested from the same donors. There were also no immediate side effects from MSC infusion and no evidence of ectopic bone formation over time. Yet, the other endpoints were not successfully met. Thus, HSC/MSC co-transplantation was not associated with accelerated engraftment. So far only in the setting of poor hematopoietic recovery, has salvage from graft failure been reported in individual patients [61, 85, 87].

Likewise, no significant difference in the incidence of acute or chronic GvHD was observed following HSC/MSC co-infusion which may well be explained by lack of appropriate inflammatory signals in the immediate post-transplant period. Indeed, in the absence of inflammation MSC are not capable of preventing or ameliorating GvHD as shown in a murine IFN-γ knockout model. Also immediately after HSCT levels of IFN-γ and TNF-α, both critical triggers of MSC activity, are low [98]. Pre-incubation of MSC with IFN-γ can compensate for this deficiency in the early transplant period. Thus, timing of MSC administration seems to be the key. Indeed, in several murine studies [98, 121], HSC/MSC co-transplantation failed to prevent GvHD whereas delayed MSC infusion seemed to effectively elicit the immunosuppressive properties of MSC [98]. Also manifest GvHD was mitigated by MSC application in a dose-dependent manner [51]. Although one needs to keep in mind the distinct immunoinhibitory mechanisms of MSC in mouse and man [83], these models do suggest that MSC might prove more useful for treatment of overt GvHD than for prevention.

7 MSC for Treatment of Steroid-Refractory Acute GvHD

For evaluation of response to MSC administration in steroid-resistant GvHD, four studies (Table 2) with a total of 289 patients and severe aGvHD III/IV in 84 % (242/289) of cases can be submitted to aggregated analyses based on the focus on aGvHD and standardized regimens for MSC-preparation from bone marrow and expansion [55, 58, 81, 101]. Although a multicenter phase II study, the European Group of Blood and Marrow Transplantation employed a consensus protocol for FCS-supported MSC-generation [58]. The three other studies employed industrially manufactured MSC. The general challenge in comparing efficacy of GvHD therapies between studies resided in the variability of endpoint definitions with regard to the scoring of clinical benefit as well as choice of timepoints for such an assessment. Also durability of response is not uniformly addressed [75, 79, 80].

In GvHD, a complete response (CR) is defined by disappearance of all symptoms. Yet, partial response (PR) may simply indicate an improvement from baseline but not necessarily a clinically meaningful benefit.

A consensus statement [80] demands that PR should signify a difference by two grades, however, this recommendation is not consistently followed or even specified. In the above studies, PR thus refers to improvement by at least one GvHD grade, mixed response (MR) describes reduction in severity of symptoms at a minimum of one affected site, and overall response (OR) summarizes the frequency of complete and partial responses. Few studies provide prospective timeframes for response evaluation and duration of follow-up, yet in most studies, best responses are documented. Aggregated calculation of OR is 65.4 % (189/289 pts.) of the above 289 high-risk patients with 47 % aGvHD grade III and 37 % grade IV [55, 58, 81, 101]. Complete responses are presented for 230 patients in three of the four studies with an aggregated CR of 44 % (101/230 pts.) [58, 81, 101]. This is a noteworthy result that compares well with other forms of second-line immunomodulatory interventions for refractory aGvHD such as treatment with TNF-α and IL-2 antibodies [4, 47, 48, 57].

One of the reasons for this favorable outcome might be the fact that in addition to 193 adults, 96 children are included in these studies. The European Group of Blood and Marrow Transplantation (EBMT) multicenter effort is the only prospective trial that includes equal numbers of children ($n = 25$) and adults ($n = 30$) clinically matched for aGvHD grade to allow for prospective comparison of age-dependent benefit from MSC within one study. In the EBMT trial 84 % OR and 68 % CR in children versus 60 % OR and 43 % CR in adults and a superior two-year overall survival in children with 45 % versus 26 % in adults ($p = 0.06$) confirms a more favorable outcome in the younger patient cohort.

One of the unresolved issues to date is the question of how many applications of MSC are required to maintain a durable response in aGvHD. Among the four studies described above, three consistently administer a minimum of eight infusions of 2×10^6 MSC/kg [55, 81, 101]. In these studies employing the commercially prepared MSC product Prochymal® an OR of 64 % is achieved compared to an OR of 71 % in the EBMT study limiting MSC application to 1–2 $\times$ 0.6–2.0 $\times 10^6$ MSC/kg in 89 % (49/55 pts.) of patients. In children, multiple infusions of Prochymal® resulted in 66 % OR compared to a considerably higher OR of 84 % in children treated in the EBMT study.

Although on first sight this seems to suggest that the commercially prepared third-party donor-derived MSC exhibit a trend towards lower efficacy, there is a variety of confounding factors in study design and endpoint assessment that may have considerable influence on such an interstudy comparison. Still, a closer look at MSC preparations seems justified. In the Prochymal® studies as well as in the EBMT multicenter trial, expansion of MSC did not exceed more than four to six passages. Yet, seeding densities may also play a role. The end product in the EBMT study is characterized according to the ISCT criteria. The Prochymal® studies submit their MSC product to additional functional immunological testing. Cryopreservation prior to infusion is one aspect in MSC preparation that is known to be critical for vitality but also with regard to the immunosuppressive MSC-mediated activity. It is this distinct difference that could contribute to discrepancies in clinical outcome, as immediately after thawing the immunosuppressive properties of MSC are severely impaired. Thus, defrosted MSC are refractory to IFN-γ which is the key signal for IDO-induction as well as for up-regulation of immunosuppressive cell surface molecules such as PD-L1. The immunoinhibitory activity of defrosted MSC is, however, fully restored if submitted to 24 h of cell culture [33]. This insight might have significant impact on the future design of MSC-facilitated studies.

8 MSC for the Treatment of De Novo Acute GvHD

Kebriaei et al. conducted the first large prospective, open-labeled multicentered phase II study in the United States, Canada, and Australia (Table 2b). Thirty-one adult patients in 16 centers with de novo grade II–IV aGvHD were enrolled, with MSC manufactured by Osiris Therapeutics, Inc., Baltimore, from bone marrow aspirates of six healthy donors. Sixteen patients received low-dose MSC (LD 2×10^6 cells/kg), and 15 received high dose (HD 8×10^6 cells/kg) infusions within 48 h from diagnosis of aGvHD and a second infusion three days later. Of note, only 32 % patients suffered from aGvHD III–IV, considerably fewer than in the studies evaluating MSC efficacy in steroid-resistant aGvHD. MSC infusions proved safe and initial response rate was high, with 24 patients in CR (14 LD-pts., 10 HD-pts.) and 7 in PR (2 LD-pts., 5 HD-pts.). Time to response was also rapid with 42 % patients achieving CR at day 7, 52 % by day 14, and 77 % at day 28. CR was not correlated to donor source, grade, or location of GvHD.

A total of 71 % of patients survived to 90 days with a significantly improved survival of responders (88 % CR vs. 14 % non-CR; $p = 0.0008$). Overall, nine patients died within 13–63 days after MSC-infusion; three patients who had achieved CR died from infections, three nonresponders died from progressive GvHD, and one nonresponding patient from relapsed malignancy or brain bleed. Three patients relapsed within a two-year follow-up period.

9 Alternative Cell Culture Supplements for Clinical-Grade MSC Products

In cell therapy, transmission of prion, viral, and other zoonotic diseases in addition to xenogenic immunization is a concern when preparing clinical-grade products supplemented with FCS. Therefore, alternative sources for expansion and maintenance of MSC have been explored. In vitro platelet lysate (PL) and to a lesser degree autologous serum (AS) proved efficacious, yielding MSC preparations with comparable surface marker profile and tri-lineage differentiation capacity to

FCS-risen MSC. With regard to their influence on T cell effector functions such as cytokine production, cytotoxicity, or proliferation, some variability in the spectrum of immunomodulatory properties and secreted mediators was observed [6, 14, 30]. There is only one study that suggests that overall PL-MSC might be less immunosuppressive than FCS-MSC. In this report PL-MSC had a weaker inhibitory influence on T and NK cell proliferation and NK cell cytotoxicity [1] which was associated with lower constitutive PGE_2-production compared to FCS-MSC.

Yet, clinical experience is sparse with few patients, 8 children and 13 adults, treated in two studies with PL-MSC for GvHD treatment thus far (Table 2b) [72, 126]. Patients enrolled in these pilot/phase I studies suffered from different degrees of steroid-refractory severe aGvHD III–IV ranging from 36 % in children [72] to 100 % in adults [126]. Accordingly in the latter study, OR was only 16 % (2/13) in the adults with steroid-refractory GvHD IV. In contrast, in the trial assessing efficacy in children [72] with slightly less severe GvHD, OR was 63 % (5/8 pts) which is more in line with the FCS-MSC studies described earlier.

In another small study with 10 aGvHD patients human autologous serum was used for MSC culture resulting in a very low CR 10 % and OR 60 % and high early toxicity and mortality (33 %) within the first 100 days post transplant [94]. Thus, these approaches to expand and activate MSC warrant further clinical evaluation. One study has already been in progress in the Netherlands since 2009 employing human plasma compared to platelet lysate for MSC expansion. In this phase I/II study patients with de novo grade II–IV aGvHD and cGvHD are included (http://www.clinicaltrials.gov; identifier: NCT00827398).

10 MSC for Treatment of Refractory Chronic GvHD

Only few studies have been conducted for treatment of cGvHD. In some of the MSC-trials for treatment of steroid-resistant aGvHD, single patients with cGvHD were enrolled (overall 17 pts.) with an aggregated OR of 47 % (8/17) [72, 87, 94, 107] (Table 2). In one small trial, MSC-mediated tissue repair after direct intra-BM injection was assessed in four sclero-dermatuos cGvHD patients. Reversal of the Th1 cells to Th2 cell ratio was observed with reported gradual improvement of symptoms in all four patients [140].

There is, however, one trial with a total of 19 patients focusing entirely on MSC application for refractory cGvHD in patients who failed six months of prior intensive immunosuppressive therapy [133]. This study is noteworthy as it provides clear definitions with regard to indication of MSC infusion, severity of GvHD, and response. Thus, the NIH consensus criteria for organ scoring and global assessment of cGvHD were used. MSC were transfused directly after preparation without intermittent cryopreservation.

As discussed above, this may be one of the reasons why in spite of relatively low MSC doses (median 0.6×10^6; range $0.2–1.4 \times 10^6$/kg) patients still experienced a considerable clinical benefit. Two of the severely ill patients had organ and four multiorgan disease. Still 14/19 patients (74 %) responded to 1–2 MSC infusions with CR in 4 patients (21 %) and PR in 10 patients (53 %). The highest clinical benefit was observed for cGvHD of the oral mucosa, GI tract, liver, and skin. Concomitantly applied immunosuppressive agent could be tapered in 5 patients and in another 5 patients immunosuppressive therapy could be stopped altogether. These encouraging results commend further evaluation of MSC for the treatment of extended cGvHD. The response profile in this study would also suggest that MSC need to be administered at a timepoint when attenuation of inflammation and tissue repair still hold a chance for facilitating clinical improvement. In contrast end-stage fibrotic disease will no longer benefit from MSC infusion.

Additional prospective studies are under way. Thus a randomized phase I/II study started in Korea in early 2012. Here, umbilical cord blood-derived MSC grown in the presence of FCS (PROMOCHEM™) are employed for the treatment of steroid-refractory aGvHD and cGvHD (http://www.clinicaltrials.gov; Indent: NCT01549665). Another study is a phase I/II randomized multicenter study in Spain which started recruitment of patients with extensive cGvHD in 2010 for treatment with MSC derived from adipose tissue (http://www.clinicaltrials.gov; Indent: NCT01222039).

Adipose tissue in future might prove to be a highly attractive source for MSC preparation due to its abundant availability and the encouraging results from the one study by Fang et al. in acute GvHD patients grade III–IV disease and a complete response in 83 % of patients (Table 2a).

11 Summary

Overall MSC hold promise in the treatment of acute and chronic GvHD. The application seems to be safe thus far with no evidence of malignant transformation. The influence of different MSC sources and various cell culture supplements in MSC generation on the regenerative and immunomodulatory properties as well as efficacy in the different clinical settings will have to be carefully explored in the future. Also, it would be desirable to accompany the clinical studies with immune-monitoring analyses to better understand the underlying mechanisms in responding and nonresponding patients. This will then provide a basis for further improving MSC therapy.

References

1. Abdelrazik H, Spaggiari GM, Chiossone L, Moretta L (2011) Mesenchymal stem cells expanded in human platelet lysate display a decreased inhibitory capacity on T- and NK-cell proliferation and function. Eur J Immunol 41:3281–3290

2. Aggarwal S, Pittenger MF (2005) Human mesenchymal stem cells modulate allogeneic immune cell responses. Blood 105:1815–1822
3. Anderson BE, McNiff JM, Jain D, Blazar BR, Shlomchik WD, Shlomchik MJ (2005) Distinct roles for donor- and host-derived antigen-presenting cells and costimulatory molecules in murine chronic graft-versus-host disease: requirements depend on target organ. Blood 105:2227–2234
4. Arora M, Burns LJ, Davies SM, Macmillan ML, Defor TE, Miller WJ, Weisdorf DJ (2003) Chronic graft-versus-host disease: a prospective cohort study. Biol Blood Marrow Transplant 9:38–45
5. Arora M, Klein JP, Weisdorf DJ, Hassebroek A, Flowers ME, Cutler CS, Urbano-Ispizua A, Antin JH, Bolwell BJ, Boyiadzis M, Cahn JY, Cairo MS, Isola L, Jacobsohn DA, Jagasia M, Klumpp TR, Lee SJ, Petersdorf EW, Santarone S, Gale RP, Schouten HC, Spellman S, Wingard JR, Horowitz MM, Pavletic SZ (2011) Chronic GVHD risk score: a Center for International Blood and Marrow Transplant Research analysis. Blood 117:6714–6720
6. Avanzini MA, Bernardo ME, Cometa AM, Perotti C, Zaffaroni N, Novara F, Visai L, Moretta A, Del Fante C, Villa R, Ball LM, Fibbe WE, Maccario R, Locatelli F (2009) Generation of mesenchymal stromal cells in the presence of platelet lysate: a phenotypic and functional comparison of umbilical cord blood- and bone marrow-derived progenitors. Haematologica 94:1649–1660
7. Baker MB, Altman NH, Podack ER, Levy RB (1996) The role of cell-mediated cytotoxicity in acute GVHD after MHC-matched allogeneic bone marrow transplantation in mice. J Exp Med 183:2645–2656
8. Ball LM, Bernardo ME, Roelofs H, Lankester A, Cometa A, Egeler RM, Locatelli F, Fibbe WE (2007) Cotransplantation of ex vivo expanded mesenchymal stem cells accelerates lymphocyte recovery and may reduce the risk of graft failure in haploidentical hematopoietic stem-cell transplantation. Blood 110:2764–2767
9. Ballen KK, Koreth J, Chen YB, Dey BR, Spitzer TR (2012) Selection of optimal alternative graft source: mismatched unrelated donor, umbilical cord blood, or haploidentical transplant. Blood 119:1972–1980
10. Barker JN, Wagner JE (2003) Umbilical-cord blood transplantation for the treatment of cancer. Nat Rev Cancer 3:526–532
11. Baron F, Lechanteur C, Willems E, Bruck F, Baudoux E, Seidel L, Vanbellinghen JF, Hafraoui K, Lejeune M, Gothot A, Fillet G, Beguin Y (2010) Cotransplantation of mesenchymal stem cells might prevent death from graft-versus-host disease (GVHD) without abrogating graft-versus-tumor effects after HLA-mismatched allogeneic transplantation following nonmyeloablative conditioning. Biol Blood Marrow Transplant 16:838–847
12. Behre G, Theurich S, Weber T, Christopeit M (2009) Reply to 'the correlation between cotransplantation of mesenchymal stem cells and higher recurrence rates in hematologic malignancy patients: outcome of a pilot clinical study' by Ning et al. Leukemia 23:178; author reply 179–180
13. Berger M, Figari O, Bruno B, Raiola A, Dominietto A, Fiorone M, Podesta M, Tedone E, Pozzi S, Fagioli F, Madon E, Bacigalupo A (2008) Lymphocyte subsets recovery following allogeneic bone marrow transplantation (BMT): CD4+ cell count and transplant-related mortality. Bone Marrow Transplant 41:55–62
14. Bernardo ME, Avanzini MA, Perotti C, Cometa AM, Moretta A, Lenta E, Del Fante C, Novara F, de Silvestri A, Amendola G, Zuffardi O, Maccario R, Locatelli F (2007) Optimization of in vitro expansion of human multipotent mesenchymal stromal cells for cell-therapy approaches: further insights in the search for a fetal calf serum substitute. J Cell Physiol 211:121–130
15. Bianco P, Riminucci M, Gronthos S, Robey PG (2001) Bone marrow stromal stem cells: nature, biology, and potential applications. Stem Cells 19:180–192
16. Boxall SA, Jones E (2012) Markers for characterization of bone marrow multipotential stromal cells. Stem Cells Int 2012:975871

17. Braun MY, Lowin B, French L, Acha-Orbea H, Tschopp J (1996) Cytotoxic T cells deficient in both functional fas ligand and perforin show residual cytolytic activity yet lose their capacity to induce lethal acute graft-versus-host disease. J Exp Med 183:657–661
18. Cahn JY, Klein JP, Lee SJ, Milpied N, Blaise D, Antin JH, Leblond V, Ifrah N, Jouet JP, Loberiza F, Ringden O, Barrett AJ, Horowitz MM, Socie G (2005) Prospective evaluation of 2 acute graft-versus-host (GVHD) grading systems: a joint Societe Francaise de Greffe de Moelle et Therapie Cellulaire (SFGM-TC), Dana Farber Cancer Institute (DFCI), and International Bone Marrow Transplant Registry (IBMTR) prospective study. Blood 106:1495–1500
19. Choi SW, Kitko CL, Braun T, Paczesny S, Yanik G, Mineishi S, Krijanovski O, Jones D, Whitfield J, Cooke K, Hutchinson RJ, Ferrara JL, Levine JE (2008) Change in plasma tumor necrosis factor receptor 1 levels in the first week after myeloablative allogeneic transplantation correlates with severity and incidence of GVHD and survival. Blood 112:1539–1542
20. Couriel DR, Hosing C, Saliba R, Shpall EJ, Anderlini P, Rhodes B, Smith V, Khouri I, Giralt S, de Lima M, Hsu Y, Ghosh S, Neumann J, Andersson B, Qazilbash M, Hymes S, Kim S, Champlin R, Donato M (2006) Extracorporeal photochemotherapy for the treatment of steroid-resistant chronic GVHD. Blood 107:3074–3080
21. Dao MA, Nolta JA (2000) CD34: to select or not to select? That is the question. Leukemia 14:773–776
22. Daubener W, Schmidt SK, Heseler K, Spekker KH, MacKenzie CR (2009) Antimicrobial and immunoregulatory effector mechanisms in human endothelial cells. Indoleamine 2,3-dioxygenase versus inducible nitric oxide synthase. Thromb Haemost 102:1110–1116
23. Dazzi F, Ramasamy R, Glennie S, Jones SP, Roberts I (2006) The role of mesenchymal stem cells in haemopoiesis. Blood Rev 20:161–171
24. Dominici M, Le Blanc K, Mueller I, Slaper-Cortenbach I, Marini F, Krause D, Deans R, Keating A, Prockop D, Horwitz E (2006) Minimal criteria for defining multipotent mesenchymal stromal cells. The International Society for Cellular Therapy position statement. Cytotherapy 8:315–317
25. Dustin ML (2001) Role of adhesion molecules in activation signaling in T lymphocytes. J Clin Immunol 21:258–263
26. Fang B, Song Y, Zhao RC, Han Q, Cao Y (2007) Treatment of resistant pure red cell aplasia after major ABO-incompatible bone marrow transplantation with human adipose tissue derived mesenchymal stem cells. Am J Hematol 82:772–773
27. Federmann B, Bornhauser M, Meisner C, Kordelas L, Beelen DW, Stuhler G, Stelljes M, Schwerdtfeger R, Christopeit M, Behre G, Faul C, Vogel W, Schumm M, Handgretinger R, Kanz L, Bethge WA (2012) Haploidentical allogeneic hematopoietic cell transplantation in adults using CD3/CD19 depletion and reduced intensity conditioning: a phase II study. Haematologica 97(10):1523–1531
28. Ferrara JL, Levine JE, Reddy P, Holler E (2009) Graft-versus-host disease. Lancet 373:1550–1561
29. Filipovich AH, Weisdorf D, Pavletic S, Socie G, Wingard JR, Lee SJ, Martin P, Chien J, Przepiorka D, Couriel D, Cowen EW, Dinndorf P, Farrell A, Hartzman R, Henslee-Downey J,Jacobsohn D, McDonald G, Mittleman B, Rizzo JD, Robinson M, Schubert M, Schultz K, Shulman H, Turner M, Vogelsang G, Flowers ME (2005) National Institutes of Health consensus development project on criteria for clinical trials in chronic graft-versus-host disease: I. Diagnosis and staging working group report. Biol Blood Marrow Transplant 11:945–956
30. Flemming A, Schallmoser K, Strunk D, Stolk M, Volk HD, Seifert M (2011) Immunomodulative efficacy of bone marrow-derived mesenchymal stem cells cultured in human platelet lysate. J Clin Immunol 31:1143–1156
31. Flomenberg N, Baxter-Lowe LA, Confer D, Fernandez-Vina M, Filipovich A, Horowitz M, Hurley C, Kollman C, Anasetti C, Noreen H, Begovich A, Hildebrand W, Petersdorf E, Schmeckpeper B, Setterholm M, Trachtenberg E, Williams T, Yunis E, Weisdorf D (2004) Impact of HLA class I and class II high-resolution matching on outcomes of unrelated donor

bone marrow transplantation: HLA-C mismatching is associated with a strong adverse effect on transplantation outcome. Blood 104:1923–1930
32. Fowler DH, Foley J, Whit-Shan Hou J, Odom J, Castro K, Steinberg SM, Gea-Banacloche J, Kasten-Sportes C, Gress RE, Bishop MR (2004) Clinical "cytokine storm" as revealed by monocyte intracellular flow cytometry: correlation of tumor necrosis factor alpha with severe gut graft-versus-host disease. Clin Gastroenterol Hepatol 2:237–245
33. Francois M, Copland IB, Yuan S, Romieu-Mourez R, Waller EK, Galipeau J (2012) Cryopreserved mesenchymal stromal cells display impaired immunosuppressive properties as a result of heat-shock response and impaired interferon-gamma licensing. Cytotherapy 14:147–152
34. Fujii N, Takenaka K, Shinagawa K, Ikeda K, Maeda Y, Sunami K, Hiramatsu Y, Matsuo K, Ishimaru F, Niiya K, Yoshino T, Hirabayashi N, Harada M (2001) Hepatic graft-versus-host disease presenting as an acute hepatitis after allogeneic peripheral blood stem cell transplantation. Bone Marrow Transplant 27:1007–1010
35. Gieseke F, Bohringer J, Bussolari R, Dominici M, Handgretinger R, Muller I (2010) Human multipotent mesenchymal stromal cells use galectin-1 to inhibit immune effector cells. Blood 116:3770–3779
36. Glucksberg H, Storb R, Fefer A, Buckner CD, Neiman PE, Clift RA, Lerner KG, Thomas ED (1974) Clinical manifestations of graft-versus-host disease in human recipients of marrow from HL-A-matched sibling donors. Transplantation 18:295–304
37. Goker H, Haznedaroglu IC, Chao NJ (2001) Acute graft-vs-host disease: pathobiology and management. Exp Hematol 29:259–277
38. Gonzalo-Daganzo R, Regidor C, Martin-Donaire T, Rico MA, Bautista G, Krsnik I, Fores R, Ojeda E, Sanjuan I, Garcia-Marco JA, Navarro B, Gil S, Sanchez R, Panadero N, Gutierrez Y, Garcia-Berciano M, Perez N, Millan I, Cabrera R, Fernandez MN (2009) Results of a pilot study on the use of third-party donor mesenchymal stromal cells in cord blood transplantation in adults. Cytotherapy 11:278–288
39. Gratwohl A, Baldomero H, Aljurf M, Pasquini MC, Bouzas LF, Yoshimi A, Szer J, Lipton J, Schwendener A, Gratwohl M, Frauendorfer K, Niederwieser D, Horowitz M, Kodera Y (2010) Hematopoietic stem cell transplantation: a global perspective. JAMA 303:1617–1624
40. Graubert TA, DiPersio JF, Russell JH, Ley TJ (1997) Perforin/granzyme-dependent and independent mechanisms are both important for the development of graft-versus-host disease after murine bone marrow transplantation. J Clin Invest 100:904–911
41. Handgretinger R, Chen X, Pfeiffer M, Mueller I, Feuchtinger T, Hale GA, Lang P (2007) Feasibility and outcome of reduced-intensity conditioning in haploidentical transplantation. Ann N Y Acad Sci 1106:279–289
42. Hao L, Sun H, Wang J, Wang T, Wang M, Zou Z (2012) Mesenchymal stromal cells for cell therapy: besides supporting hematopoiesis. Int J Hematol 95:34–46
43. Hill GR, Ferrara JL (2000) The primacy of the gastrointestinal tract as a target organ of acute graft-versus-host disease: rationale for the use of cytokine shields in allogeneic bone marrow transplantation. Blood 95:2754–2759
44. Horwitz EM, Le Blanc K, Dominici M, Mueller I, Slaper-Cortenbach I, Marini FC, Deans RJ, Krause DS, Keating A (2005) Clarification of the nomenclature for MSC: The International Society for Cellular Therapy position statement. Cytotherapy 7:393–395
45. Huang GT, Gronthos S, Shi S (2009) Mesenchymal stem cells derived from dental tissues vs. those from other sources: their biology and role in regenerative medicine. J Dent Res 88:792–806
46. Inagaki J, Nagatoshi Y, Hatano M, Isomura N, Sakiyama M, Okamura J (2008) Low-dose MTX for the treatment of acute and chronic graft-versus-host disease in children. Bone Marrow Transplant 41:571–577
47. Jacobsohn DA (2008) Acute graft-versus-host disease in children. Bone Marrow Transplant 41:215–221
48. Jacobsohn DA, Vogelsang GB (2007) Acute graft versus host disease. Orphanet J Rare Dis 2:35

49. Jiang Z, Podack E, Levy RB (2001) Major histocompatibility complex-mismatched allogeneic bone marrow transplantation using perforin and/or Fas ligand double-defective CD4(+) donor T cells: involvement of cytotoxic function by donor lymphocytes prior to graft-versus-host disease pathogenesis. Blood 98:390–397
50. Johnston LJ, Brown J, Shizuru JA, Stockerl-Goldstein KE, Stuart MJ, Blume KG, Negrin RS, Chao NJ (2005) Rapamycin (sirolimus) for treatment of chronic graft-versus-host disease. Biol Blood Marrow Transplant 11:47–55
51. Joo SY, Cho KA, Jung YJ, Kim HS, Park SY, Choi YB, Hong KM, Woo SY, Seoh JY, Cho SJ, Ryu KH (2010) Mesenchymal stromal cells inhibit graft-versus-host disease of mice in a dose-dependent manner. Cytotherapy 12:361–370
52. Karlsson H, Samarasinghe S, Ball LM, Sundberg B, Lankester AC, Dazzi F, Uzunel M, Rao K, Veys P, Le Blanc K, Ringden O, Amrolia PJ (2008) Mesenchymal stem cells exert differential effects on alloantigen and virus-specific T-cell responses. Blood 112:532–541
53. Kebriaei P, Isola L, Bahçeci E, Holland K, Rowley S, McGuirk J, Devetten M, Jansen J, Herzig R, Schuster M, Monroy R, Uberti J (2009) Adult human mesenchymal stem cells added to corticosteroid therapy for the treatment of acute graft-versus-host disease. Biol Blood Marrow Transplant 15:804–811
54. Krampera M, Glennie S, Dyson J, Scott D, Laylor R, Simpson E, Dazzi F (2003) Bone marrow mesenchymal stem cells inhibit the response of naive and memory antigen-specific T cells to their cognate peptide. Blood 101:3722–3729
55. Kurtzberg J, Prasad V, Grimley MS, Horn B, Carpenter PA, Jacobsohn D, Prockop S (2010) Allogeneic human mesenchymal stem cell therapy (Prochymal®) as a rescue agent for severe treatment resistant GVHD in pediatric patients. Biol Blood Marrow Transplant 16:S169
56. Lazarus HM, Koc ON, Devine SM, Curtin P, Maziarz RT, Holland HK, Shpall EJ, McCarthy P, Atkinson K, Cooper BW, Gerson SL, Laughlin MJ, Loberiza FR Jr, Moseley AB, Bacigalupo A (2005) Cotransplantation of HLA-identical sibling culture-expanded mesenchymal stem cells and hematopoietic stem cells in hematologic malignancy patients. Biol Blood Marrow Transplant 11:389–398
57. Lazarus HM, Rowe JM (1995) New and experimental therapies for treating graft-versus-host disease. Blood Rev 9:117–133
58. Le Blanc K, Frassoni F, Ball L, Locatelli F, Roelofs H, Lewis I, Lanino E, Sundberg B, Bernardo ME, Remberger M, Dini G, Egeler RM, Bacigalupo A, Fibbe W, Ringden O (2008) Mesenchymal stem cells for treatment of steroid-resistant, severe, acute graft-versus-host disease: a phase II study. Lancet 371:1579–1586
59. Le Blanc K, Mougiakakos D (2012) Multipotent mesenchymal stromal cells and the innate immune system. Nat Rev Immunol 12:383–396
60. Le Blanc K, Rasmusson I, Sundberg B, Gotherstrom C, Hassan M, Uzunel M, Ringden O (2004) Treatment of severe acute graft-versus-host disease with third party haploidentical mesenchymal stem cells. Lancet 363:1439–1441
61. Le Blanc K, Samuelsson H, Gustafsson B, Remberger M, Sundberg B, Arvidson J, Ljungman P, Lonnies H, Nava S, Ringden O (2007) Transplantation of mesenchymal stem cells to enhance engraftment of hematopoietic stem cells. Leukemia 21:1733–1738
62. Lee SH, Cheuh HW, Yoo KH, Sung KW, Koo HH, Kim DH, Jung CW, Choi SJ, Oh W, Yang YS (2011) Cotransplantation of third party umbilical cord blood mesenchymal stem cells to promote engraftment in pediatric recipients of unrelated donor umbilical cord blood. Biol Blood Marrow Transplant 17:S169–S170
63. Lee SJ, Klein JP, Barrett AJ, Ringden O, Antin JH, Cahn JY, Carabasi MH, Gale RP, Giralt S, Hale GA, Ilhan O, McCarthy PL, Socie G, Verdonck LF, Weisdorf DJ, Horowitz MM (2002) Severity of chronic graft-versus-host disease: association with treatment-related mortality and relapse. Blood 100:406–414
64. Li HW, Sykes M (2012) Emerging concepts in haematopoietic cell transplantation. Nat Rev Immunol 12:403–416

65. Li JM, Giver CR, Lu Y, Hossain MS, Akhtari M, Waller EK (2009) Separating graft-versus-leukemia from graft-versus-host disease in allogeneic hematopoietic stem cell transplantation. Immunotherapy 1:599–621
66. Liem LM, Fibbe WE, van Houwelingen HC, Goulmy E (1999) Serum transforming growth factor-beta1 levels in bone marrow transplant recipients correlate with blood cell counts and chronic graft-versus-host disease. Transplantation 67:59–65
67. Lin MT, Storer B, Martin PJ, Tseng LH, Gooley T, Chen PJ, Hansen JA (2003) Relation of an interleukin-10 promoter polymorphism to graft-versus-host disease and survival after hematopoietic-cell transplantation. N Engl J Med 349:2201–2210
68. Lin W, Oh SK, Choo AB, George AJ (2012) Activated T cells modulate immunosuppression by embryonic-and bone marrow-derived mesenchymal stromal cells through a feedback mechanism. Cytotherapy 14:274–284
69. Liu K, Chen Y, Zeng Y, Xu L, Liu D, Chen H, Zhang X, Han W, Wang Y, Zhao T, Wang J, Han Q, Zhao C, Huang X (2011) Coinfusion of mesenchymal stromal cells facilitates platelet recovery without increasing leukemia recurrence in haploidentical hematopoietic stem cell transplantation: a randomized, controlled clinical study. Stem Cells Dev 20:1679–1685
70. Loiseau P, Busson M, Balere ML, Dormoy A, Bignon JD, Gagne K, Gebuhrer L, Dubois V, Jollet I, Bois M, Perrier P, Masson D, Moine A, Absi L, Reviron D, Lepage V, Tamouza R, Toubert A, Marry E, Chir Z, Jouet JP, Blaise D, Charron D, Raffoux C (2007) HLA association with hematopoietic stem cell transplantation outcome: the number of mismatches at HLA-A, -B, -C, -DRB1, or -DQB1 is strongly associated with overall survival. Biol Blood Marrow Transplant 13:965–974
71. Lopez F, Parker P, Nademanee A, Rodriguez R, Al-Kadhimi Z, Bhatia R, Cohen S, Falk P, Fung H, Kirschbaum M, Krishnan A, Kogut N, Molina A, Nakamura R, O'Donnell M, Popplewell L, Pullarkat V, Rosenthal J, Sahebi F, Smith E, Snyder D, Somlo G, Spielberger R, Stein A, Sweetman R, Zain J, Forman S (2005) Efficacy of mycophenolate mofetil in the treatment of chronic graft-versus-host disease. Biol Blood Marrow Transplant 11:307–313
72. Lucchini G, Introna M, Dander E, Rovelli A, Balduzzi A, Bonanomi S, Salvadè A, Capelli C, Belotti D, Gaipa G, Perseghin P, Vinci P, Lanino E, Chiusolo P, Orofino MG, Marktel S, Golay J, Rambaldi A, Biondi A, D'Amico G, Biagi E (2010) Platelet-lysate-expanded mesenchymal stromal cells as a salvage therapy for severe resistant graft-versus-host disease in a pediatric population. Biol Blood Marrow Transplant 16:1293–1301
73. MacKenzie CR, Heseler K, Muller A, Daubener W (2007) Role of indoleamine 2,3-dioxygenase in antimicrobial defence and immuno-regulation: tryptophan depletion versus production of toxic kynurenines. Curr Drug Metab 8:237–244
74. Macmillan ML, Blazar BR, DeFor TE, Wagner JE (2009) Transplantation of ex vivo culture-expanded parental haploidentical mesenchymal stem cells to promote engraftment in pediatric recipients of unrelated donor umbilical cord blood: results of a phase I–II clinical trial. Bone Marrow Transplant 43:447–454
75. MacMillan ML, DeFor TE, Weisdorf DJ (2010) The best endpoint for acute GVHD treatment trials. Blood 115:5412–5417
76. MacMillan ML, DeFor TE, Weisdorf DJ (2012) What predicts high risk acute graft-versus-host disease (GVHD) at onset?: identification of those at highest risk by a novel acute GVHD risk score. Br J Haematol 157:732–741
77. Maeda Y, Levy RB, Reddy P, Liu C, Clouthier SG, Teshima T, Ferrara JL (2005) Both perforin and Fas ligand are required for the regulation of alloreactive CD8+ T cells during acute graft-versus-host disease. Blood 105:2023–2027
78. Majumdar MK, Thiede MA, Haynesworth SE, Bruder SP, Gerson SL (2000) Human marrow-derived mesenchymal stem cells (MSCs) express hematopoietic cytokines and support long-term hematopoiesis when differentiated toward stromal and osteogenic lineages. J Hematother Stem Cell Res 9:841–848

79. Martin PJ (2008) Study design and endpoints in graft-versus-host disease. Best Pract Res Clin Haematol 21:357–372
80. Martin PJ, Bachier CR, Klingemann HG, McCarthy PL, Szabolcs P, Uberti JP, Schuster MW, Weisdorf D, Chao NJ, Kebriaei P, Shpall EJ, Macmillan ML, Soiffer RJ (2009) Endpoints for clinical trials testing treatment of acute graft-versus-host disease: a joint statement. Biol Blood Marrow Transplant 15:777–784
81. Martin PJ, Uberti JP, Soiffer RJ, Klingemann H, Waller EK, Daly AS, Herrmann RP, Kebriaei P (2010) Prochymal improves response rates in patients with steroid-refractory acute graft versus host disease (SR-GVHD) involving the liver and gut: results of a randomized, placebo-controlled, multicenter phase III trial in GVHD. Biol Blood Marrow Transplant 16:S169–S170
82. McGuirk JP, Weiss ML (2011) Promising cellular therapeutics for prevention or management of graft-versus-host disease (a review). Placenta 32(Suppl 4):S304–S310
83. Meisel R, Brockers S, Heseler K, Degistirici O, Bulle H, Woite C, Stuhlsatz S, Schwippert W, Jager M, Sorg R, Henschler R, Seissler J, Dilloo D, Daubener W (2011) Human but not murine multipotent mesenchymal stromal cells exhibit broad-spectrum antimicrobial effector function mediated by indoleamine 2,3-dioxygenase. Leukemia 25:648–654
84. Meisel R, Zibert A, Laryea M, Gobel U, Daubener W, Dilloo D (2004) Human bone marrow stromal cells inhibit allogeneic T-cell responses by indoleamine 2,3-dioxygenase-mediated tryptophan degradation. Blood 103:4619–4621
85. Meuleman N, Tondreau T, Ahmad I, Kwan J, Crokaert F, Delforge A, Dorval C, Martiat P, Lewalle P, Lagneaux L, Bron D (2009) Infusion of mesenchymal stromal cells can aid hematopoietic recovery following allogeneic hematopoietic stem cell myeloablative transplant: a pilot study. Stem Cells Dev 18:1247–1252
86. Muller A, Heseler K, Schmidt SK, Spekker K, Mackenzie CR, Daubener W (2009) The missing link between indoleamine 2,3-dioxygenase mediated antibacterial and immunoregulatory effects. J Cell Mol Med 13:1125–1135
87. Muller I, Kordowich S, Holzwarth C, Isensee G, Lang P, Neunhoeffer F, Dominici M, Greil J, Handgretinger R (2008) Application of multipotent mesenchymal stromal cells in pediatric patients following allogeneic stem cell transplantation. Blood Cells Mol Dis 40:25–32
88. Nasef A, Zhang YZ, Mazurier C, Bouchet S, Bensidhoum M, Francois S, Gorin NC, Lopez M, Thierry D, Fouillard L, Chapel A (2009) Selected Stro-1-enriched bone marrow stromal cells display a major suppressive effect on lymphocyte proliferation. Int J Lab Hematol 31:9–19
89. Nauta AJ, Kruisselbrink AB, Lurvink E, Willemze R, Fibbe WE (2006) Mesenchymal stem cells inhibit generation and function of both CD34+-derived and monocyte-derived dendritic cells. J Immunol 177:2080–2087
90. Newman RE, Yoo D, LeRoux MA, Danilkovitch-Miagkova A (2009) Treatment of inflammatory diseases with mesenchymal stem cells. Inflamm Allergy Drug Targets 8:110–123
91. Ning H, Yang F, Jiang M, Hu L, Feng K, Zhang J, Yu Z, Li B, Xu C, Li Y, Wang J, Hu J, Lou X, Chen H (2008) The correlation between cotransplantation of mesenchymal stem cells and higher recurrence rate in hematologic malignancy patients: outcome of a pilot clinical study. Leukemia 22:593–599
92. Oevermann L, Handgretinger R (2012) New strategies for haploidentical transplantation. Pediatr Res 71:418–426
93. Perez-Simon JA, Diez-Campelo M, Martino R, Brunet S, Urbano A, Caballero MD, de Leon A, Valcarcel D, Carreras E, del Canizo MC, Lopez-Fidalgo J, Sierra J, San Miguel JF (2005) Influence of the intensity of the conditioning regimen on the characteristics of acute and chronic graft-versus-host disease after allogeneic transplantation. Br J Haematol 130:394–403
94. Perez-Simon JA, Lopez-Villar O, Andreu EJ, Rifon J, Muntion S, Campelo MD, Sanchez-Guijo FM, Martinez C, Valcarcel D, Canizo CD (2011) Mesenchymal stem cells expanded in vitro with human serum for the treatment of acute and chronic graft-versus-host disease: results of a phase I/II clinical trial. Haematologica 96:1072–1076

95. Petersdorf EW (2007) Risk assessment in haematopoietic stem cell transplantation: histocompatibility. Best Pract Res Clin Haematol 20:155–170
96. Petersdorf EW, Malkki M, Gooley TA, Martin PJ, Guo Z (2007) MHC haplotype matching for unrelated hematopoietic cell transplantation. PLoS Med 4:e8
97. Pittenger MF, Mackay AM, Beck SC, Jaiswal RK, Douglas R, Mosca JD, Moorman MA, Simonetti DW, Craig S, Marshak DR (1999) Multilineage potential of adult human mesenchymal stem cells. Science 284:143–147
98. Polchert D, Sobinsky J, Douglas G, Kidd M, Moadsiri A, Reina E, Genrich K, Mehrotra S, Setty S, Smith B, Bartholomew A (2008) IFN-gamma activation of mesenchymal stem cells for treatment and prevention of graft versus host disease. Eur J Immunol 38:1745–1755
99. Poloni A, Sartini D, Emanuelli M, Trappolini S, Mancini S, Pozzi V, Costantini B, Serrani F, Berardinelli E, Renzi E, Olivieri A, Leoni P (2011) Gene expression profile of cytokines in patients with chronic graft-versus-host disease after allogeneic hematopoietic stem cell transplantation with reduced conditioning. Cytokine 53:376–383
100. Ponte AL, Marais E, Gallay N, Langonne A, Delorme B, Herault O, Charbord P, Domenech J (2007) The in vitro migration capacity of human bone marrow mesenchymal stem cells: comparison of chemokine and growth factor chemotactic activities. Stem Cells 25:1737–1745
101. Prasad VK, Lucas KG, Kleiner GI, Talano JA, Jacobsohn D, Broadwater G, Monroy R, Kurtzberg J (2011) Efficacy and safety of ex vivo cultured adult human mesenchymal stem cells (Prochymal) in pediatric patients with severe refractory acute graft-versus-host disease in a compassionate use study. Biol Blood Marrow Transplant 17:534–541
102. Prevosto C, Zancolli M, Canevali P, Zocchi MR, Poggi A (2007) Generation of CD4+ or CD8+ regulatory T cells upon mesenchymal stem cell-lymphocyte interaction. Haematologica 92:881–888
103. Przepiorka D, Weisdorf D, Martin P, Klingemann HG, Beatty P, Hows J, Thomas ED (1995) 1994 consensus conference on acute GVHD grading. Bone Marrow Transplant 15:825–828
104. Rastegar F, Shenaq D, Huang J, Zhang W, Zhang BQ, He BC, Chen L, Zuo GW, Luo Q, Shi Q, Wagner ER, Huang E, Gao Y, Gao JL, Kim SH, Zhou JZ, Bi Y, Su Y, Zhu G, Luo J, Luo X, Qin J, Reid RR, Luu HH, Haydon RC, Deng ZL, He TC (2010) Mesenchymal stem cells: molecular characteristics and clinical applications. World J Stem Cells 2:67–80
105. Reisner Y, Hagin D, Martelli MF (2011) Haploidentical hematopoietic transplantation: current status and future perspectives. Blood 118:6006–6017
106. Remberger M, Ackefors M, Berglund S, Blennow O, Dahllof G, Dlugosz A, Garming-Legert K, Gertow J, Gustafsson B, Hassan M, Hassan Z, Hauzenberger D, Hagglund H, Karlsson H, Klingspor L, Kumlien G, Le Blanc K, Ljungman P, Machaczka M, Malmberg KJ, Marschall HU, Mattsson J, Olsson R, Omazic B, Sairafi D, Schaffer M, Svahn BM, Svenberg P, Swartling L, Szakos A, Uhlin M, Uzunel M, Watz E, Wernerson A, Wikman A, Wikstrom AC, Winiarski J, Ringden O (2011) Improved survival after allogeneic hematopoietic stem cell transplantation in recent years. A single-center study. Biol Blood Marrow Transplant 17:1688–1697
107. Ringden O, Uzunel M, Rasmusson I, Remberger M, Sundberg B, Lonnies H, Marschall HU, Dlugosz A, Szakos A, Hassan Z, Omazic B, Aschan J, Barkholt L, Le Blanc K (2006) Mesenchymal stem cells for treatment of therapy-resistant graft-versus-host disease. Transplantation 81:1390–1397
108. Rizzo JD, Wingard JR, Tichelli A, Lee SJ, Van Lint MT, Burns LJ, Davies SM, Ferrara JL, Socie G (2006) Recommended screening and preventive practices for long-term survivors after hematopoietic cell transplantation: joint recommendations of the European Group for Blood and Marrow Transplantation, the Center for International Blood and Marrow Transplant Research, and the American Society of Blood and Marrow Transplantation. Biol Blood Marrow Transplant 12:138–151
109. Salem HK, Thiemermann C (2010) Mesenchymal stromal cells: current understanding and clinical status. Stem Cells 28:585–596

110. Salmasian H, Rohanizadegan M, Banihosseini S, Rahimi Darabad R, Rabbani-Anari M, Shakiba A, Ferrara JL (2010) Corticosteroid regimens for treatment of acute and chronic graft versus host disease (GvHD) after allogenic stem cell transplantation. Cochrane Database Syst Rev, CD005565(1)
111. Saresella M, Rainone V, Al-Daghri NM, Clerici M, Trabattoni D (2012) The PD-1/PD-L1 pathway in human pathology. Curr Mol Med 12:259–267
112. Sattler C, Steinsdoerfer M, Offers M, Fischer E, Schierl R, Heseler K, Daubener W, Seissler J (2011) Inhibition of T-cell proliferation by murine multipotent mesenchymal stromal cells is mediated by CD39 expression and adenosine generation. Cell Transplant 20:1221–1230
113. Schmitz N, Eapen M, Horowitz MM, Zhang MJ, Klein JP, Rizzo JD, Loberiza FR, Gratwohl A, Champlin RE (2006) Long-term outcome of patients given transplants of mobilized blood or bone marrow: a report from the International Bone Marrow Transplant Registry and the European Group for Blood and Marrow Transplantation. Blood 108:4288–4290
114. Schonberger S, Meisel R, Adams O, Pufal Y, Laws HJ, Enczmann J, Dilloo D (2010) Prospective, comprehensive, and effective viral monitoring in children undergoing allogeneic hematopoietic stem cell transplantation. Biol Blood Marrow Transplant 16:1428–1435
115. Schonberger S, Niehues T, Meisel R, Bernbeck B, Laws HJ, Kogler G, Enzmann J, Wernet P, Gobel U, Dilloo D (2004) Transplantation of haematopoietic stem cells derived from cord blood, bone marrow or peripheral blood: a single centre matched-pair analysis in a heterogeneous risk population. Klinische Padiatrie 216:356–363
116. Schultz KR, Miklos DB, Fowler D, Cooke K, Shizuru J, Zorn E, Holler E, Ferrara J, Shulman H, Lee SJ, Martin P, Filipovich AH, Flowers ME, Weisdorf D, Couriel D, Lachenbruch PA, Mittleman B, Vogelsang GB, Pavletic SZ (2006) Toward biomarkers for chronic graft-versus-host disease: National Institutes of Health consensus development project on criteria for clinical trials in chronic graft-versus-host disease: III. Biomarker Working Group Report. Biol Blood Marrow Transplant 12:126–137
117. Shlomchik WD, Couzens MS, Tang CB, McNiff J, Robert ME, Liu J, Shlomchik MJ, Emerson SG (1999) Prevention of graft versus host disease by inactivation of host antigen-presenting cells. Science 285:412–415
118. Sioud M (2011) New insights into mesenchymal stromal cell-mediated T-cell suppression through galectins. Scand J Immunol 73:79–84
119. Spaggiari GM, Capobianco A, Abdelrazik H, Becchetti F, Mingari MC, Moretta L (2008) Mesenchymal stem cells inhibit natural killer-cell proliferation, cytotoxicity, and cytokine production: role of indoleamine 2,3-dioxygenase and prostaglandin E2. Blood 111:1327–1333
120. Spaggiari GM, Capobianco A, Becchetti S, Mingari MC, Moretta L (2006) Mesenchymal stem cell-natural killer cell interactions: evidence that activated NK cells are capable of killing MSCs, whereas MSCs can inhibit IL-2-induced NK-cell proliferation. Blood 107:1484–1490
121. Sudres M, Norol F, Trenado A, Gregoire S, Charlotte F, Levacher B, Lataillade JJ, Bourin P, Holy X, Vernant JP, Klatzmann D, Cohen JL (2006) Bone marrow mesenchymal stem cells suppress lymphocyte proliferation in vitro but fail to prevent graft-versus-host disease in mice. J Immunol 176:7761–7767
122. Sugiyama T, Nagasawa T (2012) Bone marrow niches for hematopoietic stem cells and immune cells. Inflamm Allergy Drug Targets 11:201–206
123. Tomblyn M, Chiller T, Einsele H, Gress R, Sepkowitz K, Storek J, Wingard JR, Young JA, Boeckh MJ (2009) Guidelines for preventing infectious complications among hematopoietic cell transplant recipients: a global perspective. Preface. Bone Marrow Transplant 44:453–455
124. Tu Z, Li Q, Bu H, Lin F (2010) Mesenchymal stem cells inhibit complement activation by secreting factor H. Stem Cells Dev 19:1803–1809

125. van den Brink MR, Burakoff SJ (2002) Cytolytic pathways in haematopoietic stem-cell transplantation. Nat Rev Immunol 2:273–281
126. von Bonin M, Stolzel F, Goedecke A, Richter K, Wuschek N, Holig K, Platzbecker U, Illmer T, Schaich M, Schetelig J, Kiani A, Ordemann R, Ehninger G, Schmitz M, Bornhauser M (2009) Treatment of refractory acute GVHD with third-party MSC expanded in platelet lysate-containing medium. Bone Marrow Transplant 43:245–251
127. Wagner JE, Gluckman E (2010) Umbilical cord blood transplantation: the first 20 years. Semin Hematol 47:3–12
128. Waldman E, Lu SX, Hubbard VM, Kochman AA, Eng JM, Terwey TH, Muriglan SJ, Kim TD, Heller G, Murphy GF, Liu C, Alpdogan O, van den Brink MR (2006) Absence of beta7 integrin results in less graft-versus-host disease because of decreased homing of alloreactive T cells to intestine. Blood 107:1703–1711
129. Wang H, Yan H, Wang Z, Zhu L, Liu J, Guo Z (2012) Cotransplantation of allogeneic mesenchymal and hematopoietic stem cells in children with aplastic anemia. Pediatrics 129:e1612–e1615
130. Weisdorf D, Haake R, Blazar B, Miller W, McGlave P, Ramsay N, Kersey J, Filipovich A (1990) Treatment of moderate/severe acute graft-versus-host disease after allogeneic bone marrow transplantation: an analysis of clinical risk features and outcome. Blood 75:1024–1030
131. Weisdorf D, Zhang MJ, Arora M, Horowitz MM, Rizzo JD, Eapen M (2012) Graft-versus-host disease induced graft-versus-leukemia effect: greater impact on relapse and disease-free survival after reduced intensity conditioning. Biol Blood Marrow Transplant 11:1727–1733
132. Welniak LA, Blazar BR, Murphy WJ (2007) Immunobiology of allogeneic hematopoietic stem cell transplantation. Annu Rev Immunol 25:139–170
133. Weng JY, Du X, Geng SX, Peng YW, Wang Z, Lu ZS, Wu SJ, Luo CW, Guo R, Ling W, Deng CX, Liao PJ, Xiang AP (2010) Mesenchymal stem cell as salvage treatment for refractory chronic GVHD. Bone Marrow Transplant 45:1732–1740
134. Wolf D, von Lilienfeld-Toal M, Wolf AM, Schleuning M, von Bergwelt-Baildon M, Held SA, Brossart P (2012) Novel treatment concepts for graft-versus-host disease. Blood 119:16–25
135. Yagi H, Soto-Gutierrez A, Parekkadan B, Kitagawa Y, Tompkins RG, Kobayashi N, Yarmush ML (2010) Mesenchymal stem cells: mechanisms of immunomodulation and homing. Cell Transplant 19:667–679
136. Yaniv I, Stein J (2008) Reduced-intensity conditioning in children: a reappraisal in 2008. Bone Marrow Transplant 41(Suppl 2):S18–S22
137. Zhang H, Chen J, Que W (2012) Allogeneic peripheral blood stem cell and bone marrow transplantation for hematologic malignancies: meta-analysis of randomized controlled trials. Leuk Res 36:431–437
138. Zhang X, Li JY, Cao K, Lu H, Hong M, Qian S, Wu H, Liu P, Xu W (2010) Cotransplantation of HLA-identical mesenchymal stem cells and hematopoietic stem cells in Chinese patients with hematologic diseases. Int J Lab Hematol 32:256–264
139. Zhang YN, Lie PC, Wei X (2009) Differentiation of mesenchymal stromal cells derived from umbilical cord Wharton's jelly into hepatocyte-like cells. Cytotherapy 11:548–558
140. Zhou H, Guo M, Bian C, Sun Z, Yang Z, Zeng Y, Ai H, Zhao RC (2010) Efficacy of bone marrow-derived mesenchymal stem cells in the treatment of sclerodermatous chronic graft-versus-host disease: clinical report. Biol Blood Marrow Transplant 16:403–412

Adv Biochem Eng Biotechnol (2013) 130: 163–197
DOI: 10.1007/10_2012_141

Published Online: 7 August 2012

New Cell-Based Therapy Paradigm: Induction of Bone Marrow-Derived Multipotent Mesenchymal Stromal Cells into Pro-Inflammatory *MSC1* and Anti-inflammatory *MSC2* Phenotypes

Aline M. Betancourt

Abstract Cell-based therapies (CBTs) are quickly taking hold as a revolutionary new approach to treat many human diseases. Among the cells used in these treatments, multipotent mesenchymal stromal cells, also often and imprecisely termed mesenchymal stem cells (MSC), are widely used because they are considered clinically safe, unique in their immune-modulating capabilities, easily obtained from adult tissues, and quickly expanded as well as stored. However, despite these established advantages, there are limiting factors to employing MSCs in these therapeutic strategies. Foremost is the lack of a general consensus on a definition of these cells, marring efforts to prepare homogeneous lots and more importantly complicating their in vitro and in vivo investigation. Furthermore, although one of the most profound clinical effects of MSC intravenous administration is the modulation of host immune responses, no adequate ex vivo assays exist to consistently predict the therapeutic effect of each MSC lot in the treated patient. Until these issues are addressed, this very promising and safe new therapeutic approach cannot be used to its full advantage. However, these confounding issues do present exciting opportunities. The first is an opportunity to discover unknown aspects of host immune responses because the unique effect driven by MSC infusion on a patient's immunity has not yet been identified. In addition, there is an opportunity to develop methods, tests, and tools to better define MSCs and MSC-based therapy and provide consistency in preparation and effect. To this end, my laboratory recently developed a new approach to induce uniform pro-inflammatory *MSC1* and anti-inflammatory *MSC2* phenotypes from bone marrow-derived MSC preparations. I anticipate that *MSC1* and *MSC2* provide convenient tools with which to address some of these limitations and will help advance safe and effective CBTs for human disease.

A. M. Betancourt (✉)
Center for Stem Cell Research and Regenerative Medicine,
Tulane University, New Orleans LA, USA
e-mail: alibscan@tulane.edu

Keywords Anti-inflammatory · ASC, adipose-derived multipotent stromal or mesenchymal stem cell · Bioactive factors · BM-MSC, bone marrow-derived multipotent stromal or mesenchymal stem cell · Cell-based therapy · Chemokines · Cytokines · Danger or stress responses · HSC Hematopoietic stem cells · IFN, interferon · IL, interleukin · Immune modulation or immunomodulating · Immune response · Immune suppression or immunosuppression · MSC, multipotent stromal or mesenchymal stem cell · *MSC1*, pro-inflammatory MSC phenotype · *MSC2*, anti-inflammatory MSC phenotype · Pro-inflammatory · TLR, Toll-like receptors

Abbreviations

ASC	Adipose-derived multipotent stromal or mesenchymal stem cell
BM-MSC	Bone marrow-derived multipotent stromal
COX	Cyclooxygenase
CTL	Cytotoxic T lymphocyte
ESCs	Embryonic stem cells
HLA	Human leukocyte antigen
HSCs	Hematopoietic stem cells
IDO	Indoleamine 2,3-dioxygenase
IFN	Interferon
IL	Interleukin
iNOS	Inducible nitric-oxide synthase
iPSCs	Induced pluripotent stem cells
LPS	Lipopolysaccharide
MHC	Major histocompatibility complex
MSC	Multipotent stromal or mesenchymal stem cell
MSC1	Pro-inflammatory MSC phenotype
MSC2	Anti-inflammatory MSC phenotype
NF	Nuclear factor
PGE-2	Prostaglandin E2
TGF	Transforming growth factor
TLR	Toll-like receptor
TNF	Tumor necrosis factor
Th1	T Helper cell 1
Treg	T Regulatory lymphocyte

Contents

1 Cell-Based Therapy

Cell-based therapies (CBTs) are quickly gaining ground as a promising new way to treat many human diseases. Fueling this movement is their proven efficacy in myriad ailments, growing international use, increasing approval by regulatory agencies (the U.S. Food and Drug Administration (FDA) and the European Medicines Agency), and commercial appeal [1]. However, although CBTs have the exciting potential to treat many human diseases that currently lack any established treatment, the overwhelming and unrealistic expectation by the public that these new therapies solve every ailment without consequence is driving a dangerously hastened translation of CBTs from lab bench to bedside without adequate study. In the U.S. alone, there are almost 3000 ongoing clinical trials for CBTs in the treatment of a wide range of diseases (www.Clinicaltrials.gov). Additionally there is the growing new phenomenon known as "medical tourism" whereby American patients (>500,000 in 2010) are opting to travel abroad to gain access to these new experimental medical treatments in countries practicing less stringent drug regulations for these procedures, which greatly risks patient safety (www.health-tourism.com). Further complicating matters is the fact that, because this therapeutic approach is so new, there are no adequate criteria defined for CBTs that ensure systematic testing for all aspects of these products or that can consistently predict treatment outcomes for each patient. Therefore, there is still a significant amount of investigation to be done to ensure safe and effective translation of CBTs for human disease. This review aims to outline the benefits and challenges that exist in CBTs, with the primary focus on multipotent mesenchymal stromal cell (MSC)-based therapies. It also presents a new paradigm for MSCs that I suggest may provide convenient tools to assist with tackling some of the current challenges facing the advancement of safe and effective CBTs.

1.1 Cell-Based Therapy Benefits

CBTs have their origin in the 1820s, with the first recorded successful human-to-human blood transfusion performed by Dr. James Blundell in London's Guy's Hospital. The history of CBTs is also rooted in many subsequent years of study in bone marrow and organ transplantation, tissue banking, and reproductive in vitro fertilization. CBTs are simply defined as the use of cells to treat disease [1–5]. There are two different defined cell therapy approaches to treat patients. First, cells may be harvested from a patient, manipulated or expanded, and introduced back into the same patient. This re-introduction of self-cells—the *autologous* method—is generally preferred due to the lack of required immune phenotyping or matching. A second approach involves the harvesting of cells from one or a few universal donors followed by large-scale expansion and banking of multiple doses, known as *allogeneic* cell therapy. This last approach uses cell types that do not elicit harmful immune responses upon delivery and therefore has the potential to treat hundreds of patients from a single manufacturing lot of cells. This methodology is more suited for conventional drug manufacturing practices because the product can be readily available for large-scale expansion and "off the shelf" storage. Multipotent MSC-based therapies, the primary focus of this chapter, fall into this last category due to their favorable immune status, as further discussed below.

The field of CBTs has come a long way and is once again back on track. The U.S. government is working to promote human embryonic stem cell (ESC) research [1]. Ethical issues surround ESCs because they are typically derived from fertilized human embryos. The ability to generate embryonic stem cell-like cells from adult cells or induced pluripotent stem cells (iPSCs) provides a potential way to avoid the moral and ethical controversies of using ESCs. Apart from ESCs and iPSCs, the number of different types of cells used in CBTs is growing from easily isolated blood cells including hematopoietic stem cells (HSCs) to less well-defined MSCs. I expect that these single cell type approaches will also rapidly expand in the future to include designer cells, genetically engineered cells, cell fragments, cell hybrids, substitute tissues, and the gradual emergence of complementary approaches that enhance health rather than just restore it [1].

Although CBTs can be regenerative, the field of CBTs should not be confused with the field of regenerative medicine (RM). RM is a method that replaces or regenerates cells within a tissue or organ to repair injury and restore tissue homeostasis [1]. This is unlike cell therapy, which is considered a "platform technology" because it is a therapeutic application of cells regardless of cell type or clinical indication. The distinction is that RM is an approach to treating patients; it is similar to a field of medicine, such as surgery or anatomy, not a platform technology such as CBTs. This distinction is important to avoid further confusion in the field and to drive better understanding that guides more carefully designed investigation within the respective fields.

There is evidence for the growing acceptance of CBTs in medicine, as indicated by the numerous (ca. 3000) ongoing CBT clinical trials, as well as the increasing approval by the various drug regulatory agencies. There currently are more than 8 FDA/European Medicines Agency-approved CBTs. Provenge® from Dandreon (Seattle, WA) was the first approved autologous cell-based prostrate cancer vaccine; more important than its commercial potential was its eligibility for reimbursement in the United States by the Centers for Medicare and Medicaid Services

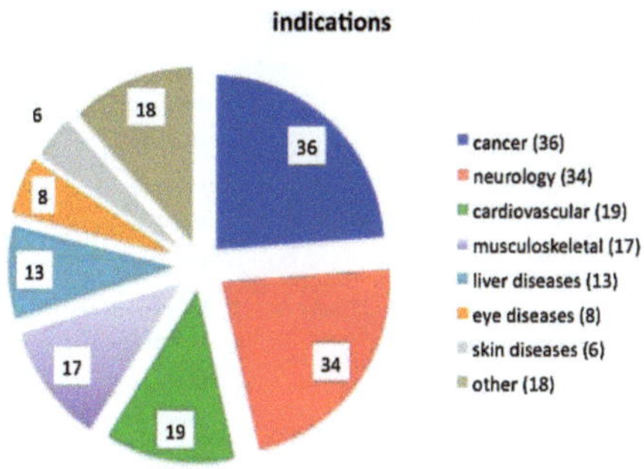

Fig. 1 Different indications testing CBTs in 2011. Adapted from [Bersenev Alexey. Cell therapy clinical trials in 2011 Hematopoiesis blog. January 3, 2011. Available: http://hematopoiesis.info/2012/01/04/cell-therapy-trials-2011]

Table 1 Potential disease conditions for CBTs. Adapted from [6]

Conditions	Predicted patients in the US (million)
Alzheimer disease	5.5
Autoimmune diseases	30
Birth defects	0.5
Burns	0.3
Cardiovascular diseases	58
Cancers	8.2
Diabetes	16
Osteoporosis	10
Parkinson disease	5.5
Spinal cord injuries	0.25

(CMS) at $93,000/treatment [1]. The New York Blood Center, Inc. (New York, NY) received approval for the first allogeneic cord blood cell product (Hemacord™) for hematopoetic reconstitution in 2011 (www.FDA.gov). Organogenesis Inc. (Cantos, MA) originally had success in 1998 with FDA approval for an allogeneic skin graft product (Apligraf®) under pre-market application (PMA) for a medical device; more recently, the company received approval for Gentuit™, a similar allogeneic cell-based product classified instead as a drug under biological license application (BLA) (www.FDA.gov). These products are made from keratinocytes and fibroblasts from human donors seeded on bovine extracellular matrices. Apligraf® is approved for use on venous leg ulcers and diabetic foot ulcers, whereas Gentuit™ is approved for use in oral soft tissue regeneration. These first few approved CBTs and their diverse indications demonstrate the vast potential of these therapies. Table 1 provides a more complete list of medical conditions and the predicted number of patients that are affected [6]. Figure 1 demonstrates the indications tested in 2011. In light of the prospective patient population and the billion-dollar market potential, it is not hard to see what is driving the quick CBT translation.

Another advantage of CBTs, apart from those already mentioned, is their unique mechanism of action. Although one might predict that with greater than 200 potential cell types in the adult body to serve as sources for CBTs, there might be 200 different mechanisms of actions for CBTs, these can largely be simplified

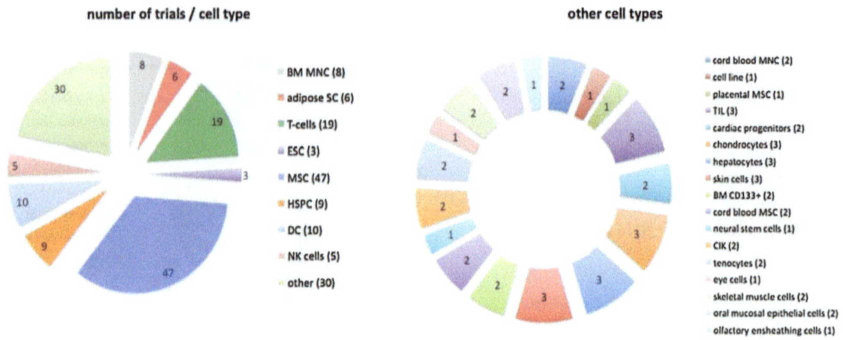

Fig. 2 Types of cells used in CBTs in 2011. Abbreviations: ESC–embryonic stem cells, MSC–multipotent mesenchymal stromal cells, HSPC–hematopoietic stem/ progenitor cells, TIL–tumor-infiltrating lymphocytes, DC–dendritic cells, BM–bone marrow, MNC–mononuclear cells, NK–natural killer cells, CIK–cytokine-induced killers, SC–stem cells. Adapted from [Bersenev Alexey. Cell therapy clinical trials in 2011. Hematopoiesis blog. January 3, 2011. Available: http://hematopoiesis.info/2012/01/04/cell-therapy-trials-2011]

Table 2 CBT treatment paradigms

	Tissue replacement	Corrective factor(s) contribution
CBT design	Cells differentiated into injured tissue cells	Cells designed to deliver corrective factors
Engraftment	Long-term (months-years)	Short-term (hours-days)
Typical cell sources	Differentiated ESCs, iPSCs, HSCs	Original or engineered MSCs, HSCs or other

or reduced to two treatment paradigms (Table 2). Figure 2 demonstrates the cell sources tested in CBTs in 2011. The first and original concept is to replace the injured tissue cells. In this paradigm, the cells administered by the therapy are expected to migrate to the site of injury and engraft, or they can be locally administered at the injury whereby they will differentiate or, if differentiated ex vivo, they will replace the damaged tissue cells and restore tissue homeostasis. The implication is that the administered cells will remain in place for the long term (months to years). In this treatment paradigm, ESCs that are pluripotent—that is, able to give rise or differentiate into more different cell types—are more coveted. iPSCs are presumably also pluripotent while circumventing the ethical issues of ESCs; thus, efforts are shifting toward their use in medicine. Adult-derived multipotent stem or progenitor cells are more limited and thus will have fewer cell types they can replace in the body. An example of this mode of action by the CBTs is given by the use of cells engineered into pancreatic beta cell-like cells to restore those lost to disease in a patient with diabetes, as well as in the recent discontinued Geron clinical trials (Menlo Park, CA). In a Phase I safety trial for spinal cord injury, Geron collaborated with Dr. Hans Keirstead (UC-Irvine) and used his technology to induce human ESCs to become a mixture of cells that include oligodendrocyte precursors [7]. Oligodendrocytes are cells in the brain and

Table 3 Benefits and challenges of CBTs

Benefits	Challenges
Treat many diseases	Lack of clear manufacturing and design guidelines by regulatory agencies
Impact millions of patients	Issues with obtaining consistent, sterility, identity, and purity
Known mode of action	Issues with defining potency and viability after isolation expansion and storage
Local not systemically targeted effect may avoid adverse effects or tolerance mechanisms seen with typical single agent drugs	Issues with ex vivo surrogate assays to consistently predict treatment outcome in patients

the central nervous system that provide myelin for insulation of nerve cells. The myelin sheath is necessary for the proper transmission of electric signals along the spinal cord that trigger muscle movement and relay stimuli signals such as touch and temperature. Damage to the myelin sheath is a common consequence of trauma to the spinal cord that leads to paralysis.

CBTs may also work to repair injury in a second treatment paradigm whereby they can release bioactive factors such as mitogens, cytokines, chemokines, extracellular matrix proteins, or even microvesicles (endosomes) in either a paracrine, endocrine, or juxtacrine manner. These factors are expected to orchestrate repair of the injured tissue(s). In this paradigm, the administered cells may naturally or artificially (e.g., through genetic manipulation) deliver sufficient essential and/or therapeutic factors. The therapeutic cells may also be designed to provide a mutated gene or missing factor. Alternatively, they may provide pro-angiogenic, anti-inflammatory, or anti-apoptotic factors that allow a window of healing time for the tissue to replace the injured cells and restore homeostasis. Compared to the previous treatment paradigm, these engineered cells—whether locally or systemically delivered—are expected to last for a short time in the host (days to weeks). The best examples of this treatment paradigm are MSCs used for their immune-modulating properties in multiple sclerosis (MS). In a pilot trial, bone marrow MSC therapy demonstrated possible benefits for the treatment of the disease in a small group of patients with MS. The procedure was observed to be well tolerated during the year the participants were followed. No serious adverse effects were encountered. Both the results of clinical scores and neurophysiological tests raise the possibility of benefit from the CBT. A Phase II/III study is planned to follow these initially successful pilot studies [8] (Table 3).

Other than their ability to replace injured tissue or provide bioactive factors for its repair, CBTs are set apart from other therapeutic strategies by the ability to deliver cells that specifically home, migrate, and even engraft at the site of injury; therefore, the adverse effects typically seen with global systemic single-acting drugs may be avoided. Lastly, another unique aspect of the CBT approach is that the studies are usually designed with a specific mode of action in mind. That is, the cells of the CBT are selected upfront to treat a specific aspect of the disease, which is different from conventional drugs selected by screening in high-throughput assays in which the mechanism of action is typically unknown [3].

Table 4 FDA guidelines for production of CBTs

Product characteristics	Tests	Issues
Safety	Infectious Contaminants Pyrogenic factors FCS and other growth medium components	Not possible to fully sterilize cells Animal products used in expansion of cells may contribute disease and/or immunogens
Identity and Purity	Viable nucleated cells (before and after cryopreservation) Karyotyping (before and after expansion) Distinct cell surface marker expression Cell morphology HLA typing or matching	Limited number of cell product for adequate testing Mutagenic potential of cell product not fully tested Adequate markers may not exist; e.g., MSCs Autologous vs. allogeneic cell product choices for each disease
Potency and Efficacy	Preclinical disease models and other surrogate ex vivo models	There are no true tests that adequately measure efficacy of the cell product prior to clinical use

1.2 Cell-Based Therapy Challenges

Although there is great promise for CBTs, they also face a major challenge: CBTs need to adhere to strict U.S. FDA regulations determining their *safety, purity, potency, identity,* and *efficacy* while meeting the demands of high-quality manufacturing processes that allow adequate delivery of viable product (Table 4) [3]. Although some of these requirements, such as sterility, can be modeled from conventional methodologies, CBTs require unique safety assessments. The use of live cells cultured with animal products may elicit allergic immune responses; they also have limited stability once grown and stored outside of the body. Additionally, when providing artificially expanded cells, there is the potential to introduce a cancerous cell in the patient. However, by far the greatest challenge for CBTs is the lack of ex vivo surrogate assays that predict any given cell lot's efficacy once reintroduced into the patient. Thus, if CBTs are quickly translated for human disease without addressing these limiting technological aspects, we can expect many iterations in this process—from clinical trials to the manufacturing process and back again—before sound standard methods and safe protocols are developed.

To ensure patient safety and address some of these issues, FDA scientists are developing laboratory techniques that will enable the agency to carefully evaluate and characterize CBTs in order to reliably predict whether they will be safe and effective. For instance, Dr. Stephen Bauer's research program "uses animal models and cell cultures to study how cells multiply and differentiate (mature into specialized cells with limited, specific functions). (They) also study the effects on cells of their microenvironment, both inside and outside of the body" (Cell and Tissue Therapy Branch, Center for Biologics Evaluation and Research [CBER], FDA). These studies will help develop testing methods that are practical and applicable to specific manufacturing steps and will help CBER to ensure the safety, consistency, and efficacy of stem cell-based products. However, *safety* is the primary concern for regulatory agencies when examining potential new treatments.

The next concern is ensuring *sterility* of each cell lot in development for CBTs. Current FDA guidelines outline the requirements for microbiological testing of aerobic and anaerobic bacteria and fungi (21 CFR 610.12). Here, too, CBTs present a unique challenge because the terminal product sterilization used for manufacturing routine drugs would kill a live cell product. Apart from other sterility issues, including decontamination from mycoplasma, viral, and other potential pyrogenic factors, there are many contributing factors that can affect cellular proliferation and survival (viability) when adhering to the required current good manufacturing practices (cGMPs). The need to maintain and expand cells in animal products, such as fetal calf serum (FCS), also presents a problem because these could potentially transfer disease or elicit unwanted immune reactions in the patient [9, 10]. For instance, one study reported that patients receiving HSC transplantations generated antibodies to FCS, although without clinical consequences [11]. As an alternative, autologous patient serum has been considered as an FCS replacement to address this potential issue [12]. However, this approach also presents the problem of insufficient sources, disease transfer, and inherent donor variability that may affect cell production [9, 13, 14]. Additionally, there are still many issues surrounding the patient's immune response that have to be considered during cell manufacturing, such as human leukocyte antigen (HLA) status, which is important when pairing donors to hosts in certain circumstances and to potentially improve engraftment of the cells used in CBTs [15, 16].

Another concern is that the expansion of the cells to generate enough doses leads to an increased chance for mutagenesis; thus, there is potential to introduce a tumorigenic cell in the CBT. To address this concern, karyotype testing should be performed once cells have been maintained in culture for significant periods to ensure chromosomal stability [17]. This is typically done by counting a minimum number of cells in a metaphase spread (>20) following Giemsa staining [18, 19]. In the case of MSC production, which like other cells can develop cytogenetic abnormalities, features of cell senesence, loss of pluripotency, genetic instability, and even transformation after long term in vitro expansion are further challenges (more than 10–20 passages in culture). MSCs are generally large and immature cells, mostly in G0/G1 phases of the cell cycle. This is probably the main cause for the difficulty in obtaining MSC metaphases for cytogenetic analysis with the standard Giemsa method. To address this shortcoming and improve the yield of metaphases, the delCanizo group recently improved on the method of Mareschi et al. by increasing colcemid concentration and exposure time without the addition of any extra growth factor or cytokines to the culture medium with some success [20]. Notably, this approach only detects gross chromosomal abnormalities and misses many potential genetic mutations; there is very little knowledge about how these abnormalities might be clinically significant or their potential risks to the patient.

Potency is the next requirement to fulfill in the manufacture of cells for CBTs. It is defined by the FDA as "the specific ability or capacity of the product, as indicated by appropriate laboratory tests or by adequately controlled clinical data obtained through the administration of the product in the manner intended, to effect a given result" (www.fda.gov). *Efficacy* generally refers to the ability of a drug or medicinal treatment to cause a functional response in the patient; it is proportional to the potency of the therapy. Effective tests to determine product potency will be required to ensure a CBT product is manufactured to the same consistent standards (www.fda.gov). With this in mind, proper characterization and understanding of cell function, or mode of action, is the most important factor in determining whether a CBT will function effectively in vivo. However, as mentioned previously, the complete characterization of some cell processes are still unknown. It is very difficult to accurately predict every consequence of a particular cell once placed within a patient. Preclinical safety studies and non-clinical testing are also likely to provide further information on possible in vivo actions of the product. Furthermore, regulatory authorities require such in vivo testing to identify safe dosages in humans, potential target organs for toxicity, and safety parameters for clinical monitoring [3]. Although limited to the beginning of clinical development, preclinical animal in vivo studies should also be sufficient to identify potential adverse affects that might then occur in the "first in man" clinical trial [3].

Next, the development of standard techniques that fulfill the FDA's requirements for *identity* and *purity* is required, which includes *viability* after cryopreservation of the product. The phenotype of the cell product is generally characterized through flow cytometry or fluorescence-activated cell sorting (FACS) analyses, which can analyze a cell's stage of proliferation, differentiation, or activation through cell surface and intracellular markers [21]. These markers are commonly used to identify individual cell populations in heterogeneous samples. Cellular morphology is also used to analyze cell populations using various microscopy techniques to determine whether cells appear true to their phenotype. For example, fibroblasts are commonly characterized by their elongated spindle-shaped appearance when cultured in vitro [22]. In addition, various stages of cellular senescence can be identified through microscopy, which can be seen visually as enlarged nuclei and multinucleated cells following extensive culture of human epithelial cells [23]. There are several microscopy techniques available to analyze cellular morphology, from fluorescent confocal microscopy to wide field live cell imaging [24].

One method to satisfy the criteria of *identity* and *purity* with greater ease is to develop a master cell bank (MCB) and a working cell bank to ensure consistency of the final product. The MCB must be fully validated and include the following information: the cell origin, standard operating procedures for all manipulations, genetic and/or phenotypic marker characterization, sterility testing, expiration dating, and complete testing of thawed/expanded cells (www.fda.gov; Rayment, 2010 #1916]. However, due to the potentially small number of available cells, alternative samples for the final product (e.g., working cell bank) may sometimes be used to minimize product loss. In terms of stability, the FDA has clear guidelines on testing that must be performed to gain product approval. Although these guidelines were originally developed for biotechnological products (e.g., proteins), they can be adapted to include cell-based products. In terms of long-term cell storage and batch testing, manufacturers must provide stability data on at least three batches that have undergone manufacturing but have not yet entered the formulation stage, with a minimum of 6 months data for products requiring long-term storage [3]. There are further challenges in terms of manufacturing an autologous therapy when compared with therapies using one distinct cell type for allogeneic treatments. Overall, there remain many challenges to overcome in the production of safe and effective CBTs.

2 MSC-Based Therapy

Among CBTs, the multipotent mesenchymal stromal cells or mesenchymal stem cells (MSCs) have great potential benefits as well as some considerable challenges in their use as CBTs for many diseases. Although they generally have a safe clinical profile, their use is constrained by the lack of a consensus among scientists of what constitutes a "mesenchymal stem cell." Recent research has been driven by the idea that these cells can be found not only in the bone marrow but also in other postnatal tissues; as such, they can be applied to repair multiple skeletal (internal supportive) tissues and non-skeletal tissues. However, some have reasonably argued that this broad definition is inaccurate and that MSCs are skeletal stem cells found only in the bone marrow and not in other tissues [2, 25, 26]. Furthermore, by adhering to a stricter definition and following established developmental tissue lineages, MSCs are restricted to a microvascular location and a mural cell identity; their primary function would be to support HSC and their microenvironment. MSCs by this account would be capable of replacing injured skeletal tissue cells and supporting hematopoietic niches but would not be capable of anything more. Continued debate and ongoing investigations will hopefully shed light on these challenging matters, thus making MSC-based CBTs safer and improving treatment outcomes.

2.1 MSC-Based Therapy Benefits

MSCs were originally described more than 30 years ago by Friedenstein et al. as plastic adherent fibroblast-like cells that could be easily isolated from the bone marrow. This group further reported MSCs to be spindle-shaped multipotential stromal precursor cells that are capable of ex vivo expansion and the formation of colony forming unit fibroblasts (CFU-F). Although initially isolated from the bone marrow, MSCs are now recognized to be found also in perivascular regions throughout the body [2, 25–29]. MSCs are widely studied due to their ease of ex vivo expansion, culture, and storage without loss of the capacity to differentiate towards mesodermal lineages such as osteoblasts, chondrocytes, and adipocytes [30–34]. Some MSCs derived from various tissue sources have also been shown to transdifferentiate into cells of ectodermal [35] and endodermal [36, 37] origins. As a result of these many potentials, initial preclinical evaluations focused on testing their potential to repair and replace injured or diseased tissues of all origins.

Despite these ambitious efforts, increasingly it is being appreciated that the clinical benefit of MSCs in CBTs is not their ability to replace the injured tissue but rather their poorly understood capacity to modulate aberrant host immune responses [30, 38, 39]. Following the initial report by the Le Blanc group that the introduction of MSCs staved off graft-versus-host disease (GvHD) in a young boy, these cells' dominant effects on the host's immunity became widely recognized [40]. Further evidence indicating that the benefit of MSCs in CBTs may be immunomodulation can be gleaned from many reports describing that although infused MSCs home to sites of injury and provide treatment benefit in widely ranging of diseases, they can rarely be detected within the repaired tissue. Ex vivo assays have confirmed some of the myriad ways that MSCs affect host immune responses. These appear to be mediated both by direct cell-to-cell contact and indirectly by the secretion of a plethora of bioactive factors [5, 41–45].

The immune-modulating effects of MSCs reported so far also include inhibition of the proliferation of activated CD8 + and CD4 + T lymphocytes and natural killer (NK) cells, recruitment and support of regulatory T cells, suppression of Th17 lymphocytes and immunoglobulin production by plasma cells, inhibition of maturation of dendritic cells (DCs), and attenuation of mast cells [41–44, 46]. MSCs secrete various inflammatory factors, including TNF-α-induced protein 6 (TNAIP6 or TSG-6), prostaglandin E2 (PGE2), human leukocyte antigen G5 (HLA-G5), hepatocyte growth factor (HGF), inducible nitric oxide synthase (iNOS), indoleamine-2,3-dioxygenase (IDO), transforming growth factor β (TGF-β), leukemia-inhibitory factor (LIF), and interleukin (IL)-10 [5, 45, 47, 48].

MSCs express low levels of HLA major histocompatibility complex (MHC) class I, do not express co-stimulatory molecules (B7-1/CD80 and -2/CD86, CD40, or CD40L), and must be induced to express MHC class II and Fas ligand, which likely allow the safe delivery of these cells in non-self (allogeneic) hosts [44, 45]. Indeed, MSCs stand alone among other types of stem cells, such as ESCs or iPSCs that are being considered in RM for their safe, non-immune provoking, allogeneic host delivery capability. This has prompted many new and established businesses to amass expanded stockpiles of MSCs ready for use in the treatment of many human diseases [4].

Apart from this, and aside from overcoming the ethical considerations of ESCs, the benefits of MSCs as a biologic therapeutic for a diverse range of clinical applications include direct homing and migration to sites of tissue injury. Homing is essentially the process by which cells migrate to and engraft in the tissue in which they will exert functional and protective effects. The molecular details of MSC homing are not fully understood, but they are typically considered to be mostly analogous to those of leukocytes. Therefore, during inflammation, the recruitment of inflammatory cells to the site requires a coordinated sequence of chemotaxis, adhesion, and accompanying signaling events (e.g., selectin-mediated rolling, cell recruitment by chemokines and cytokines, ligation of integrins, integrin-mediated adhesion to endothelia, transendothelial migration, migration and invasion into the inflamed tissue) [49, 50]. MSCs express several adhesion molecules, chemokines, and chemokine receptors including CCL12 (SDF-1) [51, 52]. The CCL12-CXCR4 axis has been the studied the most and has been shown to stimulate not only HSC engraftment but also the recruitment of MSCs and other progenitor cells to the site of tissue injury [53]. Another chemokine axis important in homing of MSCs during wound healing is the CCL21-CCR7 axis. Sasaki et al. recently demonstrated that keratinocytes within wounded skin express CCL21, and MSC homing is dependent upon their expression of the CCL21 receptor-CCR7 [54, 55].

Among the integrin-mediated homing mechanisms, podocalyxin (PODXL) a member of the CD34 family of membrane mucin-proteins, and integrin $\alpha 6$ (CD49f) are important for MSC migration and homing to injured hearts and kidneys. Notably, the homing efficiency of clinical-grade MSCs can be affected by the isolation and expansion protocols used, which most likely affect the expression of the ligands, receptors, and integrins mentioned [4].

The therapeutic potential of MSCs is currently being explored in more than 2000 phase I/II and III clinical trials, many of which have recently been completed or are underway (clinicaltrials.gov). As a result of the early clinical success reported by the LeBlanc group mentioned previously, many of these studies are investigating the use of MSC therapy to mediate HSC engraftment and reduce or eliminate graft-versus-host disease [40]. Recently, it was shown that the delivery of haploidentical MSCs into unrelated pediatric umbilical cord blood transplantation recipients could be performed safely, with no adverse effects or associated toxicity. The study demonstrated that all patients achieved neutrophil engraftment [56]. Another small study reported that haploidentical MSCs infused in conjunction with allogeneic HSC transplantation led to enhanced engraftment. In addition, all patients achieved neutrophil and platelet engraftment and 100 % donor chimerism, again without any associated toxicity [40].

Many clinical studies have analyzed the effect of MSCs on acute myocardial infarction (AMI). For example, studies by Chen et al. demonstrated that intracoronary injection of MSCs in patients with AMI resulted in significant improvement compared with controls after a 3-month follow-up [57]. In addition, significant efforts have been expended for MSC therapies by industry experts. Osiris Therapeutics Inc. (Columbia, MD, USA, http://www.osiristx.com/clinical_trials.php) is currently evaluating their proprietary adult stem cell product, Prochymal™, in clinical trials for various indications, including steroid refractory acute GvHD, diabetes, chronic obstructive pulmonary disease, and Crohn's disease. In their phase II trials for the treatment of acute GvHD with Prochymal™, 74 % of patients experienced total clinical resolution of the disease. Additionally, follow-up data for the phase II trials for the treatment of Crohn's disease for patients who had failed to respond to standard treatments, such as steroids, demonstrated a significant reduction in disease severity by day 28 with relatively low doses of Prochymal™ and a short treatment course. Athersys (Cleveland, OH, USA, http://www.athersys.com) is another company investigating the therapeutic potential of MSCs. Athersys is currently evaluating their cell-based product MultiStem™ in several clinical trials for ischemic injury (such as myocardial infarction and stroke) and inflammatory diseases with recent patent approval of MultiStem™ for GvHD.

As these clinical studies suggest, a growing interest in delivering MSCs in CBTs for many human diseases has occurred in the past few years. Many technologic advancements have been achieved in a short time from significant in vitro and in vivo approaches using MSCs in many different model systems. Although MSCs were originally slated for long-term tissue regeneration of mesenchymal tissues, such as those needed to repair arthritic joints and cartilage, the preclinical and clinical experience has not supported this potential; instead, a short-term immune-modulating role is supported as a benefit achieved by MSCs used in CBTs.

2.2 MSC-based Therapy Challenges

By far the greatest challenge in translating MSC-based therapies to the clinical setting is the lack of a consensus in defining these cells. The International Society for Cellular Therapy provided the following minimum criteria for defining the multipotent mesenchymal stromal cells (MSCs) in an effort to address this issue [17]. First, plastic adherence should occur following isolation and ex vivo culture. In addition, there should be consistent expression of CD105, CD73, and CD90, and no expression of hematopoetic cell surface markers CD34, CD45, CD11a, CD19, and HLA–DR. Lastly, under inductive in vitro culture conditions, MSCs should differentiate into osteocytes, adipocytes, and chondrocytes. Despite these attempts to make MSCs more standard, there remain many sources of conflict in their description. Because there are many different protocols for the isolation of these cells, there is no clear unique MSC marker analogous to the CD34 marker of hematopoetic cells; also, as mentioned, there is little agreement on what is a true MSC. Additionally, there is inherent variability in MSC functions inside and outside of the body given the recognized differences in age, gender, ethnicity, and health status among human donors. For instance, Gallipeau's group recently proposed that the discrepancies observed in the immune suppressive activities of MSC in clinical trials arise from intrinsic variability of each MSC donor source [58]. Furthermore, they proposed that tumor necrosis factor-α (TNF-α) and interferon-γ (IFN-γ)-activated MSC derived from normal adult volunteers suppress T cell proliferation in vitro in an inconsistent manner—an observation they linked to IFN-mediated indoleamine 2,3-dioxygenase (IDO) regulation. However, as an example of the complexity of interpreting all MSC studies, they did not address the fact that the mean age of their donors was >65 years old, which could also explain the discrepancies they found.

Another example of the difficulty of pinning down the true function of MSCs in and out of the body comes from studies describing the potential new antimicrobial activity by these cells [59, 60]. Our group reported that human bone marrow-derived MSCs were recruited to the ovarian tumor microenvironment by the pro-inflammatory antimicrobial cathelicidin peptide (hCAP18/LL37) [60–64]. Furthermore, we demonstrated that MSCs expressed LL37 and its receptor formyl receptor-like protein 1 (FRP2) in this setting. Based on these reports, the Matthay group subsequently established that LL37 expression was partly responsible for the increased bacterial clearance and survival seen in their mouse models of acute lung injury treated with MSCs [59]. More recently, however, they suggested that lipocalin 2 was responsible for bacterial clearance following MSC treatment of murine bacterial pneumonia models [60]. In our laboratory, we have not been able to demonstrate that the enhanced bacterial clearance is mediated by LL37, despite using human and murine MSCs in both in vitro and in vivo assays (unpublished results). Therefore, whether these conflicting results are due to differences in MSC donors or preparations and whether this antimicrobial activity can be explored in MSC-based CBTs remain to be determined.

A graver concern over the clinical use of MSCs comes from the observation alluded to previously that MSCs home to tumors as they do to injured tissues; once resident in the tumor microenvironment (TME), they may support tumor growth and spread [61, 64–67]. Conversely, other studies have reported that MSCs home to TMEs and diminish tumor growth [67–69]. MSCs in the TME are expected to contribute many soluble factors, such as mitogens, extracellular matrix proteins, angiogenic and inflammatory factors, and exosomes with as yet poorly defined potentials. MSCs are also expected to affect tumor-associated leukocytes either directly by cell–cell contact or indirectly by the secretion of trophic factors. MSCs are known to modulate the proliferation and differentiation of dendritic cells, monocytes/macrophages, B and T cells, NK cells, and even mast cells. There has been a great deal of debate in the field in trying to assert whether MSCs resident in the TME contribute to tumor growth and spread or prevent it—and if so, by what mechanisms. Many reasons have been advanced to explain the contradictory MSC role in cancer, including the heterogeneity of MSC preparations, the age or health of the MSC donor, and the experimental model or condition. In addition, our group has suggested that because pre-clinical and clinical experience has defined a primarily immune-modulating role for MSCs and the majority of tumor models are established in immune incompetent animals, the studies are missing an important aspect of the role of MSCs in tumors. Despite these conflicting issues—and as a result of the propensity of MSCs for the TME—most scientists agree that genetically modified MSCs that can act as "Trojan horses" and deliver anti-cancer therapeutics into the tumor stroma provide a promising new specific CBT for cancer.

Finally, the challenges remaining in translating MSC-based therapy to the clinical setting include the issues of satisfying the FDA's guidelines for cell-based products, including those mentioned previously in scaling up the production of these therapeutic cells, and the determination of whether to use allogeneic or autologous sources of cells for specific disease indications. Because there is no consensus on MSC identity and because of the challenges reported above for other CBT cells, there are no standard ways to test the potency of clinical-grade MSC products. Therefore, like other CBTs, there will be iterations of certain MSC protocols from each MSC manufacturer to the clinic and back until safe and effective therapies are implemented.

3 New MSC Paradigm: Induction into Pro-Inflammatory *MSC1* and Anti-Inflammatory *MSC2* Phenotypes

Recent studies in my laboratory were partly an attempt to resolve some of the challenges surrounding the potential use of MSCs in CBTs. These studies followed from ones in which we were attempting to study the homing and engraftment mechanisms that MSCs used in responding to stress and injury signals [70]. We recognized that Toll-like receptors (TLRs) are vital in coordinating the pro-homeostatic tissue injury responses of immune cells and possibly also that of MSCs and other stem cells of various origins. In trying to tease out the molecular details of TLR signaling within human MSCs (hMSCs), we initially observed distinct effects after stimulation of TLR3 when compared with TLR4 activation using our short-term, low-level TLR priming protocol [71]. By use of this protocol, we found that TLR3 stimulation of hMSCs supports their established immunosuppressive effects, whereas TLR4 activation of hMSCs more consistently provides a pro-inflammatory signature. From these observations, we proposed a new paradigm for MSCs that takes its cue from the monocyte literature: these heterogeneous cells can be induced to polarize into two diverse but homogeneously acting phenotypes [72].

Monocytes, when stimulated with known cytokines or agonists to their TLRs, including IFN-γ and endotoxin (LPS, TLR4-agonist), undergo polarization into a classical M1 phenotype that participates in early pro-inflammatory responses. IL-4 treatment of monocytes yields the alternative M2 phenotype that is associated with anti-inflammatory resolution responses [72]. We proposed that MSCs, like monocytes, are polarized by downstream TLR signaling into two homogenously acting phenotypes, classified as *MSC1* and *MSC2*, following this nomenclature. We reported that TLR4 agonists polarized MSCs toward a pro-inflammatory *MSC1* phenotype, whereas the downstream consequence of TLR3 stimulation of MSCs was a skewing toward an anti-inflammatory *MSC2* phenotype. This novel MSC polarization paradigm is based on the consistent but novel outcomes observed for *MSC1* when compared with *MSC2* for several parameters, including dissimilar patterns of secretion of cytokines and chemokines and differences in differentiation capabilities, extracellular matrix deposition, TGF-β signaling pathways, and Jagged, IDO and PGE-2 expression (Fig. 3) [71]. The most

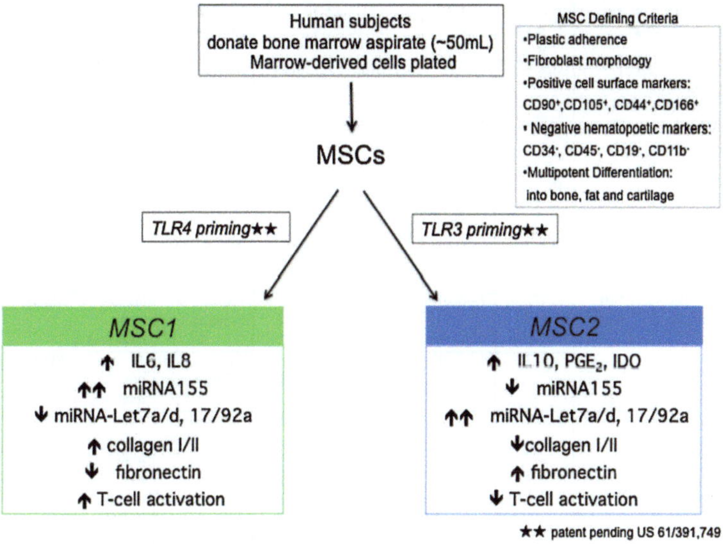

Fig. 3 Characteristics of the MSC1 and MSC2 Phenotypes. Short-term and low-level priming of TLR4 (*left* side) and TLR3 (*right* side) leads to the induction of heterogeneous hMSC preparations into a pro-inflammatory MSC1 phenotype or an antiinflammatory MSC2 phenotype. (adapted from [70, 71]

compelling outcome was the opposite effects of each cell phenotype on T-lymphocyte activation [71].

We also suggested that many of the conflicting reports on the net effect of TLR stimulation within stem cells can be resolved by taking into consideration the source of cells, their originating species, and the time and concentration of TLR agonist exposure. In line with this and the new MSC paradigm, we proposed that short-term, low-level exposure with TLR4 agonists polarizes hMSCs toward a pro-inflammatory MSC1 phenotype important for early injury responses. By contrast, the downstream consequences of TLR3 agonist exposure of hMSCs are its polarization toward an immunosuppressive MSC2 phenotype essential to later anti-inflammatory responses that help to resolve the tissue injury [71].

Although our findings that hMSCs can be pro-inflammatory challenged the current dogma, recent reports support this allegation [73–75]. For example, Romieu-Mourez et al. showed that TLR stimulation in MSCs resulted in the formation of an inflammatory site attracting innate immune cells in neutrophil chemotaxis assays and by the analyses of immune effectors retrieved from TLR-activated MSC microenvironments within mice. As mentioned, we found differences among the phenotypes in the secretion of cytokines and chemokines, as well as differences in differentiation capabilities, ECM deposition, TGFβ signaling pathways, Jagged expression, IDO and PGE2 expression, and their polar opposite effects on T-lymphocyte activation by *MSC1* and *MSC2*. We provided further support for TLR3-mediated elevated secretion of CCL10 (IP10), CCL5 (RANTES), and IL10 because this effect could be specifically inhibited by dominant-

Animal Disease Model	MSC-based Therapy	MSC Dose (cells)	Treatment Frequency (Time of treatment)	Disease Impact	Length of study	Adverse Effects
1. LPS-induced Acute Lung Injury (ALI) (BalbC and C57BL/6J, *n*=12)	MSCs	0.5X10^6	1X (24hrs post-disease onset)	Mostly anti-inflammatory	1 week post-treatment	**NONE**
	MSC1	0.5X10^6	1X (24hrs post-disease onset)	Pro-inflammatory	1 week post-treatment	**NONE**
	MSC2	0.5X10^6	1X (24hrs post-disease onset)	Anti-inflammatory	1 week post-treatment	**NONE**
2. Streptozotocin-Induced Diabetes and neuropathic pain (C57BL/6J, *n*=30)	MSCs	1-3X10^6	3X (given in 10-day intervals post-disease onset)	Mostly anti-inflammatory	70 days post-treatment	**NONE**
	MSC1	1-3X10^6	3X (given in 10-day intervals post-disease onset)	Pro-inflammatory	70 days post-treatment	**NONE**
	MSC2	1-3X10^6	3X (given in 10-day intervals post-disease onset)	Anti-inflammatory	70 days post-treatment	**NONE**
3. Immune-incompetent human tumor xenografts (Balb *scid* and *nude* *n*=60)	MSCs	0.5X10^6	3X (given weekly post-disease onset)	Mostly anti-inflammatory	>120 days post-treatment	**NONE**
	MSC1	0.5X10^6	3X (given weekly post-disease onset)	Pro-inflammatory	>120 days post-treatment	**NONE**
	MSC2	0.5X10^6	3X (given weekly post-disease onset)	Anti-inflammatory	>120 days post-treatment	**NONE**
4. Immune-competent MOSEC (C57/BL6J *n*=20)	MSCs	0.5X10^6	3X (given weekly post-disease onset)	Mostly anti-inflammatory	>70 days post-treatment	**NONE**
	MSC1	0.5X10^6	3X (given weekly post-disease onset)	Pro-inflammatory	>70 days post-treatment	**NONE**
	MSC2	0.5X10^6	3X (given weekly post-disease onset)	Anti-inflammatory	>70 days post-treatment	**NONE**

Fig. 4 Human MSC-based therapy of murine disease models. Please NOTE that for all of the data presented MSCs represent conventionally prepared human MSCs, MSC1 are defined as the hMSCs incubated for 1hr with 10 ng/mL LPS and washed prior to delivery. MSC2 are defined as the hMSCs incubated for 1hr with 1 mg/mL poly(I:C) and washed prior to delivery (patent-pending US 61/391,749)

negative TLR3 expression and not TLR4-dominant negative expression. However, we found that the enhanced IL6 and IL8 expression after TLR priming was downstream of both TLR3 and TLR4 activation, and that the secretion of other soluble mediators was indirectly affected by these because no direct effect was noted by the dominant negative strategy. We pointed out that all the siRNA-driven TLR3 inhibition strategies we attempted were unsuccessful because the double-stranded RNAs used as the interfering agents are most likely also acting as the agonists for the targeted TLR3 receptor. In support of this, the inhibition of the expression of TLR3 and TLR4 receptors by nucleofection with knockdown plasmids reduced NF-κB-driven luciferase expression by 90 %, along with the effect on the soluble mediators. Importantly, we also observed that hMSC migration is affected by both the stimulant and the time it is exposed to it. Whereas TLR-priming promoted hMSC migration, the equivalent short-term exposure with TNFα and CCL5 did not promote migration. Conversely, long-term TLR priming

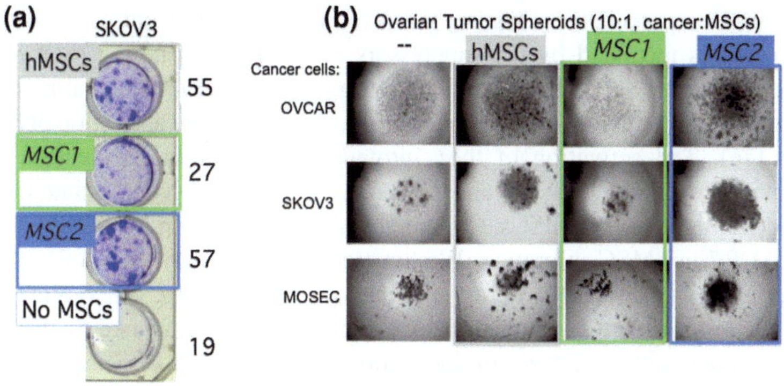

Fig. 5 MSC1 reduce tumor growth whereas MSC2 favor tumor growth. **a** Data demonstrates that there are distinct effects on colony forming units (CFU) after coculture of different human cancer cell lines with untreated conventional MSCs (hMSCs), MSC1, or MSC2. Methods: CFU assays were performed by culturing human tumor cells (200 cells/well) mixed with hMSCs, MSC1, or MSC2 (2 cells/well) at a ratio of 10 cancer cells per 1 MSC and plated in 24-well plates in growth medium supplemented with 10% FBS as indicated in figure. Cultures were grown for 14 days at 37°C in a humidified atmosphere of 5% carbon dioxide balance air. Growth medium was changed every 3-4 days. Colonies were visualized by staining with a crystal violet solution (0.5% crystal violet/10% ethanol). The resulting colonies were enumerated by the colony counting macro in Image J software, SKOV3- ovarian cancer cell lines. Micrographs of the stained plates are shown. Colony counts are at right (n=8). **b** Data demonstrates that there are distinct effects on tumor spheroids after coculture of different cancer cell lines with unprimed MSCs, MSC1, or MSC2. Methods: Tumor spheroids were formed by culturing tumor cells (2000 cells/well) mixed without any other cells (–) or with hMSCs, MSC1, or MSC2 (20 cells/well) at a ratio of 10 cancer cells per 1 MSC and plated over 1.5% agarose in 96-well plates in growth medium supplemented with 10% FBS as indicated in figure. Cultures were grown for 14 days at 37°C in a humidified atmosphere of 5% carbon dioxide balance air. Growth medium was changed every 3-4 days. Micrographs shown represent 20X magnified field of the 96-well plate. Cancer cell lines used are: OVCAR-human ovarian cancer, SKOV3-human ovarian cancer cell lines, and MOSEC-murine ovarian surface epithelium carcinoma cells. Data indicate distinct effects by MSC1 and MSC2 on cancer cell growth and spread

inhibited hMSC migration but was effective for TNFα- and CCL5-mediated migration [71].

We also contended that short-term, low-level TLR priming mimics the gradient of danger signals endogenous MSCs encounter and respond to at a distance from the site of injury that draws them to the appropriate target. Further, once the hMSCs arrive at the site spilling large amounts of these danger signals, migration pathways need to be turned off and the reparative programs turned on. Transfection of hMSCs with the dominant negative-expressing TLR3 and TLR4 plasmids diminished migration by 50 % in unstimulated hMSCs, as expected [71]. However, poly(I:C) or LPS stimulation of these transfected cells resulted in further enhancement of migration when compared with unstimulated controls. We speculated that specific TLR3 or TLR4 receptor inhibition by the transfected dominant negative-expressing plasmids de-repressed chemokine or other chemotactic receptors' inhibition downstream from these receptors while potentiating alternative poly(I:C) or LPS receptors. One potential mechanism that we proposed in a recent study to explain this finding is mediated by the suppressors of cytokine signalling (SOCS)1 and SOCS3 within hMSCs. TLR3 stimulation triggers a JAK/STAT signaling cascade indirectly by its induction of type I interferons, resulting in the activation of SOCS 1 and 3 [76]. The activation of these proteins modulates the expression of the chemokine receptors CXCR4 and CXCR7 by altering CXCR4- and CXCR7-dependent migration of hMSCs. Our study suggests a new role for SOCS, CXCR4, and CXCR7 in hMSC migration. We hypothesize that polarization of hMSCs by TLR priming also affects their programming towards tri-lineage differentiation, and that the various reported contrasting effects might also be explained by differences of source, amount, and time of incubation with the TLR agonists during the induction periods.

We further suggested that specific TLR activation affects many aspects guiding stem cell fates, but unfortunately a consensus on the effect of TLR stimulation and tri-lineage differentiation of stem cells is not possible because some of the experimental details of others' studies were not always included. Apart from the effects on differentiation, the TLR priming protocol affected the ability of hMSCs to deposit ECM, another established classical function of these cells. Unlike unprimed hMSCs and TLR4-primed hMSCs that deposited more collagen, TLR3-primed hMSCs deposited more fibronectin. To help explain these results, it has been proposed that differences in TGFβ expression are an established component of mechanisms that control ECM deposition and are also linked to immune modulation [71].

Finally, we speculated that only an immunosuppressive phenotype has been recognized for current heterogeneous MSC preparations because of the manner in which they are isolated from the host and the way they are expanded in ex vivo culture. We reasoned that the default, more progenitor-like MSC phenotype must be an immunosuppressive one to avoid profound and deleterious consequences from a pro-inflammatory *MSC1* phenotype in the context of the HSCs that MSCs maintain and support within the progenitor/stem cell niches both of these cells share. We envision that circulating or quiescent stem/progenitor cells are equipped to respond to environmental cues but must not be actively engaging immune cells or repair cells while circulating throughout the body or maintaining HSCs in the bone marrow niche. In a manner analogous to the immature state maintained for monocytes, dendritic cells, and other immune cells until a response is needed, MSCs are immunosuppressive until a pro-inflammatory role is essential to promote tissue repair. We also surmise that TLR4 priming is not the optimal way to induce the *MSC1* phenotype. It is likely that a combination of other factors, such as interferons or contact with other pro-inflammatory cells and their microenvironments (along the lines of that reported by Romieu-Mourez et al.), will more readily induce the *MSC1* phenotype [75]. However, we propose that induction of MSCs, regardless of the protocol used to isolate them into the *MSC1* and *MSC2* phenotype, is one way to make the conventionally prepared mixed pool of cells behave in a more consistent and predictable manner.

Table 5 Benefits and Challenges of MSC-based therapy

Benefits	Challenges
Safety	Lack of definition
Autologous and allogeneic donors	Donor variability
Ease of expansion and storage without loss of capabilities	These two above issues hinder reproducible manufacture of MSCs
Homing and migration toward injured tissues	Issues with potency and viability parameters to test after isolation, expansion, and storage
Local not systemically targeted effect may avoid adverse effects or tolerance	Issues with ex vivo surrogate assays that predict therapeutic effect of MSC lots
Short-term engraftment precludes long-term mutagenic cell potential	Issues with the MSC potential to support tumor growth and spread

3.1 MSC1 *and* MSC2 *in Preclinical Models of Disease*

In an effort to begin to translate these findings into CBTs, my laboratory tested the effect of *MSC1* and *MSC2* therapy in several animal disease models. We have succeeded in inducing more than eight young (<30 yrs old) donors induced into these phenotypes with very consistent and predictable inflammatory responses (Table 5). We include some of the data from two of our recently completed studies.

3.1.1 Anti-Tumor Effect by *MSC1*-Based Therapies

Our first study describes the effects of *MSC1* and *MSC2* in tumor growth and spread [67, 77]. We reasoned that these different, more uniform phenotypes may help resolve some of the controversies surrounding the role of MSCs in cancer. To this end, we extended our work on MSCs and ovarian cancer by adding these new MSC phenotypes (Fig. 3). Our initial in vitro experiments demonstrated that *MSC1* co-culture with various cancer cells diminished their capacity to form colonies in contrast to growth promoting MSC- or *MSC2*-co-cultures (Fig. 4). This effect remained constant even when tested by 3D tumor spheroid models. In this study, we only tested cancer cells derived from solid organ tumors and not from leukemia or other blood-related malignancies. We also used MSC-to-cancer cell ratios of 1:10 throughout the study to more closely resemble the proportions that might be achieved in the clinic with MSC-based therapies; this was different than the 1:1 ratios used by other MSC and cancer studies (e.g., [78–81]).

MSCs targeted to cancers are expected to contribute many bioactive factors once resident in the TME, such as mitogens, extracellular matrix proteins, angiogenic, and inflammatory factors, as well as exosomes or microvescicles. MSCs are also expected to affect tumor-associated leukocytes either directly by cell–cell contact or indirectly by these secreted factors [67]. We previously reported that there were differences among several of these secreted factors following the induction of MSCs into *MSC1* and *MSC2* [71]. Co-cultures of these MSC phenotypes with the cancer cells also reflected distinct effects on the secreted

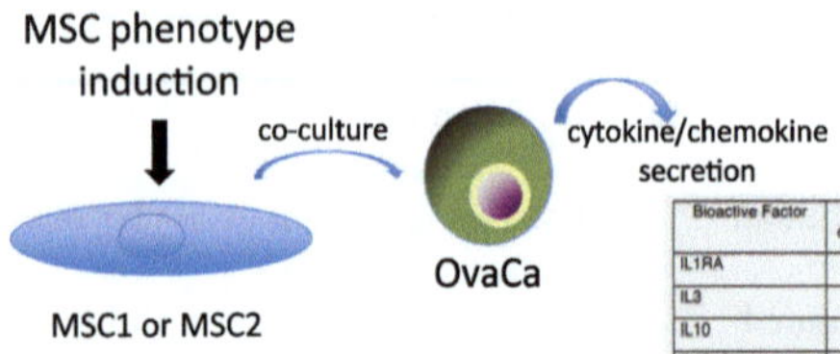

Bioactive Factor	Contact dependent effect	*MSC1*	*MSC2*
IL1RA	-	↓	↑↑
IL3	-	↑↑	↓
IL10	+	-	↑
IL12p40	-	↓	↑
IL17	+	↑↑	↑
CXCL1 (Groα)	-	↓	↑↑
CXCL10 (IP10)	+	↓	↑↑
CCL5 (RANTES)	+	-/↓	↑↑
MIG	+	↑	-
MIP1β	+	↑	↑↑
GM-CSF	+	↑↑	↓
HGF	+	↓	-
LIF	-	↑↑	↓
TRAIL	+	↑↑	↓

Fig. 6 Ovarian cancer cells co-cultured with MSC1 differ from MSC2 co-cultures in their secretion of bioactive factors. SKOV3 ovarian cancer cells were plated on 24-well plates until they reached 50-70% confluence. MSC1, MSC2, (25,000 cells/insert) or medium control were then added into 0.4mM (no cell-cell contact) or 8mM transwell inserts and the cocultures were allowed another 72hr prior to collecting the conditioned medium and testing by Bio-Plex Cytokine Assays following the manufacturer's instructions (Human Group I & II; Bio-Rad, Hercules, CA). *Arrows* represent relative normalized changes compared with the SKOV3 alone control. Biofactor levels that were different between the MSCs grown in 0.4mM (no cell-cell contact) versus 8mM transwell inserts are represented by "+." Those biofactor levels that were similar in both sample groups are represented by "-." Data are representative of triplicate experiments

factors (Fig. 5). Both contact-dependent and -independent effects were observed. Increases in the levels of CCL5 (RANTES) by the pro-tumor *MSC2* groups are in agreement with previous reports [71, 79]. *MSC1* treatment groups had elevated levels of IL17, GM-CSF, and TRAIL that would suggest an overall inflammatory and pro-apoptotic effect by these cells. *MSC2* treatment groups had elevated levels of IL1RA, IL10, and most chemokines tested; in contrast to *MSC1* treated groups, this suggests a net tumor supportive immunosuppressive effect by these cells [78]. However, we note the limitation of this type of analysis because it represents only one time point and sampling site.

We also investigated the effect that co-culture with *MSC1* and *MSC2* phenotypes might have on cancer cell migration and invasion (Fig. 6). However, although we measured fewer migrating and invading cells for the *MSC1* sample groups compared to the other MSC sample groups, we could not attribute this difference to decreased expression of activated MMP2. Additionally, we have not been able to detect significant levels of either the zymogen or active forms of MMP9 in MSC phenotype in vitro cultures or co-cultures with cancer cells. These results are intriguing given the documented importance of MMP2 and MMP9 in tumor spread and invasion [82]. Further studies are needed to investigate this complex tumor process and how the MSCs might affect it.

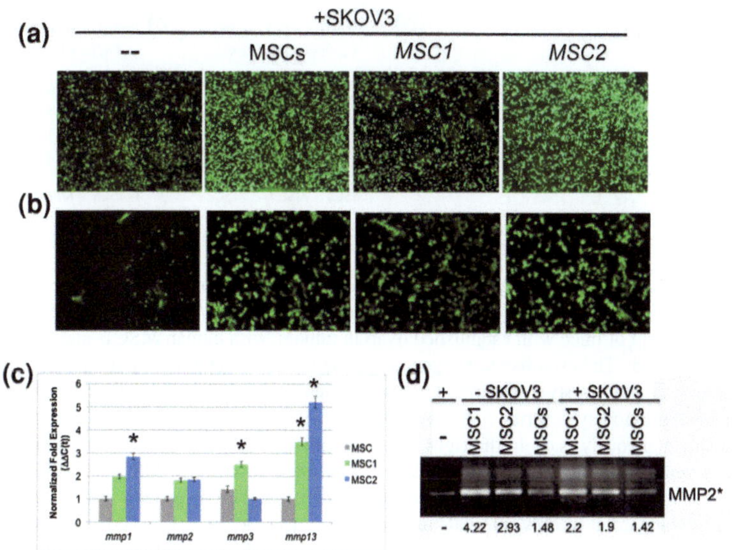

Fig. 7 Migration and Invasion of Cancer Cells following MSC phenotype co-culture. Transwell migration and matrigel invasion assays were performed with 3mM Falcon fluoroblok transwell inserts as described previously (10, 17, 23). MSCs were added at a 10:1 ratio of SKOV3 to MSC. These were co-cultured on traditional 2D dishes 72hr prior to placing the dissociated cells within the transwell inserts. Representative micrographs of **a** transwell migrating and **b** matrigel invading cells were visualized and obtained on an inverted fluorescence microscope (A. 100X and B. 200X, Olympus, MetaMorph analysis software). Data are representative of duplicates in at least three independent experiments **c** Representative bar graph of quantitative real-time PCR (qPCR) assays carried out as previously described (24). Gene expression of mmps among the MSC samples is expressed by the normalized cumulative threshold method (DDC(t)). *$P<0.05$ versus the normalized values for MSC. Statistically significant differences were not measured among the other samples. Samples were run in triplicate for at least four different MSC donors **d** Representative micrograph following gelatin zymography of the condition medium from MSC-SKOV3 co-cultures (1:10) or SKOV3 and MSC samples cultured alone as indicated for 72hr. Bands are of pro-MMP2 (72 kDa) and active MMP2* (62 kDa). The numbers below micrograph are the fold changes relative to SKOV3 alone sample obtained following densitometric analysis (Image J). Data are representative of at least three independent experiments

The effect that *MSC1* and *MSC2* had on human ovarian tumors (SKOV3 and OVCAR) in immune-compromised mice was investigated as before [61]. As was observed by the in vitro assays, *MSC1* treatment yielded smaller tumors when compared with MSCs and *MSC2* treated tumors in this animal model (data not shown). Because immune modulation is an important clinical effect of MSCs, we also used an immune-competent mouse model of ovarian cancer that has been useful in similar studies [83–85]. With this approach, we consistently observed that the *MSC1* treatment groups had smaller tumors without any detectable metastasis and accumulated little to no ascites when compared to the MSCs- or *MSC2*-treated groups (Fig. 7). Upon staining of the collected ascites, it was evident that there were large tumor aggregates or spheroids present in the MSCs- and *MSC2*-treatment groups but not in the *MSC1* ones.

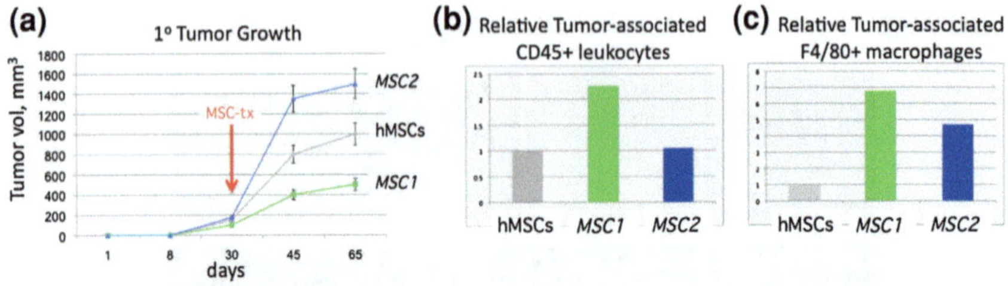

Fig. 8 MSC1 attenuate tumor growth whereas MSC2 favor tumor growth and metastasis. The data show differences in tumor volume, CD45+leukocyte and F4/80+ macrophage recruitment after the treatment of mice with established ovarian tumors, with human MSC1- and MSC2-based therapies. Methods The established syngeneic mouse model for epithelial ovarian cancer used is based upon a spontaneously transformed mouse ovarian surface epithelial cell (MOSEC) line ID8 that has been previously described. 4-6 week-old female mice (n>10 mice/MSC-treatment) were injected subcutaneously (s.c.) in the right hind leg with 1 X107 MOSEC cells. At approximately 4 weeks a single dose of labeled human conventional MSCs (hMSCs), MSC1 or MSC2 (1X106/ per mouse) were injected i.p. as indicated by *red arrow* **a** Tumor growth was measured with callipers as standard at weekly intervals until day of mouse sacrifice (Day 65). Harvested tumors and metastasis were weighed, counted and processed for flow cytometry and immunohistochemical analysis (IHC, Coffelt et al., 2009). Metastasis was found only in MSC2-treated mice (data not shown). MSCs were detected by flow cytometry and IHC. All MSC-treated samples had similar detectable MSCs within the tumor tissue-trending towards more MSC1 and MSC2 measured than hMSCs: approximately 15-25 cells counted per 200X field after 24hr of MSC-treatment and 2-5 cells at time of tissue harvest (day 65, data not shown). Sectioned tumor sample slides were stained with murine CD45 **b** or F4/80 **c** antibodies and the number of positively stained immune cells per 200X field were scored as described previously , Coffelt et al., 2009. Data are expressed as average cells counted in 4 fields/slide relative to hMSC sample. Data indicate in vivo stability and predictably distinct effects on MOSEC tumors by the MSC1 and MSC2 therapy

We also found interesting differences in the effect of the MSC therapies on the tumor-associated leukocytes (Figs. 7 and 8). These treatments appeared to distinctly affect CD45 + cells, F4/80 + cells, other monocytes, and mast cells. Tumor-associated macrophages (TAMs) are known to be educated from tumor-eradicating cells to tumor-promoting cells with F4/80 expression, potentially changing from one population to the other [86–88]. It will be interesting to determine in future studies whether tumor-associated MSCs and TAMs directly affect each other and can be "re-educated" from one form to the other following this interaction. Notably, macrophages, mast cells (MCs), and MSCs also affect ECM proteins important to tumor growth and spread [82, 89–91]. We observed very interesting correlations between MSC treatment, MC degranulation, and collagen deposition, which may help to explain some of the tumor effects we found. These differences may be explained by direct in vivo interactions between the MSCs and MCs, which was recently reported and that would be present in the TME [92]. MSCs, macrophages, and MCs seem to share many properties affecting the secretion of bioactive factors and recruited leukocytes within tumors that we hope to identify in future studies [93–98].

We also believe that subtle differences in the net effect of the MSC-based therapies on tumor immunity, growth, and spread will be learned from the study of other solid tumor and leukemia models, as well as other immune competent strains of mice. We propose here again that iterations of the therapy to determine the optimal dose, frequency, and timing of the MSC-based therapy will need to be done for each cancer. Despite these challenges, the proclivity of MSCs for the tumor microenvironment makes them ideally suited for the directed delivery of anti-cancer payloads. We also suggest that an ideal therapeutic approach for the complex pathology of cancer may be a complementary one that employs conventional methods to target the cancer cells combined with MSC-based therapies that target the TME. Finally, the new *MSC1*- and *MSC2*-therapy approach we have identified provides a convenient tool with which to begin to dissect the contribution of MSCs to tumors, and may help resolve some of the controversies surrounding the use of MSCs in CBTs of many human diseases.

3.1.2 Anti-inflammatory *MSC2*-based therapies attenuate diabetic neuropathic pain

In contrast to the *MSC1* effects in tumor therapy, we have also shown that *MSC2* can be used as consistent anti-inflammatory CBTs [99]. Our recent studies indicate that *MSC2*-treated diabetic mice had significant improvement in symptoms of painful diabetic peripheral neuropathy (pDPN) compared to the vehicle and conventional MSC treatment groups when evaluated by established tactile allodynia (von Frey's) and thermal hypoalgesia (Hardgreave's) behavioral assays. These observations, along with the changes found in immune-modulating factors from the blood sera of the *MSC2*-treated diabetic mice, suggest that the anti-inflammatory properties of the *MSC2* CBT most likely contribute to the improvements in the pDPN symptoms we observed for the STZ-induced diabetic mice. These findings also provide strong evidence for exploring *MSC2*-based therapy as an improved treatment option for pDPN.

pDPN is a serious complication of diabetes, but the mechanisms behind it are not completely understood [100–106]. Because many reports have implicated inflammatory mechanisms in pDPN, we felt that testing the anti-inflammatory properties of *MSC2* therapy was a logical next step [104, 106–110]. As mentioned above, preclinical and clinical experience with MSC-based therapy suggests it has mainly an anti-inflammatory effect [111]. Additionally, conventional MSCs used in the treatment of several murine models of diabetic peripheral neuropathy provided encouraging results [112, 113]. Furthermore, similar to these studies, the C57BL/6 J STZ-treated mice we studied developed tactile allodynia and heat hypoalgesia [114–118]. These are features of insensate neuropathy seen in advanced cases of DPN and correlate with pDPN [119]. We found that throughout this study, all STZ-treated mice remained diabetic (Fig. 9). Further, we suggested that given the improvement seen in the diabetic animals treated with CBT despite uncontrolled glucose levels, better results may be expected in a model where

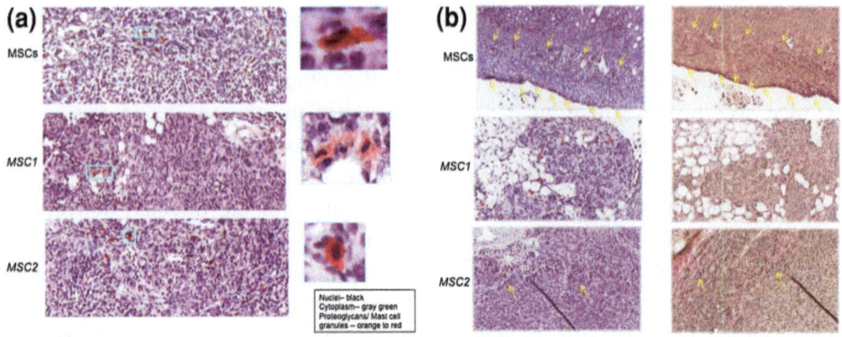

Fig. 9 Differences found in tumor-associated immune cells in MSC-treated tumor groups **a** MOSEC tumors were established in C57BL/6 mice for 4 weeks. MSCs, MSC1, or MSC2 (1X106 in 0.5mL HBSS) were infused IP and the mice were harvested after 65 days. Tumors were excised, fixed, and cut into 5 mM sections by standard methods (7). Sections were processed for safranin O proteoglycan staining (www.ihcworld.com). Representative micrographs of several MSC-treated tumor sections are included from images obtained from the Aperio ScanScope (200X, Aperio, Vista, CA). The expected color for each tissue element is described in the inset on the lower right hand side. 400X images are included in boxed insets. Data are representative of three independent experiments with at least 6 mice per treatment group **b** Co-localization of tumor associated mast cells with collagen. MOSEC tumors were established in C57BL/6 mice for 4 weeks. MSCs, MSC1, or MSC2 (1X106 in 0.5mL HBSS) were infused IP and the mice were harvested after 65 days. Tumors were excised, fixed, and cut into 5 mM sections by standard methods (7). Sections were processed for Verhoeff-Van Gieson (VVG) elastic fiber/collagen staining (left panels) or for safranin O proteoglycan staining (right panels, www.ihcworld.com). Representative micrographs of several MSC-treated tumor sections are included from images obtained from the Aperio ScanScope (40X, Aperio, Vista, CA). *yellow arrows* indicate comparable sections among the tumor tissue sections. Data are representative of three independent experiments with at least 6 mice per treatment group

animals are treated with the *MSC2* therapy and their glucose levels are controlled by conventional methods.

Additionally in this study, we found that although conventional MSCs predictably attenuated the immune response in an STZ-induced diabetic murine model, *MSC2* treatment did so to a greater extent (Fig. 10). Also, treating pDPN with the cell-based anti-inflammatory *MSC2* yielded even greater improvement in the symptoms of pDPN than conventional MSC-based therapy (Fig. 11). The animals that received MSC treatment showed significant improvement in thermal hypoalgesia but not in tactile allodynia. Studies have shown improvements in animal models of neuropathy after treatment with MSCs, but used different strains of animals and measured different endpoints [112, 113]. Another difference observed from our study was the use of autologous MSCs; in our study, we used allogeneic human MSCs, which may help explain the difference in behavioral responses. Nonetheless, we were surprised that the MSC-treated mice did not show significant improvement in the evaluation for these mechanical stimuli. This finding could be due to the type of nerve fibers tested in each of the behavioral assays, as well as study conditions. The von Frey assay for mechanical allodynia

	Vehicle	MSCs	*MSC2*	Control
Number per Treatment Group	5	6	6	2
Average Pre Study Weights (gm)	23.54 ± 0.695	20.97 ± 0.692	20.59 ± 0.472	24.43 ± 0.695
Average Post Study Weights (gm)	21.33 ± 0.925†	20.82 ± 0.723	20.84 ± 0.901	29.70 ± 0.768*
Average Pre Study Glucose (gm/dL)	372.8 ± 43.064	417.6 ± 32.979	325.6 ± 32.344	159 ± 3**
Average Post Study Glucose (gm/dL)	466.6 ± 61.20	495 ± 23.679	498.6 ± 37.823	159.5 ± 11.5**

Fig. 10 Characteristics of treatment groups (mean + SE) used in the study. MSCs, conventional mesenchymal stem cells; MSC2, anti-inflammatory mesenchymal stem cells. † When compared to pretreatment weight, post-treatment weight varied significantly (p=0.04). *When compared to streptozotocin treated groups, the control group weight was significantly more (p=0.0002). **When compared to streptozotocin treated groups, the control group blood glucose was significantly less (p=0.0001)

stimulates large myelinated A-beta fibers sensitive to light touch and small unmyelinated C fibers involved in the pain response [120]. In the Hargreave's assay for thermal hypoalgesia, pain and temperature sensing C and A-delta fibers are stimulated [120–123]. At day 40, the MSC treated mice demonstrated improvement in mechanical allodynia after decreases from baseline on days 10 and 20.

We also found that the levels of the anti-inflammatory cytokine IL-10 were greater in the *MSC2* treatment arm compared with the MSC group. IL-10 is known to decrease levels of IL-1, IL-6, and TNF-α [124]. This cytokine does not have a long duration; by modifying IL-10 with polyethylene glycol, Soderquist and colleagues have shown that the therapeutic duration and magnitude of the anti-inflammatory cytokine can be increased [125]. Because the half-life of IL-10 is short, the time interval for measurement may not have been optimal for this investigation. Because IL-10 does influence a number of pro-inflammatory cytokines, it is reasonable to speculate that IL-10 involvement in the inflammatory process contributed to the effects seen with the CBTs.

In other diabetic-induced rodent model reports, they observed increased levels of pro-inflammatory markers [126–129]. For example, elevated levels of IL-1β, IL-6 and TNFα were found in spinal cord tissue of STZ-treated rats, whereas increased IL-6, cyclo-oxygenase (COX-2), iNOS, and TNF-α were found in their blood [126, 127]. Many of the inflammatory markers we evaluated in this study have been measured in painful disease states [104, 106, 108, 126, 127]; therefore, their attenuation likely contributed to the improvement in the symptoms of pDPN we measured.

The pro-inflammatory cytokine IL-6, which can be decreased by the anti-inflammatory cytokine IL-10, was found to be significantly less expressed in the *MSC2*-treated mice compared with the MSC-treated mice. IL-6 plays a significant role in pain; after sciatic nerve injury in an animal model, IL-6 receptors are found to be elevated [130]. In addition, IL-6 is seen in greater concentration at the

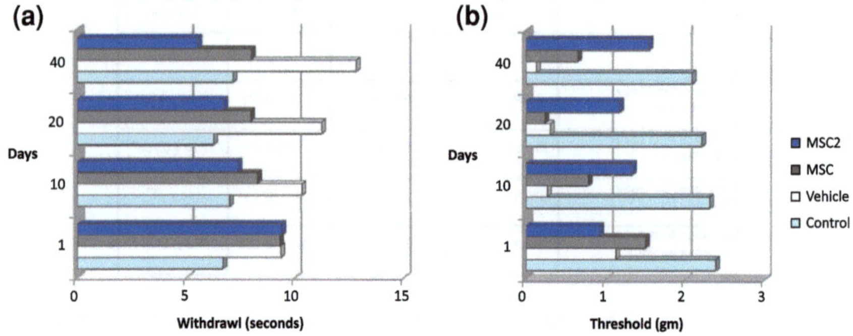

Fig. 11 The effect of MSCıtreatments of streptozotocin (STZ)ıinduced diabetic mice **a** Hargreave's heat hypoalgesia behavioral assays. Results of thermal sensitivity. MSCs, conventional mesechymal stem cells; MSC2, anti-inflammatory mesenchymal stem cells. At baseline streptozotocin treated groups had significantly higher thresholds than the control. At days 10,20,40 the control, MSCs, and MSC2 groups had significantly lower thresholds when compared to the vehicle group. At day 40, MSC2 varied significantly from MSCs and control **b** von Frey's mechanical allodynia behavioral assays. Microfilaments of various weights (gm) were applied to the hindpaw. Mice with mechanical allodynia are not capable of withstanding increased pressure from the microfilaments. At baseline, there were no statistically significant differences between the three treatments. After the 2nd and 3rd treatments, a statistically significant ability to tolerate increases in weight of microfilaments was found between MSC2 and both MSC and placebo. This ability to withstand the increases is an indication of improvement in mechanical allodynia

surgical sites of elective surgeries that are considered severe [131, 132]. IL-1α and IL-1β were also both lower in *MSC2*-treated mice when compared with MSC-treated mice. IL-1β has been implicated in chronic pain conditions and is elevated in the cerebral spinal fluid of patients with chronic pain [133, 134]. We suggest that the overall dampened immune state of the *MSC2*-treated mice may help explain the improvement in their behavioral assays.

Overall, these results demonstrate the beneficial therapeutic effects on pDPN by the anti-inflammatory *MSC2*-based therapy. Moreover, our results show greater modification of the inflammatory state and vast improvement of symptoms, which call for the continued investigation of *MSC2*-based therapy in such diseases. We aim to continue to evaluate the potential of *MSC2* to be developed as an off-the-shelf product and used as an improved anti-inflammatory CBT for pDPN.

4 Concluding Remarks

CBTs are a promising new approach to treat many human diseases, although many challenges remain before we can immediately and safely translate them to the clinic. Multipotent mesenchymal stromal cells (MSCs) are prevalent among the nearly 3000 ongoing clinical trials testing new CBTs. MSCs are increasingly considered to be safe, are unique in their ability to modulate host immune responses, are obtained from *adult* tissues and thus avoid the ethical issues of embryonic cell sources, and are quickly expanded and stored. However, despite these established advantages, one of the greatest challenges to using MSCs in the clinic is the fact that there is no general consensus on how to clearly define these cells; this significantly confounds efforts to study them in vitro and in vivo, as well as hinders efforts to prepare homogeneous lots. Furthermore, although one of their most profound clinical effects upon intravenous administration is the modulation of host immune responses, we do not yet have adequate ex vivo assays that consistently predict each MSC lot's therapeutic consequence in the treated patient. My laboratory recently developed a new approach for the induction of MSCs into uniform and consistently acting pro-inflammatory *MSC1* or anti-inflammatory *MSC2* phenotypes that we suggest provide convenient experimental tools to address some of these challenges. The continued investigation of these cells and others used for CBTs will ensure safe and effective therapy of human disease.

Acknowledgments This work was supported by the National Institutes of Health grant 1P20RR20152-01, Department of Defense OC073102 Concept Award and research support from the Tulane Cancer Center and the Center for Stem Cell Research and Regenerative Medicine.

References

1. Mason C et al (2011) Cell therapy industry: billion dollar global business with unlimited potential. Regen Med 6(3):265–272
2. Bianco P (2011) Back to the future: moving beyond "mesenchymal stem cells". J Cell Biochem 112(7):1713–1721
3. Rayment EA, Williams DJ (2010) Concise review: mind the gap: challenges in characterizing and quantifying cell- and tissue-based therapies for clinical translation. Stem Cells 28(5):996–1004
4. Salem HK, Thiemermann C (2010) Mesenchymal stromal cells: current understanding and clinical status. Stem Cells 28(3):585–596
5. Singer NG, Caplan AI (2011) Mesenchymal stem cells: mechanisms of inflammation. Annu Rev Pathol 6:457–478
6. Perry D (2000) Patients' voices: the powerful sound in the stem cell debate. Science 287(5457):1423
7. Sharp J, Keirstead HS (2007) Therapeutic applications of oligodendrocyte precursors derived from human embryonic stem cells. Curr Opin Biotechnol 18(5):434–440
8. Scolding N (2011) Adult stem cells and multiple sclerosis. Cell Prolif 44(Suppl 1):35–38
9. Mannello F, Tonti GA (2007) Concise review: no breakthroughs for human mesenchymal and embryonic stem cell culture: conditioned medium, feeder layer, or feeder-free; medium with fetal calf serum, human serum, or enriched plasma; serum-free, serum replacement nonconditioned medium, or ad hoc formula? All that glitters is not gold! Stem Cells 25(7):1603–1609
10. Martin MJ et al (2005) Human embryonic stem cells express an immunogenic nonhuman sialic acid. Nat Med 11(2):228–232
11. Sundin M et al (2007) No alloantibodies against mesenchymal stromal cells, but presence of anti-fetal calf serum antibodies, after transplantation in allogeneic hematopoietic stem cell recipients. Haematologica 92(9):1208–1215

12. Lepperdinger G et al (2008) Controversial issue: is it safe to employ mesenchymal stem cells in cell-based therapies? Exp Gerontol 43(11):1018–1023
13. Abdallah BM et al (2006) Inhibition of osteoblast differentiation but not adipocyte differentiation of mesenchymal stem cells by sera obtained from aged females. Bone 39(1):181–188
14. Zouboulis CC et al (2008) Human skin stem cells and the ageing process. Exp Gerontol 43(11):986–997
15. Giacomini M, Baylis F, Robert J (2007) Banking on it: public policy and the ethics of stem cell research and development. Soc Sci Med 65(7):1490–1500
16. Duquesnoy RJ (2008) Clinical usefulness of HLAMatchmaker in HLA epitope matching for organ transplantation. Curr Opin Immunol 20(5):594–601
17. Dominici M et al (2006) Minimal criteria for defining multipotent mesenchymal stromal cells. The international society for cellular therapy position statement. Cytotherapy 8(4): 315–317
18. Steidl C et al (2005) Adequate cytogenetic examination in myelodysplastic syndromes: analysis of 529 patients. Leuk Res 29(9):987–993
19. Yang W et al (2010) FISH analysis in addition to G-band karyotyping: utility in evaluation of myelodysplastic syndromes? Leuk Res 34(4):420–425
20. Muntion S et al (2012) Optimisation of mesenchymal stromal cells karyotyping analysis: implications for clinical use. Transfus Med 22(2):122–127
21. Sergent-Tanguy S et al (2003) Fluorescent activated cell sorting (FACS): a rapid and reliable method to estimate the number of neurons in a mixed population. J Neurosci Methods 129(1):73–79
22. Dobra K et al (2000) Differentiation of mesothelioma cells is influenced by the expression of proteoglycans. Exp Cell Res 258(1):12–22
23. Walen KH (2005) Budded karyoplasts from multinucleated fibroblast cells contain centrosomes and change their morphology to mitotic cells. Cell Biol Int 29(12):1057–1065
24. Stephens DJ, Allan VJ (2003) Light microscopy techniques for live cell imaging. Science 300(5616):82–86
25. Bianco P et al (2010) "Mesenchymal" stem cells in human bone marrow (skeletal stem cells): a critical discussion of their nature, identity, and significance in incurable skeletal disease. Hum Gene Ther 21(9):1057–1066
26. Sacchetti B et al (2007) Self-renewing osteoprogenitors in bone marrow sinusoids can organize a hematopoietic microenvironment. Cell 131(2):324–336
27. da Silva Meirelles L, Caplan AI, Nardi NB (2008) In search of the in vivo identity of mesenchymal stem cells. Stem Cells 26(9):2287–2299
28. Feng J et al. (2011) Dual origin of mesenchymal stem cells contributing to organ growth and repair. Proc Natl Acad Sci U S A 108(16):6503–6508
29. Zwezdaryk KJ et al (2007) Erythropoietin, a hypoxia-regulated factor, elicits a pro-angiogenic program in human mesenchymal stem cells. Exp Hematol 35(4):640–652
30. Pittenger MF et al (1999) Multilineage potential of adult human mesenchymal stem cells. Science 284(5411):143–147
31. Jaiswal N et al (1997) Osteogenic differentiation of purified, culture-expanded human mesenchymal stem cells in vitro. J Cell Biochem 64(2):295–312
32. Bruder SP, Jaiswal N, Haynesworth SE (1997) Growth kinetics, self-renewal, and the osteogenic potential of purified human mesenchymal stem cells during extensive subcultivation and following cryopreservation. J Cell Biochem 64(2):278–294
33. Digirolamo CM et al (1999) Propagation and senescence of human marrow stromal cells in culture: a simple colony-forming assay identifies samples with the greatest potential to propagate and differentiate. Br J Haematol 107(2):275–281
34. Phinney DG et al (1999) Plastic adherent stromal cells from the bone marrow of commonly used strains of inbred mice: variations in yield, growth, and differentiation. J Cell Biochem 72(4):570–585

35. Kopen GC, Prockop DJ, Phinney DG (1999) Marrow stromal cells migrate throughout forebrain and cerebellum, and they differentiate into astrocytes after injection into neonatal mouse brains. Proc Natl Acad Sci U S A 96(19):10711–10716
36. Sun Y et al (2007) Differentiation of bone marrow-derived mesenchymal stem cells from diabetic patients into insulin-producing cells in vitro. Chin Med J (Engl) 120(9):771–776
37. Ju S et al (2010) In vivo differentiation of magnetically labeled mesenchymal stem cells into hepatocytes for cell therapy to repair damaged liver. Invest Radiol 45(10):625–633
38. Prockop DJ (2003) Further proof of the plasticity of adult stem cells and their role in tissue repair. J Cell Biol 160(6):807–809
39. Prockop DJ (2009) Repair of tissues by adult stem/progenitor cells (MSCs): controversies, myths, and changing paradigms. Mol Ther 17(6):939–946
40. Le Blanc K et al (2004) Treatment of severe acute graft-versus-host disease with third party haploidentical mesenchymal stem cells. Lancet 363(9419):1439–1441
41. Abdi R et al (2008) Immunomodulation by mesenchymal stem cells: a potential therapeutic strategy for type 1 diabetes. Diabetes 57(7):1759–1767
42. Nemeth K, Mayer B, Mezey E (2009) Modulation of bone marrow stromal cell functions in infectious diseases by toll-like receptor ligands. J Mol Med 27:551–589
43. Uccelli A, Moretta L, Pistoia V (2008) Mesenchymal stem cells in health and disease. Nat Rev Immunol 8(9):726–736
44. Aggarwal S, Pittenger MF (2005) Human mesenchymal stem cells modulate allogeneic immune cell responses. Blood 105(4):1815–1822
45. Bunnell BA, Betancourt AM, Sullivan DE (2010) New concepts on the immune modulation mediated by mesenchymal stem cells. Stem Cell Res Ther 1(5):34
46. Nemeth K et al (2010) Bone marrow stromal cells use TGF-beta to suppress allergic responses in a mouse model of ragweed-induced asthma. Proc Natl Acad Sci U S A 107(12):5652–5657
47. Gur-Wahnon D et al (2009) The induction of APC with a distinct tolerogenic phenotype via contact-dependent STAT3 activation. PLoS One 4(8):e6846
48. Krampera M et al (2006) Regenerative and immunomodulatory potential of mesenchymal stem cells. Curr Opin Pharmacol 6(4):435–441
49. Honczarenko M et al (2006) Human bone marrow stromal cells express a distinct set of biologically functional chemokine receptors. Stem Cells 24(4):1030–1041
50. Ringe J et al (2007) Towards in situ tissue repair: human mesenchymal stem cells express chemokine receptors CXCR1, CXCR2 and CCR2, and migrate upon stimulation with CXCL8 but not CCL2. J Cell Biochem 101(1):135–146
51. Wynn RF et al (2004) A small proportion of mesenchymal stem cells strongly expresses functionally active CXCR4 receptor capable of promoting migration to bone marrow. Blood 104(9):2643–2645
52. Sordi V et al (2005) Bone marrow mesenchymal stem cells express a restricted set of functionally active chemokine receptors capable of promoting migration to pancreatic islets. Blood 106(2):419–427
53. Lapidot T (2001) Mechanism of human stem cell migration and repopulation of NOD/SCID and B2mnull NOD/SCID mice. The role of SDF-1/CXCR4 interactions. Ann N Y Acad Sci 938:83–95
54. Sasaki M et al (2008) Mesenchymal stem cells are recruited into wounded skin and contribute to wound repair by transdifferentiation into multiple skin cell type. J Immunol 180(4):2581–2587
55. Inokuma D et al (2006) CTACK/CCL27 accelerates skin regeneration via accumulation of bone marrow-derived keratinocytes. Stem Cells 24(12):2810–2816
56. Macmillan ML et al (2009) Transplantation of ex vivo culture-expanded parental haploidentical mesenchymal stem cells to promote engraftment in pediatric recipients of unrelated donor umbilical cord blood: results of a phase I-II clinical trial. Bone Marrow Transplant 43(6):447–454

57. Chen SL et al (2004) Improvement of cardiac function after transplantation of autologous bone marrow mesenchymal stem cells in patients with acute myocardial infarction. Chin Med J (Engl) 117(10):1443–1448
58. Romieu-Mourez R, Coutu DL, Galipeau J (2012) The immune plasticity of mesenchymal stromal cells from mice and men: concordances and discrepancies. Front Biosci (Elite Ed) 4:824–837
59. Krasnodembskaya A et al (2011) Antibacterial effect of human mesenchymal stem cells is mediated in part from secretion of the antimicrobial peptide LL-37. Stem Cells 28(12):2229–2238
60. Gupta N et al. (2012) Mesenchymal stem cells enhance survival and bacterial clearance in murine Escherichia coli pneumonia. Thorax 67(6):533–539
61. Coffelt SB et al (2009) The pro-inflammatory peptide LL-37 promotes ovarian tumor progression through recruitment of multipotent mesenchymal stromal cells. Proc Natl Acad Sci U S A 106(10):3806–3811
62. Coffelt SB, Scandurro AB (2008) Tumors sound the alarmin(s). Cancer Res 68(16):6482–6485
63. Coffelt SB et al (2009) Leucine leucine-37 uses formyl peptide receptor-like 1 to activate signal transduction pathways, stimulate oncogenic gene expression, and enhance the invasiveness of ovarian cancer cells. Mol Cancer Res 7(6):907–915
64. Coffelt SB et al (2008) Ovarian cancers overexpress the antimicrobial protein hCAP-18 and its derivative LL-37 increases ovarian cancer cell proliferation and invasion. Int J Cancer 122(5):1030–1039
65. Spaeth E et al (2008) Inflammation and tumor microenvironments: defining the migratory itinerary of mesenchymal stem cells. Gene Ther 15(10):730–738
66. Spaeth EL et al (2009) Mesenchymal stem cell transition to tumor-associated fibroblasts contributes to fibrovascular network expansion and tumor progression. PLoS One 4(4):e4992
67. Waterman RS, Betancourt AM (2012) The role of mesenchymal stem cells in the tumor microenvironment. Tumor Microenvironment and Myelomonocytic Cells. InTech
68. Klopp AH et al. (2010) Dissecting a Discrepancy in the Literature: Do Mesenchymal Stem Cells Support or Suppress Tumor Growth? Stem Cells
69. Klopp AH et al (2011) Concise review: dissecting a discrepancy in the literature: do mesenchymal stem cells support or suppress tumor growth? Stem Cells 29(1):11–19
70. Tomchuck SL et al (2008) Toll-like receptors on human mesenchymal stem cells drive their migration and immunomodulating responses. Stem Cells 26(1):99–107
71. Waterman RS et al (2010) A new mesenchymal stem cell (MSC) paradigm: polarization into a pro-inflammatory MSC1 or an Immunosuppressive MSC2 phenotype. PLoS One 5(4):e10088
72. Verreck FA et al (2006) Phenotypic and functional profiling of human proinflammatory type-1 and anti-inflammatory type-2 macrophages in response to microbial antigens and IFN-gamma- and CD40L-mediated costimulation. J Leukoc Biol 79(2):285–293
73. Cassatella MA et al (2011) Toll-like receptor-3-activated human mesenchymal stromal cells significantly prolong the survival and function of neutrophils. Stem Cells 29(6):1001–1011
74. Krampera M (2011) Mesenchymal stromal cells: more than inhibitory cells. Leukemia 25(4):565–566
75. Romieu-Mourez R et al (2009) Cytokine modulation of TLR expression and activation in mesenchymal stromal cells leads to a proinflammatory phenotype. J Immunol 182(12):7963–7973
76. Tomchuck SL, Henkle SL, Coffelt SB, Betancourt AM (2012) Toll-like receptor 3 and suppressor of cytokine signaling proteins regulate CXCR4 and CXCR7 expression in bone marrow-derived human multipotent stromal cells. Manuscript submitted (in press)
77. Waterman RS, Henkle SL, Betancourt AM (2012) Mesenchymal stem cell 1 (MSC1)-based therapy attenuates tumor growth whereas MSC2-treatment promotes tumor growth and metastasis. Manuscript submitted (in press)

78. Djouad F et al (2003) Immunosuppressive effect of mesenchymal stem cells favors tumor growth in allogeneic animals. Blood 102(10):3837–3844
79. Karnoub AE et al (2007) Mesenchymal stem cells within tumour stroma promote breast cancer metastasis. Nature 449(7162):557–563
80. Khakoo AY et al (2006) Human mesenchymal stem cells exert potent antitumorigenic effects in a model of Kaposi's sarcoma. J Exp Med 203(5):1235–1247
81. Shinagawa K et al (2010) Mesenchymal stem cells enhance growth and metastasis of colon cancer. Int J Cancer 127(10):2323–2333
82. Gialeli C, Theocharis AD, Karamanos NK (2010) Roles of matrix metalloproteinases in cancer progression and their pharmacological targeting. FEBS J 278(1):16–27
83. Cubillos-Ruiz JR, Rutkowski M, Conejo-Garcia JR (2010) Blocking ovarian cancer progression by targeting tumor microenvironmental leukocytes. Cell Cycle 9(2):260–268
84. Nesbeth Y et al (2009) CCL5-mediated endogenous antitumor immunity elicited by adoptively transferred lymphocytes and dendritic cell depletion. Cancer Res 69(15):6331–6338
85. Roby KF et al (2000) Development of a syngeneic mouse model for events related to ovarian cancer. Carcinogenesis 21(4):585–591
86. Condeelis J, Pollard JW (2006) Macrophages: obligate partners for tumor cell migration, invasion, and metastasis. Cell 124(2):263–266
87. Baay M et al (2011) Tumor cells and tumor-associated macrophages: secreted proteins as potential targets for therapy. Clin Dev Immunol 2011:565187
88. Sica A, Bronte V (2007) Altered macrophage differentiation and immune dysfunction in tumor development. J Clin Invest 117(5):1155–1166
89. Murdoch C et al (2008) The role of myeloid cells in the promotion of tumour angiogenesis. Nat Rev Cancer 8(8):618–631
90. Navi D, Saegusa J, Liu FT (2007) Mast cells and immunological skin diseases. Clin Rev Allergy Immunol 33(1–2):144–155
91. Ingman WV et al (2006) Macrophages promote collagen fibrillogenesis around terminal end buds of the developing mammary gland. Dev Dyn 235(12):3222–3229
92. Brown JM et al (2011) Bone marrow stromal cells inhibit mast cell function via a COX2-dependent mechanism. Clin Exp Allergy 41(4):526–534
93. Bianchi G et al (2011) Immunosuppressive cells and tumour microenvironment: focus on mesenchymal stem cells and myeloid derived suppressor cells. Histol Histopathol 26(7):941–951
94. Coffelt SB, Hughes R, Lewis CE (2009) Tumor-associated macrophages: Effectors of angiogenesis and tumor progression. Biochim Biophys Acta 1796(1):11–18
95. Coussens LM, Werb Z (2002) Inflammation and cancer. Nature 420(6917):860–867
96. Fidler IJ (2003) The pathogenesis of cancer metastasis: the 'seed and soil' hypothesis revisited. Nat Rev Cancer 3(6):453–458
97. Rabinovich GA, Gabrilovich D, Sotomayor EM (2007) Immunosuppressive strategies that are mediated by tumor cells. Annu Rev Immunol 25:267–296
98. Strausberg RL (2005) Tumor microenvironments, the immune system and cancer survival. Genome Biol 6(3):211
99. Waterman RS, Morgenweck J, Nossaman BD, Scandurro AE, Scandurro SA, Betancourt AM (2012) Anti-inflammatory mesenchymal stem cells (MSC2) attenuate symptoms of painful diabetic peripheral neuropathy. Manuscript submitted (in press)
100. Bailey CJ (2007) Treating insulin resistance: future prospects. Diab Vasc Dis Res 4(1):20–31
101. Dominiczak MH (2003) Obesity, glucose intolerance and diabetes and their links to cardiovascular disease. Implications for laboratory medicine. Clin Chem Lab Med 41(9):1266–1278
102. Kampoli AM et al (2011) Potential pathogenic inflammatory mechanisms of endothelial dysfunction induced by type 2 diabetes mellitus. Curr Pharm Des 17(37):4147–4158
103. Niehoff AG et al (2007) C-reactive protein is independently associated with glucose but not with insulin resistance in healthy men. Diabetes Care 30(6):1627–1629

104. Purwata TE (2011) High TNF-alpha plasma levels and macrophages iNOS and TNF-alpha expression as risk factors for painful diabetic neuropathy. J Pain Res 4:169–175
105. Sjoholm A, Nystrom T (2005) Endothelial inflammation in insulin resistance. Lancet 365(9459):610–612
106. Uceyler N et al (2007) Differential expression of cytokines in painful and painless neuropathies. Neurology 69(1):42–49
107. Chopra K et al (2011) Sesamol suppresses neuro-inflammatory cascade in experimental model of diabetic neuropathy. J Pain 11(10):950–957
108. Doupis J et al (2009) Microvascular reactivity and inflammatory cytokines in painful and painless peripheral diabetic neuropathy. J Clin Endocrinol Metab 94(6):2157–2163
109. Pabreja K et al (2011) Minocycline attenuates the development of diabetic neuropathic pain: possible anti-inflammatory and anti-oxidant mechanisms. Eur J Pharmacol 661(1–3):15–21
110. Valsecchi AE et al (2011) The soy isoflavone genistein reverses oxidative and inflammatory state, neuropathic pain, neurotrophic and vasculature deficits in diabetes mouse model. Eur J Pharmacol 650(2–3):694–702
111. Newman RE et al (2009) Treatment of inflammatory diseases with mesenchymal stem cells. Inflamm Allergy Drug Targets 8(2):110–123
112. Kim BJ, Jin HK, Bae JS (2011) Bone marrow-derived mesenchymal stem cells improve the functioning of neurotrophic factors in a mouse model of diabetic neuropathy. Lab Anim Res 27(2):171–176
113. Shibata T et al (2008) Transplantation of bone marrow-derived mesenchymal stem cells improves diabetic polyneuropathy in rats. Diabetes 57(11):3099–3107
114. Cameron NE et al (2001) Effect of the hydroxyl radical scavenger, dimethylthiourea, on peripheral nerve tissue perfusion, conduction velocity and nociception in experimental diabetes. Diabetologia 44(9):1161–1169
115. Drel VR et al (2006) The leptin-deficient (ob/ob) mouse: a new animal model of peripheral neuropathy of type 2 diabetes and obesity. Diabetes 55(12):3335–3343
116. Drel VR et al (2007) Evaluation of the peroxynitrite decomposition catalyst Fe(III) tetra-mesitylporphyrin octasulfonate on peripheral neuropathy in a mouse model of type 1 diabetes. Int J Mol Med 20(6):783–792
117. Ilnytska O et al (2006) Poly(ADP-ribose) polymerase inhibition alleviates experimental diabetic sensory neuropathy. Diabetes 55(6):1686–1694
118. Stevens MJ et al (2007) Nicotinamide reverses neurological and neurovascular deficits in streptozotocin diabetic rats. J Pharmacol Exp Ther 320(1):458–464
119. Obrosova IG (2009) Diabetic painful and insensate neuropathy: pathogenesis and potential treatments. Neurotherapeutics 6(4):638–647
120. Beiswenger KK, Calcutt NA, Mizisin AP (2008) Epidermal nerve fiber quantification in the assessment of diabetic neuropathy. Acta Histochem 110(5):351–362
121. Herrmann DN et al (1999) Epidermal nerve fiber density and sural nerve morphometry in peripheral neuropathies. Neurology 53(8):1634–1640
122. Muller KA et al (2008) Abnormal muscle spindle innervation and large-fiber neuropathy in diabetic mice. Diabetes 57(6):1693–1701
123. Sinnreich M, Taylor BV, Dyck PJ (2005) Diabetic neuropathies. Classification, clinical features, and pathophysiological basis. Neurologist 11(2):63–79
124. Fiorentino DF et al (1991) IL-10 inhibits cytokine production by activated macrophages. J Immunol 147(11):3815–3822
125. Soderquist RG et al (2010) PEGylation of interleukin-10 for the mitigation of enhanced pain states. J Biomed Mater Res A 93(3):1169–1179
126. Bishnoi M et al (2011) Streptozotocin-induced early thermal hyperalgesia is independent of glycemic state of rats: role of transient receptor potential vanilloid 1(TRPV1) and inflammatory mediators. Mol Pain 7:52
127. Kumar A, Negi G, Sharma SS (2011) JSH-23 targets nuclear factor-kappa B and reverses various deficits in experimental diabetic neuropathy: effect on neuroinflammation and antioxidant defence. Diabetes Obes Metab 13(8):750–758

128. Luo L et al (2011) Sildenafil improves diabetic vascular activity through suppressing endothelin receptor A, iNOS and NADPH oxidase which is comparable with the endothelin receptor antagonist CPU0213 in STZ-injected rats. J Pharm Pharmacol 63(7):943–951
129. Silva DC et al (2011) Pectin from Passiflora edulis shows anti-inflammatory action as well as hypoglycemic and hypotriglyceridemic properties in diabetic rats. J Med Food 14(10):1118–1126
130. Cui JG et al (2000) Possible role of inflammatory mediators in tactile hypersensitivity in rat models of mononeuropathy. Pain 88(3):239–248
131. Cruickshank AM et al (1990) Response of serum interleukin-6 in patients undergoing elective surgery of varying severity. Clin Sci (Lond) 79(2):161–165
132. Holzheimer RG, Steinmetz W (2000) Local and systemic concentrations of pro- and anti-inflammatory cytokines in human wounds. Eur J Med Res 5(8):347–355
133. Alexander GM et al (2005) Changes in cerebrospinal fluid levels of pro-inflammatory cytokines in CRPS. Pain 116(3):213–219
134. Watkins LR, Maier SF, Goehler LE (1995) Immune activation: the role of pro-inflammatory cytokines in inflammation, illness responses and pathological pain states. Pain 63(3):289–302

Adv Biochem Eng Biotechnol (2013) 130: 199–208
DOI: 10.1007/10_2012_154

Published Online: 7 August 2012

Interactions Between Mesenchymal Stem Cells and Dendritic Cells

Grazia Maria Spaggiari and Lorenzo Moretta

Abstract Mesenchymal stem or stromal cells (MSC) are considered a promising new therapeutic strategy for the treatment of several pathological conditions. Due to their immunomodulatory properties, they are currently employed in clinical trials aimed at preventing or treating steroid-resistant acute graft-versus-host disease (GvHD), a frequent complication of allogeneic hematopoietic stem cell transplantation (HSCT). In addition, the use of MSC has been proposed for the treatment of autoimmune diseases. A number of recent studies have focused on the influence of MSC on dendritic cell (DC) function. DCs play a critical role in initiating and regulating immune responses by promoting antigen-specific T cell activation. Moreover, they are involved in efficient cross-talk with different cells of the innate immune system. DC are the most effective antigen-presenting cells and prime na T cells to initiate adaptive immune responses including those against allogeneic cells or self-antigens. Thus, alteration of DC generation or function may greatly contribute to the inhibition of T cell responses. In this context, MSC were shown to interfere with DC maturation from monocytes or $CD34^{+}$ hemopoietic precursors thus further confirming their role in immune regulation and their usefulness in cell-based therapies.

Keywords DC cytoskeleton rearrangement · DC maturation · DC phenotype · Dendritic cell differentiation · Dendritic cells · Hematopoietic stem cell transplantation · Immune synapse · MSC licensing · MSC-mediated immunosuppression · Notch-2 · PGE2 · Tolerogenic DC · Treg cells · Unrestricted somatic stem cells

G. M. Spaggiari (✉)
Department of Experimental Medicine, University of Genoa,
Via L.B. Alberti 2, Genoa 16132, Italy
e-mail: gra_ma@yahoo.com

L. Moretta
Scientific Direction, Giannina Gaslini Institute,
Via G. Gaslini, 5, 16147 Genoa, Italy
e-mail: lorenzomoretta@ospedale-gaslini.ge.it

List of abbreviations

Ag	Antigen
DC	Dendritic cells
GM-CSF	Granulocyte macrophage-colony stimulating factor
GvHD	Graft-versus-Host Disease
HGF	Hepatocyte growth factor
HO-1	Haemoxygenase-1
HSCT	Hematopoietic stem cell transplantation
IDO	Indoleamine 2,3-dioxygenase
IFN-γ	Interferon-γ
IL	Interleukin
LPS	Lipopolysaccharide
MHC	Major histocompatibility complex
MLR	Mixed lymphocyte reactions
MSC	Mesenchymal stem cells
NK cells	Natural killer cells
NO	Nitric oxide
PGE2	Prostaglandin E2
TGF-β	Transforming growth factor-β
TNF-α	Tumor necrosis factor-α

Contents

1 Introduction

Mesenchymal stem cells (MSC) are known for their characteristic of being multipotent stem cells, capable of forming bone, cartilage, and other mesenchymal tissues [20]. In particular, in vitro experiments demonstrated that clonal MSC can differentiate into different lineages including not only osteoblasts, chondrocytes, and adipocytes but also muscle cells, cardiomyocytes, and neural precursors. Moreover, MSC are a component of the bone marrow stroma that have been shown to support hematopoiesis by providing suitable cytokines and growth factors [15]. More recently, another function has been ascribed to MSC: a strong immunosuppressive effect on cells of both innate and adaptive immunity including T and B cells, natural killer (NK) cells, and dendritic cells (DC; [18], [24]). The mechanisms underlying the MSC-mediated inhibitory effect are only in part understood. Both contact-dependent mechanisms and soluble factors are thought to be involved in the induction of MSC-mediated immunosuppression. The first step in these interactions usually involves cell-to-cell contact mediated by adhesion molecules. Subsequently, several soluble factors appear to be involved in MSC-mediated immunoregulation, produced constitutively by MSC or released as a result of the interaction with other cell types. Examples of these molecules are indoleamine 2,3-dioxygenase (IDO) and nitric oxide (NO), which are released by MSC only after triggering by interferon-γ (IFN-γ). Other soluble factors, including transforming growth factor-β (TGF-β), hepatocyte growth factor (HGF), prostaglandin E2 (PGE2), hemoxygenase-1(HO-1), interleukin (IL)-6, and soluble HLA-G5, are constitutively produced by MSC. In addition, the production of some of these molecules can be increased by cytokines, such as IFN-γ, released by cells interacting with MSC [6, 12].

In the setting of allogeneic hematopoietic stem cell transplantation (HSCT), MSC have been brought to the clinic mainly to promote hematopoietic engraftment and for prevention/treatment of graft-versus-host disease (GvHD) [7]. The use of MSC for clinical purposes takes advantage of their poor immunogenicity in vitro, in pre-clinical studies, and in human studies, which supported the possible infusion of MSC from allogeneic donors in adoptive immunotherapy. The therapeutic potential of MSC is currently being explored in a number of phase I/II and II clinical trials [23]. So far, most of the data reported in these studies have accounted for the safety of infusion of culture-expanded allogeneic MSC, together with sustained hematopoietic engraftment after HSCT, reduced incidence of GvHD [3], and, in the case of steroid-resistant acute GvHD, markedly improved survival rate of MSC-infused patients [13].

DC play a critical role in initiating and regulating immune responses by promoting antigen (Ag)-specific T cell activation [4, 5]. In addition, as revealed by recent studies, they can efficiently interact with and trigger or modulate cells of the innate immune system [9, 16, 17]. DC are the most effective antigen-presenting cells and prime naïve T cells to initiate adaptive immune responses including proliferative responses to allogeneic cells, that can be tested in vitro in mixed lymphocyte reactions (MLR). Some DC reside in an immature state in peripheral tissues and are highly specialized in Ag uptake. Immature DC (iDC) display low levels of major histocompatibility complex (MHC) and co-stimulatory molecules (CD80, CD86) at their surface. Remarkably, in mature DC, these molecules become highly upregulated. Immature DC can be rapidly recruited at the site of inflammation where Ag capture and processing primarily occur. After Ag cleavage into peptides, peptide loading on MHC molecules and migration to T cell areas of the draining lymph nodes, DC undergo complete maturation. Mature DC (mDC) lose their uptake capability while they acquire the ability to stimulate T cells including unpolarized Th0 or polarized Th1 or Th2 responses and also tolerogenic T cells.

In 2005 [1], Aggarwal and Pittenger first reported that bone marrow-derived MSC could influence the outcome of an ongoing inflammatory immune response by altering the cytokine secretion profile of peripheral blood DC resulting in a shift from a pro-inflammatory immune response towards an anti-inflammatory or tolerant cell environment. In particular, they showed that MSC could decrease tumor necrosis factor-α (TNF-α) secretion by CD1c+ myeloid DC subset while inducing increases of IL-10 production by BDCA-4^{+} plasmacytoid DC [1]. In addition, Beyth et al. [8] showed that MSC could inhibit T cells indirectly, by contact-dependent induction of regulatory or aberrant antigen-presenting cells, including dendritic cells, characterized by T cell suppressive properties.

Since then, a series of studies has been reported investigating the MSC-mediated inhibition of DC differentiation and function. However, in spite of the general evidence that MSC can inhibit the generation of functional DC, data on specific aspects of such inhibitory effects are contradictory, possibly reflecting differences in experimental protocols. In this chapter, we offer an overview of the most relevant data regarding DC-MSC interactions in humans.

2 MSC Inhibit Dendritic Cell Differentiation

A large body of evidence accounts for the ability of MSC to strongly inhibit DC generation from both monocytes and CD34^{+} cell precursors. Indeed, MSC have been shown to affect the acquisition of DC-specific markers when added to monocytes induced to differentiate towards DC with granulocyte macrophage-colony stimulating factor (GM-CSF) and IL-4 [10, 19, 22]. In particular, monocyte-derived cells obtained in the presence of MSC failed to express CD1a while still maintaining the monocyte marker CD14. Moreover, upon stimulation with lipopolysaccharide (LPS), which normally induces full DC maturation, cells expressed lower levels of CD80 and CD86 co-stimulatory molecules and of the DC maturation marker CD83 as compared to control cells. In agreement with an altered phenotype, cells did not display typical DC morphology (i.e., a veiled appearance and an abundant cytoplasm), but rather developed macrophage morphology [10]. On the other hand, different results were obtained by van den Berk et al. [25], who showed that cord blood-derived MSC (also called unrestricted somatic stem cells, USSC) did not interfere with DC differentiation from monocytes. Thus, immature DC generated in the presence of MSC lost CD14 and acquired normal levels of CD40, CD86, CD209, and HLA-DR. The different origin of MSC (cord blood vs. bone marrow) may explain these different results suggesting that MSC of different origin/site of isolation may display different functional properties.

It is of note that MSC-mediated inhibition of DC differentiation was not accompanied by cell loss. Indeed, Jiang et al. [10] reported that cell viability was not affected by co-culture with MSC, and the cell recovery of these co-cultures was comparable to that of control cultures. However, although no evidence existed of a pro-apoptotic effect induced by MSC on monocytes, MSC prevented monocytes from entering the G1 phase of the cell cycle with a progressive number of cells accumulating in the G0 phase [21]. Although monocytes do not require DNA synthesis and cell division to become functional DC, they must enter the cell cycle. MSC could arrest monocytes in G0 rendering them unable to stimulate allogeneic T cells. Downregulation of cyclin D2 expression was shown to be primarily responsible for cell cycle arrest.

An important aspect of the inhibition exerted by MSC is the reversibility of the effect. Jiang et al. [10] reported that inhibition of DC differentiation from monocytes was reversible. Thus, upon removal of MSC and addition of fresh cytokines, monocyte-derived cells acquired the DC phenotype, that is, loss of CD14 and acquisition of CD1a and CD83. Nauta et al. [19] showed that when MSC were removed after 2 days from monocyte cultures, cells downregulated CD14 but did not express CD1a, suggesting that, in this experimental setting, inhibition was only partially reversible and that an early conditioning with MSC was essential for the inhibitory effect. The discrepancy of the results may indeed reflect differences in the experimental settings. Jiang et al. performed their experiments under transwell culture conditions, however, Nauta et al. performed co-cultures in which cells were in direct contact.

Other groups investigated the effect of MSC on DC differentiation by using $CD34^+$ cells as DC precursors. Nauta et al. [19] demonstrated that MSC could prevent DC generation from $CD34^+$ cells derived from umbilical cord blood. In these studies, they also showed that MSC inhibited the differentiation of dermal/interstitial DC, by blocking the transition of $CD14^+CD1a^-$ intermediate prescursors to the $CD14^-CD1a^+$ differentiation stage, whereas they did not prevent the generation of $CD14^-CD1a^+$ Langerhans cells. Moreover, the $CD14^+CD1a^-$ subset expressed low levels of CD80, CD86, CD83, and CD40 after cell stimulation with LPS. In contrast, Li et al. [14] reported that MSC could inhibit the differentiation of both dermal/interstitial and Langerhans cells. In addition, they showed that MSC could also inhibit the proliferation of DC precursors by inducing a threefold decrease of their proliferation rate. Also in this case, variability in results may be explained in part by the fact that different DC precursors were used in these experiments (i.e., cord blood- vs. adult bone marrow-derived $CD34^+$ cells).

Regarding the functional capabilities of the phenotypically abnormal DC obtained in the presence of MSC, all studies reported that their ability to stimulate allogeneic T cell proliferation in MLR was strongly impaired as compared to control DC. In addition, these abnormal DC, generated either from monocytes or $CD34^+$ cells, produced very low levels of IL-12 upon stimulation with LPS [10, 22] or CD40L [19]. Moreover, it was shown that LPS-induced phosphorylation of p38, a kinase involved in an intracellular signaling pathway positively regulating IL-12 secretion, was greatly reduced in the presence of MSC [10]. However, so far, it is still poorly defined whether cells generated in the presence of MSC are simply DC with an impaired function or rather "educated" DC with regulatory activity. In the study by Li et al. [14], secondary allostimulation of T cells by DC generated in the presence of MSC induced the generation of FoxP3-expressing alloantigen-specific T cells. This finding implies that MSC may promote differentiation of tolerogenic DC, capable of stimulating expansion of Treg cells. In this context, future studies aimed at further characterizing MSC-conditioned DC will help to clarify this point. In this context, it is possible that, depending on the DC/MSC source or experimental settings, different types of "nonclassical" DC may be generated capable of finally exerting immunomodulatory/anti-inflammatory effects.

3 Effects of MSC on the Final Maturation of DC

Another relevant question concerning DC-MSC interactions is whether MSC can interfere not only with the early generation of DC from their precursors but also with later stages of differentiation, such as the progression from immature to mature DC. Several groups investigated this point in their studies and obtained contradictory results. In some cases, MSC were shown to moderately suppress LPS-induced maturation of monocyte-derived DC. Thus, the resulting cells displayed decreased ability to stimulate allogeneic T cell proliferation in MLR, associated with lower levels of IL-12 production and IFN-γ induction compared to control mDC [10]. On the other hand, Spaggiari et al. [22] demonstrated that MSC failed to interfere with LPS-induced maturation of DC. Indeed, DC induced to mature in the presence of MSC displayed normal phenotype, with adequate levels of surface CD80, CD86, and CD83 molecules and were even more efficient stimulators in MLR than mDC obtained under standard conditions. In an interesting study by Aldinucci et al. [2], a new pathway of MSC-mediated regulation of DC function was proposed. Immature DC, stimulated with LPS in the presence of MSC, became unable to form active immune synapses with lymphocytes, despite their expression of a mature phenotype and a normal IL-12/IL-10 production profile. In addition, MSC-treated DC retained endocytic activity and podosome-like structures, typical of immature DC. The inability of DC to establish synapses was associated with alteration of the cytoskeleton rearrangement, consisting of absence of actin redistribution, which normally occurs in iDC upon stimulation by LPS. As a consequence, DC while undergoing some sort of differentiation retained features of immaturity, thus becoming unable to activate alloreactive T cells efficiently.

An opposite effect, promoting rather than inhibitory on LPS-induced maturation of iDC, seems to be exerted by USSC [25]. Immature DC stimulated with LPS in the presence of USSC displayed higher migratory capacity in response to CCL21 chemokine than control DC. Accordingly they expressed significantly higher levels of its specific receptor CCR7. Also IL-12 production was increased in cells that had undergone maturation in the presence of USSC. Interestingly, even in the absence of LPS, USSC could positively contribute to DC maturation by significantly increasing expression of CD80 and CD83 markers. However, in this study, DC were not analyzed for their capability of stimulating T cell response, thus it was not demonstrated whether these MSC-treated DC would be efficient antigen-presenting cells.

Few studies investigated the possible effects of the presence of MSC on fully mature DC. Zhao et al. [26] demonstrated that DC with regulatory activity could be generated by culturing fully differentiated, LPS-stimulated mDC with MSC. After conditioning with MSC, these cells acquired the ability to suppress T cell proliferation in MLR by producing TGF-β. Moreover, they could promote the generation of $CD4^{+}CD25^{+}FoxP3^{+}$ Treg cells from $CD4^{+}CD25^{-}$ T cells. Jiang et al. [10] showed that culture of mDC with MSC reverted DC to the phenotypic profile of an immature state, characterized by decreased surface expression of HLA-DR, CD80, and CD86 molecules.

4 Mechanisms Involved in the MSC-Mediated Inhibitory Effect: Cell Contact Versus Soluble Factors

Various molecular pathways appear to be involved in MSC-mediated immune regulation, including IFN-γ, IL-1β, TGF-β, IDO, IL-6, PGE2, HGF, TNF-α, NO, HO-1, and HLA-G5, most of which are strictly related and reciprocally activating. It is now largely accepted that the immunosuppressive activity is not a constitutive property of MSC, but depends on a process of activation or "licensing" to be acquired [11]. Activation of MSC is mostly consequent to cell exposure to inflammatory cytokines, such as IFN-γ, TNF-α, and IL-1α/β, which are produced by different cell types following induction of an inflammatory or immune response. These soluble activators can induce changes in both MSC phenotype and gene expression, thus allowing cells to act as immune regulators. As for other cells of the innate immune system, the inhibitory effect can be mediated by the activity of soluble factors produced by activated MSC and/or by cell contact involving specific receptor/ligand interactions not completely elucidated so far.

In the case of dendritic cells, most studies support the idea of a major involvement of soluble factors. Indeed, blocking of their activity or their production by using specific inhibitors could significantly, if not completely, restore DC differentiation and function. In this context, IL-6 and M-CSF were shown to be partially involved in the MSC-mediated inhibition of DC differentiation from monocytes, although only partial restoration of the DC phenotype (i.e., loss of CD14 but lack of expression of CD1a marker) could be obtained by using anti-IL-6 and anti-M-CSF neutralizing antibodies.

Another important MSC product, PGE2, was demonstrated as playing a major role in the inhibitory effect [22]. Indeed, PGE2 levels were strongly increased in the supernatants of monocyte-MSC co-cultures as compared to those of monocytes alone. Moreover, the selective inhibition of cyclooxigenase-2 activity and thereby of PGE2 synthesis almost completely reverted the inhibitory effect as confirmed by the restoration of both DC phenotype and function. Notably, this effect was achieved in spite of the presence of high levels of IL-6 in co-culture supernatants, thus suggesting that PGE2 and not IL-6 was predominantly involved in the inhibitory effect. Other authors reported a substantial role played by cell-to-cell interactions. Li et al. [14] showed that, in co-culture experiments performed using the transwell chamber system to separate MSC and $CD34^+$ DC precursors physically, inhibition of DC differentiation was significantly prevented. Accordingly, the inhibitory mechanism proposed was an MSC-induced expression and subsequent signaling through the Notch-2 receptor in $CD34^+$-derived DC. The inhibition of Notch-2 signaling resulted in complete restoration of DC phenotype and function. It should be noted that, in this case, cell targets of the MSC-mediated inhibition were bone marrow-derived $CD34^+$ precursors and not peripheral blood monocytes. Thus, it is conceivable that different mechanisms may be responsible for the interference with distinct differentiation pathways.

The modality of the inhibitory effect may also depend on the differentiation stage of immune cells. Indeed, Aldinucci et al. [2] demonstrated that the alteration of LPS-induced cytoskeleton rearrangement in differentiated immature DC was contact-dependent and partially mediated by V-CAM and N-cadherin molecules expressed on the MSC cell surface.

5 Concluding Remarks

The regulation of DC function represents an important strategy in the design of innovative therapeutic protocols aimed at suppressing pathological immune responses, such as GVHD and autoimmune disorders. In this context, the suppressive effect that MSC can exert on immune cells, including DC, reveals a promising therapeutic strategy. Most studies addressing the interaction between MSC and DC have demonstrated that MSC are capable of inhibiting DC at multiple levels. Indeed, cells generated in the presence of MSC from DC progenitors, either monocytes or $CD34^+$ cells, do not display the proper DC phenotype and have impaired function as compared to control DC. However, there are no converging conclusions on the immunoregulatory phenomenon, possibly as a result of different experimental settings (including DC progenitor or MSC sources employed, cell-to-cell ratios used, and the consideration of only one single MSC-derived mediator for the inhibitory effect with no comparison with others). Moreover, so far, it has not been clarified what kind of cells are generated upon interaction with MSC, whether functionally impaired DC, characterized by lower capacity of efficiently stimulating T cell proliferation and by altered cytokine profile, or regulatory DC, capable of generating $CD4^+CD25^+FoxP3^+$ Treg cells. Future studies aimed at clarifying these points will contribute to better knowledge of MSC biology and, it is hoped, to the optimal use of MSC in clinical practice.

References

1. Aggarwal S, Pittenger MF (2005) Human mesenchymal stem cells modulate allogeneic immune responses. Blood 105:1815–1822
2. Aldinucci A, Rizzetto L, Pieri L et al (2010) Inhibition of immune synapse by altered dendritic cell actin distribution: a new pathway of mesenchymal stem cell immune regulation. J Immunol 185:5102–5110
3. Ball LM, Bernardo ME, Roelofs H et al (2007) Cotransplantation of ex vivo expanded mesenchymal stem cells accelerates lymphocyte recovery and may reduce the risk of graft failure in haploidentical hematopoietic stem-cell transplantation. Blood 110:2764–2767
4. Banchereau J, Steinman RM (1998) Dendritic cells and the control of immunity. Nature 392:245–252
5. Banchereau J, Briere F, Caux F et al (2000) Immunobiology of dendritic cells. Annu Rev Immunol 18:767–811
6. Bassi EJ, de Almeida DC, Moraes-Vieira PM et al (2012) Exploring the role of soluble factors associated with immune regulatory properties of mesenchymal stem cells. Stem Cell Rev Rep 8:329–342
7. Battiwalla M, Hematti P (2009) Mesenchymal stem cells in hematopoietic stem cell transplantation. Cytotherapy 11:503–515
8. Beyth S, Borovsky Z, Mevorach D et al (2005) Human mesenchymal stem cells alter antigen-presenting cell maturation and induce T-cell unresponsiveness. Blood 105:2214–2219
9. Cooper MA, Fehniger TA, Fuchs A et al (2004) NK cell and DC interactions. Trends Immunol 25:47–52
10. Jiang X-X, Zhang Y, Liu B et al (2005) Human mesenchymal stem cells inhibit differentiation and function of monocyte-derived dendritic cells. Blood 105:4120–4126
11. Krampera M (2011) Mesenchymal stromal cell "licensing": a multistep process. Leukemia 25:1408–1414
12. Krampera M, Cosmi L, Angeli R et al (2006) Role for interferon-gamma in the immunomodulatory activity of human bone marrow mesenchymal stem cells. Stem Cells 24:386–398
13. Le Blanc K, Frassoni F, Ball LM et al (2008) Mesenchymal stem cells for treatment of steroid-resistant severe, acute graft versus-host disease: a phase II study. Lancet 371:1579–1586
14. Li Y-P, Paczesny S, Lauret E et al (2008) Human mesenchymal stem cells license adult $CD34^+$ hemopoietic progenitor cells to differentiate into regulatory dendritic cells through activation of the Notch pathway. J Immunol 180:1598–1608
15. Méndez-Ferrer S, Michurina TV, Ferraro F et al (2010) Mesenchymal and haematopoietic stem cells form a unique bone marrow niche. Nature 466:829–834
16. Moretta A (2002) Natural killer cells and dendritic cells: rendezvous in abused tissues. Nat Rev Immunol 2:957–965
17. Moretta L, Ferlazzo G, Bottino C et al (2006) Effector and regulatory events during natural killer-dendritic cell interactions. Immunol Rev 214:219–228
18. Nauta AJ, Fibbe WE (2007) Immunomodulatory properties of mesenchymal stromal cells. Blood 110:3499–3506
19. Nauta AJ, Kruisselbrink AB, Lurvink E et al (2006) Mesenchymal stem cells inhibit generation and function of both CD34+-derived and monocyte-derived dendritic cells. J Immunol 177:2080–2087
20. Pittenger MF, Mackay AM, Beck SC et al (1999) Multilineage potential of adult human mesenchymal stem cells. Science 284:143–147
21. Ramasamy R, Fazekasova H, Lam EW et al (2007) Mesenchymal stem cells inhibit dendritic cell differentiation and function by preventing entry into the cell cycle. Transplantation 83:71–76

22. Spaggiari GM, Abdelrazik H, Becchetti F et al (2009) Mesenchymal stem cells inhibit dendritic cell maturation and function by selectively interfering with the generation of immature DCs: central role of MSC-derived prostaglandin E2. Blood 113:6576–6583
23. Tolar J, Le Blanc K, Keating A et al (2010) Concise review: hitting the right spot with mesenchymal stromal cells. Stem Cells 28:1446–1455
24. Uccelli A, Moretta L, Pistoia V (2008) Mesenchymal stem cells in health and disease. Nat Rev Immunol 8:726–736
25. van den Berk LC, Roelofs H, Huijs T et al (2009) Cord blood mesenchymal stem cells propel human dendritic cells to an intermediate maturation state and boost interleukin-12 production by mature dendritic cells. Immunology 128:564–572
26. Zhao Z-G, Xu W, Sun L et al (2012) Immunomodulatory function of regulatory dendritic cells induced by mesenchymal stem cells. Immunol Invest 41:183–198

Adv Biochem Eng Biotechnol (2013) 130: 209–266
DOI: 10.1007/10_2012_150

Published Online: 19 September 2012

MSC and Tumors: Homing, Differentiation, and Secretion Influence the Therapeutic Potential

Naomi D'souza, Jorge Sans Burns, Giulia Grisendi, Olivia Candini, Elena Veronesi, Serena Piccinno, Edwin M. Horwitz, Paolo Paolucci, Pierfranco Conte and Massimo Dominici

Abstract Mesenchymal stromal/stem cells (MSC) are adult multipotent progenitors with fibroblast-like morphology able to differentiate into adipocytic, osteogenic, chondrogenic, and myogenic lineages. Due to these properties, MSC have been studied and introduced as therapeutics in regenerative medicine. Preliminary studies have also shown a possible involvement of MSC as precursors of cellular elements within tumor microenvironments, in particular tumor-associated fibroblasts (TAF). Among a number of different possible origins, TAF may originate from a pool of circulating progenitors from bone marrow or adipose tissue-derived MSC. There is growing evidence to corroborate that cells immunophenotypically defined as MSC are able to reside as TAF influencing the tumor microenvironment in a potentially bi-phasic and obscure manner: either promoting or inhibiting growth depending on tumor context and MSC sources. Here we focus on relationships between the tumor microenvironment, cancer cells, and MSC, analyzing their diverse ability to influence neoplastic development. Associated activities include MSC homing driven by the secretion of various mediators, differentiation

N. D'souza · J. S. Burns · G. Grisendi · O. Candini · E. Veronesi · S. Piccinno · PF. Conte · M. Dominici (✉)
Department of Oncology, Hematology and Respiratory Diseases,
University-Hospital of Modena and Reggio Emilia,
Via del Pozzo, 71, 41100 Modena, Italy
e-mail: massimo.dominici@unimore.it

S. Piccinno · P. Paolucci
Department of Mother and Child, Modena,
University-Hospital of Modena and Reggio Emilia,
Via del Pozzo, 71, 41100 Modena, Italy

E. M. Horwitz
Division of Oncology, The Children's Hospital of Philadelphia,
Philadelphia PA, USA

E. M. Horwitz
Department of Pediatrics, Perelman School of Medicine,
The University of Pennsylvania, Philadelphia PA, USA

towards TAF phenotypes, and reciprocal interactions with the tumor cells. These are reviewed here with the aim of understanding the biological functions of MSC that can be exploited for innovative cancer therapy.

Keywords MSC · Microenvironment · Tumor stroma · TAF · TRAIL

Abbreviations

5-FC	5-Fluorocytosine
AAV	Adeno associated viruses
APC	Antigen presenting cells
Ang-1	Angiopoietin 1
bFGF	Basic fibroblast growth factor
BM	Bone marrow
BrdUrd	Bromodeoxyuridine
c-fibronectin	Cellular fibronectin
CSC	Cancer stem cells
CTL	Cytotoxic T Lymphocytes
DC	Dendritic cells
DKK-1	Dickkopf-related protein-1
DL	Death ligand
ECM	Extracellular matrix
EGF	Epidermal growth factor
EMT	Epithelial to mesenchymal transition
EndMT	Endothelial to mesenchymal transition
EPC	Endothelial progenitor cells
ESC	Embryonic stem cell
ET-1	Endothelin-1
ETBR	Endothelin binding receptor
FAK	Focal adhesion kinase
FAP	Fibroblast activation protein
FPRL-1	Formyl peptide receptor like-1
FSP	Fibroblast specific protein
G-CSF	Granulocyte-colony-stimulating factor
GM-CSF	Granulocyte-macrophage-colony-stimulating factor
hAD-MSC	Human adipose-derived mesenchymal stromal/stem cells
hBM-MSC	Human bone marrow-derived mesenchymal stromal/stem cells
HCC	Hepatocarcinoma cancer
HGF	Hepatocyte growth factor
HLA	Human leukocyte antigen
HLA-DR	Human leukocyte antigen and its ligand
hMSC	Human mesenchymal stromal/stem cells
HO-1	Heme oxygenase-1
HSC	Hematopoietic stem cells
IDO	Indolamin 2, 3-dioxygenase

IFN-γ	Interferon-gamma
IGF	Insulin growth factor
IL	Interleukin
LIF	Leukemia inhibitory factor
MCP-1/CCL5	Monocyte chemotactic protein-1
M-CSF	Macrophage-colony-stimulating factor
MDSCs	Myeloid derived suppressor cells
miRNA	microRNA
MMP	Matrix metalloproteinases
MSC	Mesenchymal stromal/stem cells
NF-κB	Nuclear factor-kappa B
NHL	non-Hodgkin lymphoma
NK cells	Natural killer cells
NO	Nitric oxide
NSCLC	Non-small cell lung carcinoma
OI	Osteogenesis imperfecta
PAI-1	Plasminogen activator inhibitor type-1
PAI-2	Plasminogen activator inhibitor type-2
PDGF	Platelet derived growth factor
p-fibronectin	Plasma fibronectin
PGE2	Prostaglandin E2
PIFA	Platinum-induced polyunsaturated fatty acids
PlGF	Placental growth factor
ROS	Reactive oxygen species
SCID	Severe combined immunodeficieny
SDF-1/CXCL12	Stromal derived factor–1
sHLA-G5	Soluble human leukocyte antigen G5
SL-1	Stromelysin-1
STC1	Stanniocalcin-1
S-TRAIL	Secretable form of TRAIL
TAF/CAF	Tumor-associated fibroblasts/carcinoma associated fibroblasts
TAM	Tumor-associated macrophages
TGF-β	Transforming growth factor-beta
Th1 cells	T Helper 1 cells
Th2 cells	T Helper 2 cells
TIMPs	Tissue inhibitor of metalloproteinases
TLS	Tertiary lymphoid structures
Tn-C	Tenascin-C
TNF	Tumor necrosis factor
TNF-α	Tumor necrosis factor-alpha
tPA	Tissue plasminogen activator
TRAIL	Tumor necrosis factor apoptosis inducing ligand
Tregs	T regulatory cells
Tsp-1	Thrombospondin-1

uPA	Urokinase plasminogen activator
uPAR	Urokinase plasminogen activator receptor
VEGF	Vascular endothelial growth factor
WAT	White adipose tissue
α-SMA	Alpha-smooth muscle actin

Contents

1 What's Inside A Tumor?

Cancer cell development and survival is a multifactorial process, involving genetic mutation of normal cells as well as physiological changes within both cancer cells and the body's defense mechanisms [111]. It is a disease that is not only dependent on qualities intrinsic to the tumor cells themselves, but also on extrinsic factors, such as the immune and endocrine systems, stroma, vasculature, and metabolism, all of which play key roles in the development, proliferation, and evolution of cancer. As with normal tissues, tumors are composed of two discrete but interactive compartments, namely, parenchyma and stroma [221], wherein tumor cells themselves are the parenchyma, and the stroma is composed of a mixture of non malignant cells, nonhematopoietic cells, and connective tissue elements such as blood and lymphatic vessels, fibroblasts, and inflammatory cells. The stroma also includes a specialized extracellular matrix (ECM), a basement membrane composed of a variety of extracellular macromolecules, including collagens, fibronectin, fibrin, various proteoglycans, and hyaluronan whose interactions regulate many cellular processes, including cell survival, proliferation, differentiation, cell migration, tissue specificity, epithelial polarity, functionality, and structural support [147]. The parenchyma and the stroma are not independent; rather they continuously cross-talk and interact with each other, so that intrinsic factors are capable of recruiting the extrinsic ones, and the availability of extrinsic factors determines the intrinsic cellular activity.

Under normal circumstances, the equilibrium between cell renewal and cell reduction is tightly governed through connections between parenchyma progenitor cells and the microenvironment to carry out the tissue remodeling or respond to stress caused by tissue injury. In contrast, cancer cells show abnormal responses to normal physiological regulators of cell growth, and they constantly send remodeling signals for the stroma to be reorganized in an activated form to permit tumor growth. A useful concept is that to some degree a solid tumor behaves like a wound that does not heal [76].

2 Tumor Microenvironment, Its Main Components and Their Role in Tumor Progression

During wound healing, inflammation protects defined damaged tissue by recruiting cells, cytokines, and chemokines that work in a self-limiting approach to heal the wound while also isolating the area until normal tissue function is restored. Immune cells can also potentially inhibit tumor growth and progression by recognition and rejection of malignant cells, a process referred to as immunosurveillance [34]. However, immunity fails to subside and this unresolved inflammation can result in tumor cell growth, survival, and angiogenesis. The tumor microenvironment (see Fig. 1) largely consists of (1) cytokines and growth factors (2) proteolytic enzymes such as matrix metalloproteinases (MMPs) and tissue inhibitor of metalloproteinases (TIMPs) (3) extracellular matrix (ECM) proteins (4) immune cells including regulatory and cytotoxic T cells, myeloid-derived suppressor cells (MDSCs) (5) endothelial cells, and (6) fibroblasts. Tumors also produce large amounts of mediators, including cytokines such as TNF-α and TGF-β, as well as cytotoxic molecules, proteases, MMPs, interleukins, and interferons [54]. Therefore, for a better understanding of the role of the stroma, it is important to dissect the critical components regulating tumorigenic processes that affect tumor growth, angiogenesis, desmoplasia, lymphanogenesis, inflammation, and immune escape.

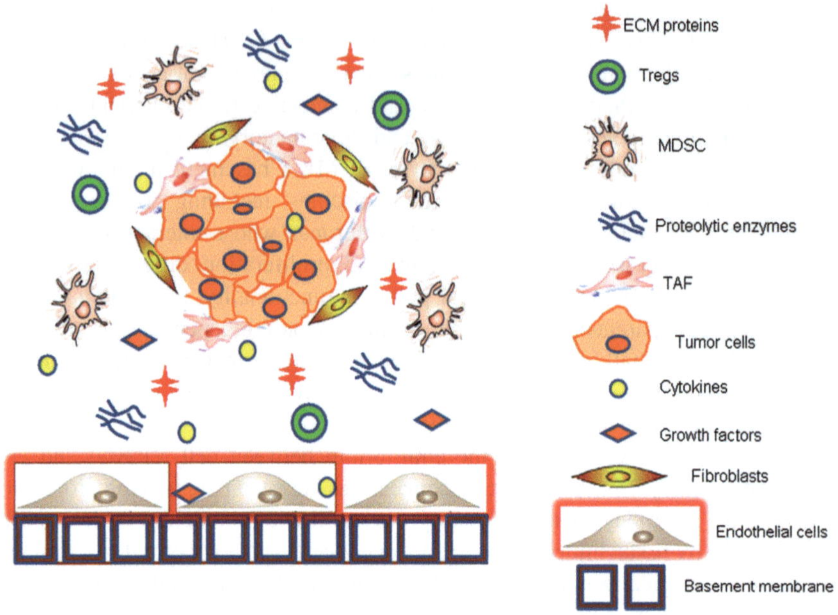

Fig. 1 *Tumor complexity.* The tumor microenvironment contains several mediators such as cytokines, growth factors, ECM proteins, proteolytic enzymes, fibroblasts, MDSCs, Tregs, endothelial cells, and tumor cells, all of which regulate tumorigenesis by affecting tumor growth, angiogenesis, inflammation, and escape

2.1 Cytokines and Growth Factors of the Tumor Microenvironment

Cytokines and growth factors are extensively involved in intercellular communication particularly in the cross-talk between tumor cells and stroma. Rather than provide a totally comprehensive description of this complex scenario, we focus here on cytokines and growth factors known to have a role in tumor biology that are also likely to be influenced or expressed by mesenchymal progenitor cells.

2.1.1 Hepatocyte Growth Factor

Hepatocyte growth factor (HGF) belongs to the plasminogen family and is tethered to ECM in a precursor form. It binds to the high-affinity receptor c-met, and its overexpression or constant oncogenic c-met signaling can influence proliferation, invasion, and metastasis [30]. Produced by mammary stromal cells, HGF may have a profound effect on developing mammary tumors [206]. Moreover, HGF provides a co-stimulatory signal to the Wnt pathway upregulated in colon carcinogenesis [331]. Recently, Li et al. demonstrated that malignant pleural mesothelioma produced fibroblast growth factor-2 (FGF-2) and platelet-derived growth factor-AA (PDGF-AA), and that these growth factors stimulated tumor-associated fibroblasts (TAF) to produce HGF, thus promoting tumor progression through a malignant cytokine network [186].

2.1.2 Interferon-γ

Interferon-γ (IFN-γ) seems to be one of most significant cytokines preventing and suppressing the development of cancers. Initially, it was believed that CD4+ T helper cell type 1 (Th1) lymphocytes, CD8+ cytotoxic lymphocytes, and NK cells exclusively produced IFN-γ. However, there is now evidence that other cells, such as B cells, NK-T cells, and antigen-presenting cells (APCs) also secrete IFN-γ [291]. IFN-γ is important for its immunomodulatory effects and is controlled by IL-12 and IL-18, whereas negative regulators include IL-4, IL-10, and transforming growth factor-beta (TGF-β) which are largely produced by tumor cells. Notably, IFN-γ and its receptors are found in normal tissues as well as in tumors, yet tumor cells within the tumor microenvironment produce very low levels of INF-γ during early tumor stages and as a result may escape its inhibitory effects. Beyond decreased expression of IFN-γ, an alternative mechanism to avoid INF-γ mediated growth inhibition includes an alteration of either of its receptors as observed during breast cancer promotion [100]. Although INF-γ may ordinarily serve to restrict cancer growth, selection for cells tolerating low levels within the tumor stroma may ultimately mean it serves to enhance tumor cell proliferation specifically. Hence, the outcome between the anti- and pro-tumorigenic functions of IFN-γ are likely to be context dependent, reflecting tumor specificity, microenvironmental factors, and signal intensity [348].

2.1.3 Interleukin-1β

Whereas IL-1α is mainly active in cell-associated forms, its close family member IL-1β is secreted by activated macrophages and is considered the more significant pleiotropic cytokine involved in inflammatory and immune responses that in turn influence cell proliferation, differentiation, and apoptosis [13]. Cancer cells directly produce IL-1 or can induce cells within the tumor microenvironment to do so with constitutive IL-1β protein production documented in human and animal cancer cell lines derived from sarcomas as well as ovarian and urothelial cell carcinomas [66]. Solid tumors in which IL-1β has been shown to be upregulated include breast, colon, lung, head and neck cancers, melanomas, and patients with IL-1β-producing tumors have a generally worse prognosis [184]. Polymorphisms in IL-1β have been linked to gastric cancer [190]. IL-1β may exhibit autocrine behavior by stimulating the tumor cell itself to invade and proliferate, or exert paracrine effects on stromal cells in the tumor microenvironment [363]. IL-1β could induce expression of metastatic genes such as matrix metalloproteinases (MMPs) and stimulate nearby cells to produce angiogenic proteins and growth factors such as vascular endothelial growth factor (VEGF), IL-8, IL-6, tumor necrosis factor alpha (TNF-α), and TGF-β [184] thus promoting tumor growth. Recent studies have indicated an important role for IL-1 in tumor angiogenesis and its modulation of tumor immunity can influence metastatic potential [41].

2.1.4 Interleukin-6

Interleukin-6 (IL-6) acts as both a pro-inflammatory and an anti-inflammatory cytokine and is secreted by T cells and macrophages to stimulate immune responses. It serves as a pro-angiogenic factor that encourages cancer cell proliferation while also inhibiting apoptosis [111]. It has also been implicated in chronic inflammation, and tumor growth and development [287]. In a variety of pre-clinical models, IL-6 has also been shown to promote tumorigenicity, angiogenesis, and metastasis [170]. In humans, IL-6 production seems to be upregulated in the tumor microenvironment and particularly by fibroblasts during cancer progression in colorectal cancer [38]. Antibody therapy targeting the IL-6 pathway is currently being explored [117].

2.1.5 Interleukin-8

Interleukin-8 (IL-8) is a chemokine produced by both macrophages and epithelial cells originally classified as a neutrophil chemoattractant and now reported to play an important role in tumor progression and metastasis in a variety of human cancers, including lung cancers. IL-8 activity in tumors and microenvironment may contribute to cancer progression through its potential function in the regulation of angiogenesis, cancer cell growth and survival, tumor cell migration, leukocyte infiltration, and modification of immune responses [187]. Within the tumor microenvironment, the cancer cells themselves secrete IL-8 in an autocrine or paracrine manner, an observation common to breast cancer, gastric cancer, colon cancer, cervical cancer, and pancreatic cancer [46] making its measurement in peritumoral fluids useful for diagnosis and prognosis [171].

2.1.6 Interleukin-10

Interleukin-10 (IL-10) is an anti-inflammatory cytokine produced by T and B cells and monocytes. IL-10 has been reported to encourage cell survival directly and suppress effector T cells, depending upon experimental conditions and the presence of regulatory T cells (Tregs) within the microenvironment [285]. It is essential for proper downregulation of inflammation and prevention of carcinogenesis in a mouse model of microbial-induced colitis [349]. In contrast, others have proposed that IL-10 induces effective anti-tumor immune surveillance [220].

IL-10 may also hinder angiogenesis within the tumor microenvironment and thus IL-10 may be acting as an anti-tumor agent that additionally reduces inflammation.

2.1.7 Platelet-Derived Growth Factor

Platelet-derived growth factor (PDGF) can be secreted not only from the alpha granules of platelets but also from a number of different cell types of mesenchymal origin, for example, fibroblasts and smooth muscle cells, and by some cells of neuroectodermal origin, such as the oligodendrocytes [91]. It participates in biological events ranging from embryogenesis to growth and from development to wound healing, atherosclerosis, and cancer. The role of PDGF in malignancies involves both autocrine and paracrine stimulation of cells within the tumor. Increasing attention has been focused on the paracrine effects of PDGF on cells of the stroma and vessel compartment of tumors. Many common tumors, such as carcinomas of the lung, breast, and colon, express PDGF ligands and its receptors whereas the presence of activated PDGF receptors in tumor homogenates derived from receptor-negative tumor cell lines [253] implicated that PDGF mediates much of its influence on angiogenesis, metastasis, and other processes via the tumor stroma [321].

2.1.8 Stromal Derived Factor-1

This cytokine is generally associated with hematopoietic stem cell homing into bone marrow stroma [295]. SDF-1, a growth factor for B cell progenitors and a chemotactic factor for T cells and monocytes, is essential for B cell lymphopoiesis and bone marrow myelopoiesis and mice lacking expression of either CXCR4 or SDF-1 die perinatally due to virtual absence of bone marrow hematopoiesis [353]. Regarding the tumor microenvironment, SDF-1 was produced by breast carcinoma-associated fibroblasts promoting proliferation of tumor cells expressing its receptor [237]. Moreover, the level of SDF-1 expression in serum has been associated with poor survival in breast cancer patients [35] and the SDF-1α/CXCR4 signaling axis was also deregulated in several malignancies, promoting cancer cell migration and metastasis, as observed for cancers of ovarian, prostate, breast, pancreatic, lung, and colorectal and hepatic tissues as well as multiple myeloma [236], suggesting a broadly relevant role for SDF-1 in cancer–stroma interactions in primary tumors and metastatic sites [281].

2.1.9 Transforming Growth Factor-β

The three isoforms of Transforming growth factor-β (TGF-β) in humans, TGF-β1, TGF-β2, and TGF-β3 are pivotal immunoregulatory cytokines produced within tumor microenvironments. They inhibit T cell growth and cytolytic T lymphocyte

(CTL) induction, differentiation, and cytokine production, as well as antigen-presenting cell functions [183]. TGF-β inhibits macrophage activation and decreases their production of inflammatory cytokines such as IL-6, TNF-α, and IL-1β, which may impede inflammation-associated cancer development. It plays an indispensable, yet complex, role in carcinogenesis and progression [201]. Tumor cells release TGF-β which also maintains infiltrating NK cells in an anergic state; as a result NK cells are unable to secrete IFN-γ to kill tumor cells, even after activation with IL-2. However, a dual role of TGF-β in tumorigenesis has also been revealed in studies with transgenic expression of TGF-β. Cui et al. used a model of murine skin multistage carcinogenesis with transgenic expression of TGF-β. Early during tumor formation malignant cells were responsive to TGF-β inhibiting tumor formation, however, at later stages when malignant cells lost their responsiveness to TGF-β, it apparently stimulated tumor growth [56]. This transition correlated to resistance of TGF-β by malignant cells with an increase of tumor aggressiveness and metastatic features.

2.1.10 Tumor Necrosis Factor

Tumor necrosis factor (TNF) family of cytokines mainly produced by activated macrophages have the primary role of regulating immune cells. Although TNF-α is the isoform predominantly investigated, polymorphisms enhancing TNF-β expression have been implicated in osteosarcoma progression [233]. However, several receptors of the TNF family are also "death receptors" signaling apoptosis in a variety of cells [193]. Induction of TNF in response to stimuli induces a cascade of other inflammatory cytokines, chemokines, growth factors, and endothelial adhesins that recruit and activate a range of cells at the site of infection or tissue damage. TNF plays an essential role in several models of cancer, and is a critical inflammatory mediator [150]. Although initially identified as an anticancer agent, TNF treatment in experimental ovarian cancer models promoted peritoneal adhesion and solid tumor formation, and TNF has now been shown to be involved in cellular transformation, tumor promotion, and induction of metastasis [297]. Also Mueller et al. demonstrated that primary carcinoma-associated fibroblasts (CAF) from colorectal liver metastasis express several inflammatory, tumor-enhancing factors and monocyte-chemoattractant protein (MCP)-1 and both these molecules were intensely induced by TNF-α [219]. Tumor cell-derived TNF-α may promote angiogenesis indirectly through stimulation of endothelial progenitor monocytes recruited to the tumor microenvironment [185]. Therefore, despite a role for TNF in stimulating immune responses it may also regulate communication between tumor and stromal cells to influence tumor progression positively.

2.1.11 Vascular Endothelial Growth Factor

Vascular endothelial growth factor-A (VEGF-A) was originally identified from bovine pituitary follicular cells, however, its secretion is now noted in fibroblasts, smooth muscle cells, hypertrophic chondrocytes, and osteoblasts [249] and it remains a principal angiogenic member of the VEGF family synthesized by human cells, including placental growth factor (PLGF), VEGF-B, VEGF-C, and VEGF-D. VEGF-A stimulates vasculogenesis and angiogenesis as well as the recruitment of both leukocytes and vascular endothelial cells. Solid tumors cannot grow beyond a limited size without an adequate blood supply, thus cancer cell secretion of angiogenic factors is considered fundamental for extensive growth and metastasis. There is considerable evidence demonstrating that VEGF-A is the principal inducer of tumor blood vessels and its levels are upregulated in a large number of tumor types [247]. The tumor microenvironment is an important site for VEGF release with stromal elements recruited as additional sources of VEGF [322]. Inhibition of VEGF activity resulted in growth suppression in a wide variety of tumor types in preclinical contexts and several therapeutic agents have been designed to inhibit VEGF-induced angiogenesis including monoclonal antibodies that block VEGF (i.e., Bevacizumab) and small molecule inhibitors of the VEGFR-2 tyrosine kinase (e.g., Sorafinib, Sunitinib). The addition of Bevacizumab to standard chemotherapy prolonged progression-free survival and overall survival in patients with advanced non small cell lung cancer (NSCLC) and colon cancer [282]. However, overall, early expectations have not been met, with several studies highlighting that inhibitors of VEGF signaling may ultimately invoke selection for an invasive metastatic phenotype, in part by creating an increasingly hypoxic microenvironment [264].

2.2 Proteolytic Enzymes

The ability of tumor cells to digest the extracellular matrix (ECM) by secreting proteolytic enzymes correlates well with their tissue invasiveness [28]. The invasive phenotype is accomplished through a combination of cytoskeletal changes to form F-actin rich protrusions called invadopodia and localized matrix degrading proteolytic activity at focal adhesion points recruiting proteases of multiple catalytic types: aspartic, cysteine, serine, threonine, and metallo, as well as lysosomal proteases and cathepsins, which penetrate connective tissue barriers, induce vascular and stromal activation [31], and ultimately remodel the surrounding normal tissue [299, 334].

2.2.1 Matrix Metalloproteinases

Matrix metalloproteinases (MMPs) are the major class of endopeptidases, characterized by utilization of a metal ion to polarize a water molecule for hydrolytic reactions, that contributes to the complete set of protease genes within the tumor degradome [324]. MMPs are not constitutively expressed, but are regulated by complex factors including cytokines, cell–cell and cell–matrix interactions, MMP inhibitors [e.g., tissue inhibitors of metalloproteinases (TIMPs)], and by the processing of their pro-enzyme form. MMPs involved in tissue remodeling in most cancers degrade ECM components and promote growth and invasion of cancer cells through the interaction of ECM molecules with integrins [347]. Under normal disease-free conditions, MMPs tend to be expressed at low levels controlled by TIMPs but during disease pathogenesis their relative levels rise, causing MMP activation. Tumor stroma assisted progression and invasion of colorectal cancer in a multistep process involving multiple interactions between tumor cells and the surrounding stroma mediated by many proteins including MMPs and TIMPs [9]. It was first thought that MMPs were expressed by the tumor cells, but subsequent work has shown that many of the MMPs present in the tumor microenvironment are actually expressed by the stromal cells of the host. For example, in breast cancer several MMPs have all been localized to the stromal compartment [337]. Intercellular communication within the tumor microenvironment is critical for regulating the expression of MMPs that can contribute to multiple stages of cancer progression [154]. The broad spectrum strategy of blocking MMP activity with synthetic MMP inhibitors, such as hydroxymates had disappointing outcomes in clinical trials. Greater appreciation of the complexity involved emerged with observations that some MMP family members behave as tumor-suppressor enzymes [62].

2.2.2 Urokinase Plasminogen Activator

Urokinase plasminogen activator (uPA) is another critical protease involved in tumor proliferation, invasive migration, and metastasis [57]. It has two specific inhibitors: plasminogen activator inhibitor type-1 (PAI-1) and type-2 (PAI-2). Activation of plasminogen to serine protease plasmin is mediated by uPA and tissue plasminogen activator [29]. The plasminogen activators are linked to degradation and remodeling of normal and cancer tissue and the surrounding ECM. Overexpression of uPA and its receptor uPAR are observed in a variety of cancers and elevated uPA and uPAR levels in tumor tissue and blood are associated with poor prognosis. uPA and uPAR are expressed both by cancer and stromal cells, depending on cancer type. In breast and colon cancer, uPA is expressed primarily by supporting stromal cells, in prostate cancer by macrophages, and in skin squamous cell carcinoma it is expressed autonomously by the cancer cells; the latter also express uPAR. Expression of uPAR was also found in macrophages of breast, colon, and prostate cancer, therefore several approaches have been developed to target the uPA/PAI-1 system in cancer aiming to reduce tumor invasion and metastases [59, 289].

2.3 *Extracellular Matrix and Its Proteins*

Extracellular matrix provides a dynamic 3D structure composed of specialized fibrous proteins and proteoglycans whose interactions regulate many cellular functions and loss of ECM-mediated control in the tumor microenvironment is a characteristic feature of malignant progression [195]. Altered ECM composition and abnormal topography deregulate stromal cell homeostasis, generating an angiogenic and inflammatory tumor microenvironment.

2.3.1 Collagens

Collagens, the most abundant proteins in ECM, provide a structural support for cells and contribute to tumor cell migration and proliferation. Increased deposition of collagen can influence ECM rigidity, altering cell morphogenesis and polarity with changes in tumor cell microRNA (miRNA) expression [227]. Let-7 g, a known tumor suppressor miRNA, downregulated COL1A2 and inhibited hepatocarcinoma cancer (HCC) cell migration and growth [143]. Cartilage, a naturally avascular tissue, provides an ECM composition including fragments of type IV, type XV, and type XVIII collagen that have been studied for their potential to reduce angiogenesis and metastasis [250]. The production of type I collagen by tumors depends on the stage of tumor progression and increased collagen synthesis by pancreatic stellate cells accompanied their malignant phenotype [14]. In contrast, type I collagen was able to inhibit growth and malignant transformation in human glioma cells [127]. In neuroblastoma, type I collagen biosynthesis was a helpful marker for studying specific patterns of trans-differentiation associated with the loss of malignant potential [61]. Thus tumor cell differentiation state and the fact that proteolytic collagen fragments known as matricryptins may serve as bioactive regulators of angiogenesis and tumor growth [271] complicate straightforward relationships between collagen expression and outcome.

2.3.2 Fibronectins

Fibronectins are glycoproteins with a fundamental role in blood vessel morphogenesis during embryonic development and pathological angiogenesis. Their expression is barely detectable in the normal adult vasculature [328], yet abundant during pathological angiogenesis in various diseases such as cancer. Fibronectin is commonly classified into two forms, plasma fibronectin (p-fibronectin), a soluble form produced by hepatocytes that circulates in blood at high concentrations, and cellular fibronectin (c-fibronectin), produced in tissues where it is incorporated in a fibrillar matrix. In addition to promoting adhesion and signaling through cell surface receptors, the fibronectin matrix functions as a fibrillar scaffold for the assembly of other matrix proteins. Often expressed by endothelial cells and TAF, fibronectin provides a platform for angiogenic signaling by increasing the bioavailability of soluble angiogenic factors and co-operating with their transmembrane receptors. It connects cells with collagen fibers in the ECM via cell surface integrins causing a reorganization of the cell's cytoskeleton that facilitates cell movement. Fibronectin expression was found to be increased in non-small cell lung carcinoma (NSCLC; [119]), an observation supported by earlier studies showing that the adhesion of lung carcinoma cells to fibronectin enhances tumorigenicity and confers resistance to apoptosis induced by standard chemotherapeutic agents [274]. Nevertheless, context remains very important for determing the influence of ECM proteins. Loss of fibronectin from the cell surface was closely associated with malignant transformation of cells and the overexpression of fibronectin in human fibrosarcoma cells suppressed their motility and growth potential [50].

2.3.3 Integrins

Integrins are surface receptor proteins that mediate cell–matrix and cell–cell adhesion. Multiple heterodimer combinations allow specific cross-talk with oncogenes and growth factor receptors on both tumor and tumor-associated cells. ECM interactions provide the necessary traction for cell motility and invasion, assisting in matrix remodeling by directing protease localization [63]. Integrins are known to play a critical role in the movement of virtually all motile cell types, including T cells, macrophages, fibroblasts, and epithelial cells during wound healing and development. The involvement of integrin receptors in the migration of both normal and transformed cells was implicated by their localization to invadopodia [272], membrane protrusions, and also enriched MMPs, tyrosine kinase signaling machinery, actin, and actin-associated proteins [17]. Integrins can activate MMP synthesis at the transcriptional level, compartmentalize the proteases to the cell surface membrane, and promote the activation of pro-MMPs. For example, the $\alpha V\beta 3$ integrin can promote tumor invasion and metastasis through recruitment and activation of MMP-2 [115] whereas disruption of MMP-2-$\alpha V\beta 3$ binding inhibited angiogenesis and tumor growth [300]. TIMPs detected in association with $\alpha V\beta 3$ expression in tumors can compete with MMP-2 to bind integrin $\alpha V\beta 3$, serving as a natural inhibitor of MMP-2 activity to modulate angiogenesis [33].

Regarding the uPA protease, integrins can enhance expression of the uPA and its receptor uPAR and govern the spatial localization of uPA/uPAR to the leading edge of migrating cells. As co-receptors of uPAR signaling, integrins co-operate with uPAR to transduce multiple signals that contribute to tumor-related events. Several integrins including $\alpha V\beta 3$, $\alpha V\beta 5$, and $\alpha V\beta 1$ are involved in angiogenesis and antagonists of these integrins can block tumor-induced angiogenesis in multiple animal models with clinical benefit against solid tumors [347]. $\alpha V\beta 3$ integrin is strongly correlated with tumor growth and metastasis in breast, prostate, pancreatic, glioblastoma, cervical, and ovarian cancers [84]. Integrins are effectors of signaling cascades including MAP kinase, Jun, NFκB, and β-catenin as well as direct downstream targets of Src-family kinases, focal adhesion kinase (FAK), and protein kinase B [344]. Interactions with these signaling cascades allow integrins to modulate cell survival, proliferation, cell migration, and invasion.

2.3.4 Laminins

Laminins are cell adhesion proteins in the ECM that form web-like structures to resist tensile forces in the basal lamina. They play an important role in the architecture of the basement membrane and interact with cell surface receptors to regulate additional functions such as development and differentiation [52]. Laminin-5 is expressed in hepatocellular carcinoma (HCC) stromal cells and its expression was associated with the metastatic phenotype of HCC [104]. Laminin-8 ($\alpha 4\beta 1\gamma 1$) supports cell migration and may be associated with tumor invasion [125]. Knockout of laminin-8 $\alpha 4$ chain is characterized by abnormal blood vessel maturation [317]. Fujita et al. have identified three laminins, namely laminin-2 ($\alpha 2\beta 1\gamma 1$), laminin-8, and laminin-10, as new breast carcinoma angiogenic markers whose expression was increased in the walls of ductal breast carcinoma blood vessels [94]. The laminin-binding integrin $\alpha 3\beta 1$ is a candidate anti-cancer target for breast cancers [311].

2.3.5 Tenascin-C

Among several ECM components, Tenascin-C (Tn-C) was elevated in the stromal microenvironment of epithelial cancers and was shown to decrease the formation of cell adhesion complexes thereby promoting proliferation and migration [133]. Its expression is elevated in embryonic tissues and high Tn-C expression has been found to correlate with lymph node metastasis and poor prognosis in several cancers [207]. Tn-C also binds multiple additional proteins including integrins, a variety of proteoglycans, adhesion molecules, and fibronectin and collagens thus affecting tissue architecture, tissue resilience, and cellular responses relevant in angiogenesis, metastasis, and the stem cell niche [328]. Recently, Oskarsson et al. demonstrated that Tn-C supported initiating breast cancer cells during the establishment of lung metastases, implicating Tn-C in the survival and outgrowth of disseminated cancer cells [240]. Tn-C is abundantly produced by carcinoma cells and incorporated into the vascular basement membrane [25] presenting a therapeutic target that may eliminate disease with minimal adverse effects [241].

2.4 Metabolic Milieu: Warburg Effect

Under normal conditions, cells can limit proliferation by inhibiting the uptake of nutrients from their environment unless stimulated to do so by growth factors. In contrast, cancer cells overcome this growth factor dependence by acquiring genetic mutations that functionally alter receptor-initiated signaling pathways. There is growing evidence that some of these pathways constitutively activate the uptake and metabolism of nutrients (particularly glucose) to meet or exceed the bioenergetic demands of cell growth and proliferation [330]. The best-characterized metabolic phenotype observed in tumor cells is the "Warburg effect" [169]. In the 1920s, Otto Warburg observed and suggested that unlike normal tissue cells that relied primarily on mitochondrial oxidative phosphorylation to generate the energy needed for cellular processes, cancer cells convert glucose into lactate by a process of aerobic glycolysis even in the presence of sufficient oxygen to support mitochondrial oxidative phosphorylation. This metabolic shift demands that proliferating cells implement an abnormally high rate of glucose uptake to meet their increased energy, biosynthesis, and redox needs. To support proliferation, cancer cells must strike a balance between energy production and macromolecule biosynthesis. One well-defined mechanism by which cancer cells establish the Warburg effect is via transcriptional upregulation of glycolytic enzymes. Elevated expression of glucose transporters and glycolytic enzymes is found in numerous cancers and may contribute to tumor progression [21].

Although the Warburg effect metabolic phenotype was initially identified in cancer tissue, it is now well appreciated that rapidly dividing normal tissues, such as embryonic stem cells (ESC) and lymphocytes, employ aerobic glycolysis to meet their energetic and biosynthetic requirements during expansion. These observations support the notion that aerobic glycolysis is a preferred metabolic program for rapid cellular expansion. Work on epithelial cancer microenvironments revealed that tumor associated fibroblasts (TAF) acquire a unique metabolic program incorporating many elements of the cancer metabolic phenotype, including the Warburg effect [266]. In this model, cancer cells induce a metabolic or oxidative stress on neighboring TAF resulting in their co-acquisition of a glycolytic phenotype. Subsequently, TAF provide nutrients (lactate or pyruvate) to tumor cells that drive anabolism for cellular growth. Interesting work from Coller and colleagues demonstrated that "quiescent" fibroblasts maintain a heightened anabolic program despite undergoing replicative arrest [182]. MSC can secrete stanniocalcin-1 (STC1) that serves to reduce intracellular reactive oxygen species (ROS) in adjacent cells, thereby protecting tumor cells from ROS-induced apoptosis [231]. Thus a combination of catabolic programs in normal cells and release of redox pressure survival factors within the tumor parenchyma can play a pivotal role in supporting the anabolic program of cancer.

2.5 *Cells of the Tumor Microenvironment*

2.5.1 Lymphocytes

Tumors develop in a complex and dynamic microenvironment and within this tumor microenvironment all immune cell types may be found. Histopathological analyses of human tumors have provided evidence that variable numbers of infiltrating immune cells are found in different tumors of the same type, and are found in different locations within and around a tumor. Lymphocytes are not randomly distributed but are located in specific areas, whereas NK cells are found in the stroma but are not in contact with tumor cells. B cells are mostly found in the invasive margin of growing tumors and in tertiary lymphoid structures (TLS) that are adjacent to tumor beds [65]. T cells, particularly CD8+ T cells, may be located in the invasive margin but can also be in the tumor core which is similar to secondary follicles in lymph nodes that contain naive T cells and memory T cells, B cells, and mature dendritic cells (DC). TLS may be sites in which tumor-controlling primary and/or secondary immune responses are generated. It has also been reported that T cells are found only in the invasive margin in liver metastases of colon cancer [118].

However, not all T cells inside the tumor microenvironment are anti-tumor effectors. For instance, a subpopulation of CD4+ T cells expressing CD25 and the master transcription factor Foxp3, termed regulatory T cells (Tregs), play a role in promoting tumor growth and progression by inhibiting the immune response against cancer [349]. These cells suppress the activation of effector immune cells that are specific for self-antigens, limiting autoimmunity and inflammation under physiologically normal conditions. Studies of murine tumors indicate that Tregs inhibit the immune response to tumors, and depletion of these cells promotes rejection of several murine tumor cell lines including melanoma, fibrosarcoma, and myeloma. In non-small cell lung cancer (NSCLC) and ovarian cancer, tumor-associated T cells contained increased proportions of CD25+ Tregs and it was further shown that these cells secrete the immunosuppressive cytokine TGF-β [338]. Somasundaram et al. demonstrated the role of TGF-β-mediated immunosuppression on CD25+ Tregs in colorectal cancer [303]. Patients with metastatic renal cell cancer were treated with chemotherapy drugs like Sunitinib or Bevacizumab and revealed a decrease in circulating Tregs after two or three treatment cycles with significantly longer overall survival than patients with no decrease in circulating Tregs [2].

2.5.2 Macrophages

Macrophages are a heterogeneous population of innate bone marrow-derived myeloid cells that can phagocytose small particles and apoptotic cells once recruited into tissues by a variety of inflammatory and immune stimuli [49]. Also functioning as antigen-presenting cells (APC), macrophages express class I and class II HLA-molecules and co-stimulatory/inhibitory molecules to instruct T cells and are characterized by the expression of transcriptional factors, cell surface markers, the production of cytokines, and their function in vitro [123]. Broadly divided into two main classes, M1 or classically activated macrophages are stimulated by bacterial products and Th1 cytokines [e.g. IFN-γ and lipopolysaccharides], whereas M2 or alternatively activated macrophages differentiate in microenvironments rich in Th2 cytokines (e.g., IL-4, IL-13, IL-10, and glucocorticoid hormones). M1 macrophages express high levels of IL-12 and low levels of IL-10, inflammatory and immunostimulating cytokines favoring a Th1 cell-mediated adaptive immune response. They also secrete reactive oxygen species (ROS) and nitrogen intermediates cytotoxic towards neoplastic cells hence facilitating anti-tumor immunity. Counterbalancing M2 macrophages produce high levels of IL-10 and low levels of IL-12 to suppress the Th1 cell response and may promote tumor progression. They also have high scavenging activity and produce several growth factors that activate the process of tissue repair [6, 244].

Macrophages may represent up to 50% of the tumor mass and can disrupt the balance of pro- and anti-angiogenic signaling by the secretion of various cytokines, growth factors, chemokines, and matrix-degrading enzymes, which are directly involved in the endothelial cell function and facilitate endothelial cell migration via extracellular matrix remodeling [67]. Depending on the mode of their activation, macrophages may promote tumor growth and suppress local immunity or attack tumor cells and sustain tumor immunity [123]. Tumor-associated macrophages (TAM) originate from the bone marrow as immature monocytes that circulate in the blood and extravasate into tumors, where they start to differentiate into TAM. Monocytes can differentiate to macrophages or TAM in response to macrophage colony-stimulating factor (M-CSF), produced by tumor cells in ovarian, breast, and endometrial cancers where M-CSF production is correlated with a poor prognosis [6]. TAM can promote tumor progression through multiple non-immune mechanisms including enhancement of angiogenesis, promoting tumor cell invasion and metastasis, and protect tumor cells from chemotherapy-induced apoptosis [244]. TAM are associated with a poor prognosis in breast, lung, and pancreatic cancers [49]. Targeting of macrophages in tumors is considered a promising therapeutic strategy; novel agents aimed at inhibiting the recruitment, activation, and pro-angiogenic functions of macrophages have great potential for the improvement of current anti-tumor therapeutics [67]. In addition to depletion of TAM, their "re-education" as anti-tumor effectors is under clinical investigation and, it is hoped, will contribute to the success of conventional anti-cancer treatments [6].

2.5.3 Myeloid-Derived Suppressor Cells

Myeloid-derived suppressor cells (MDSCs) in the tumor stroma represent a heterogeneous population of myeloid cells that are expanded and activated in response to growth factors and cytokines released by tumors. MDSCs themselves secrete VEGF and cytokines IL-6, TNF, and IL-1β, key regulators in promoting the progression of tumor development. Activated MDSCs accumulate in lymphoid organs and tumors where they exert T cell immunosuppression through nitric oxide, reactive oxygen species (ROS), and TGF-β secretion while also promoting Treg induction and favoring anti-inflammatory responses [349]. MDSCs and Tregs accumulate at the tumor site and maintain an immune tolerance to overcome tumor immunity [75]. Found within both human and murine tumors, MDSCs are far more abundant than Tregs and constitute the major components of the leukocyte infiltrate in a wide variety of solid tumors including breast, prostate, cervical, and ovarian cancers. Consistent with a role in promotion of cancer progression and metastasis, MDSCs density is correlated with poor prognosis [288]. MDSCs would normally differentiate after migration, but the cytokines and cellular factors found within the tumor microenvironment prevent their differentiation and instead encourage their expansion and activation of the immature myeloid cell population [96]. MDSCs are susceptible to chemotherapeutic agents such as Gemcitabine, which decreases the number of MDSCs and improves anti-tumor responses induced by immunotherapy in animal models of lung cancer and mammary tumors [262].

2.5.4 Endothelial Cells

Endothelial cells play crucial roles in the tumor microenvironment, directly interacting with tumor cells and generating new blood vessels. This process defined by Judah Folkman as neoangiogenesis, is required for tumor growth and metastasis [85]. Tumor-driven chaotic processes modulated by ECM proteins and growth factors such as VEGF often lead to highly abnormal tumor blood vessels and although targeting angiogenesis serves as a useful adjuvant therapy, the therapeutic effect has not met expectations [77].

Disappointing outcomes for VEGF therapy invite a broader appreciation of the complexity of endothelial cell regulation and behavior, including the concept of "vasculogenic mimicry," in which tumor cells transdifferentiate into vessel-forming endothelial-like cells [162]. The therapy met a critical reception [203] yet there is increasing evidence that at least in some tumor types, for example, Ewing sarcoma [327] or glioblastoma, cancer stem cells (CSC) may express a multipotentiality that allows them to differentiate into cells that contribute to the inner walls of tumor vasculature [333].

Given that tumor blood vessels often show abnormal leakiness, irregular flow, and structurally abnormal basement membranes, it is consistent that molecular and phenotypic differences between normal endothelial cells and tumor endothelial cells were found [307]. Only more recently has there begun to be an understanding of how these differences might be acquired and microvesicle transfer from tumor cells [152] and cytokine cross-talk between cells in the tumor microenvironment [90] are likely mechanisms in addition to influence from the tumor cell ECM [124, 134]. One of outcomes of VEGF inhibition is that it serves to complement conventional chemotherapy and there is increasing appreciation that this can come from anti-

VEGF agents leading to normalization of tumor vasculature, which in turn can improve the metabolic profile of the tumor microenvironment [105].

Lymphatic endothelial cells lining lymphatic vessels are major components of the tumor microenvironment and tumor-associated lymphangiogenesis is also continually regulated by a complicated cytokine network [235]. Lymphatic endothelial cells express an endothelin binding receptor (ETBR) making them proliferatively responsive to endothelin (ET)-1. In addition, VEGF-A, -C, and -D and hypoxia strongly influence lymphatic differentiation and agents blocking these factors can interfere with ET-1 mediated formation of lymphatic vessels that could otherwise ultimately serve as important conduits for tumor cells to metastasize [101].

2.5.5 Tumor-Associated Fibroblasts

Tumor-associated fibroblasts (TAF) or carcinoma-associated fibroblasts (CAF) are the most prominent cell type within the tumor stroma of many cancers, most notably breast and pancreatic carcinoma where they are relatively absent in corresponding normal tissue. They may be recruited regionally or from circulating populations [212, 305] and play a critical role in tumor remodeling, tumor growth and metastasis, and structural matrix formation [210]. TAF generally favor the transition of non-tumorigenic cells towards tumorigenic clones [318]. The presence of fibroblast populations within human tumors is associated with poor outcome and an increase in metastatic potential [306]. Chronic inflammation may be mediated by cytokines including IL-1β, IL-6, and IL-8 [54]; (see Fig. 2). The TAF population differs from a normal fibroblastic phenotype because of its rich source of tumor growth-promoting factors, pro-angiogenic factors, and expression of myofibroblastic characteristics. TAF are often characterized by increased expression of pathology-associated or ''activated'' fibroblast markers such as fibroblast-specific protein (FSP) and fibroblast activation protein (FAP). TAF also express markers of malignant tumors, including stromelysin-1 (SL-1), thrombospondin-1 (Tsp-1), and Tn-C. TAF also express pro-tumorigenic growth factors including HGF, members of the epidermal growth factor (EGF), fibroblast growth factor (FGF), and Wnt families plus cytokines, such as SDF-1α and IL-6, all which contribute to cancer progression as previously stated in breast, prostate, ovarian, pancreatic, and colon cancers [128].

TAF can also produce a variety of ECM-degrading enzymes that release latent angiogenic factors such as bFGF, VEGF, and TGF-β, rendering them bioavailable to their receptors on endothelial cells [148]. Finally, TAF can also produce chemoattractants that recruit pro-angiogenic macrophages, neutrophils, and other myeloid cells, thereby orchestrating tumor angiogenesis [265], directly stimulating recruitment of endothelial precursor cells via secretion of CXCL12 [239]. TAF secreting TGF-β1 prevented destructive inflammatory responses that might otherwise disrupt tumor growth and progression [308]. It has also been demonstrated that TAF express markers of fibrovascularization such as α-smooth muscle actin (α-SMA), desmin, and vimentin in vivo and ex vivo following co-culture with tumor cells or their

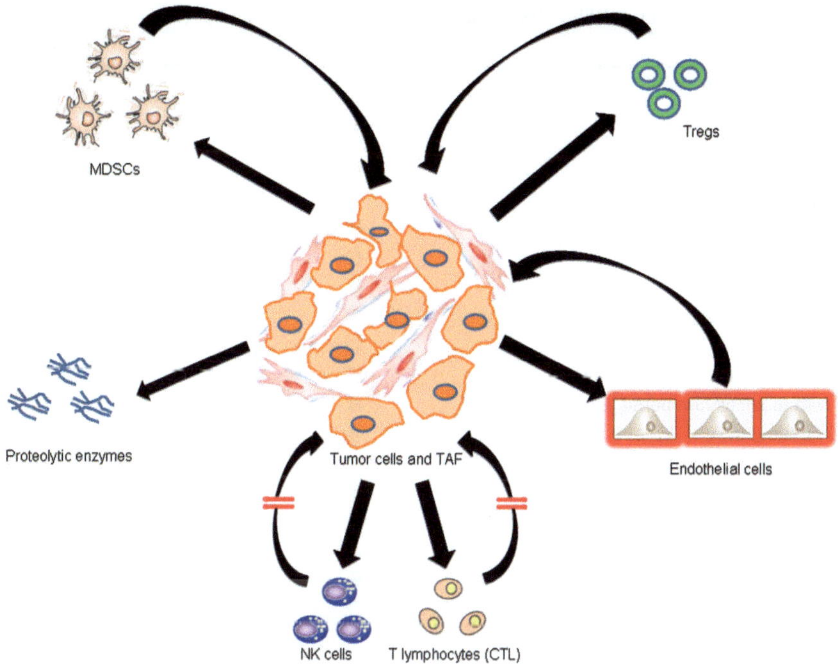

Fig. 2 *Interactions between tumor cells and TAF.* TAF are the most prominent cell population within the tumor stroma and play critical roles in tumor growth and metastasis. TAF secrete various cytokines and growth factors that may favor a transition from pre-tumor cells to tumorigenic clones. TAF also chemoattract pro-angiogenic macrophages, MDSCs, and Tregs that further orchestrate angiogenesis and promote tumor progression. TAF secrete several cytokines such as TGF-β which prevents inflammatory responses by NK and T lymphocytes

conditioned media. The importance of TAF in promoting tumorigenesis has been well established in multiple tumor models (see Table 1); [243]. TAF extracted from human tumors facilitate the growth of human breast and ovarian cancers when co-injected into immunosuppressed mice. This appears to involve multiple mechanisms, including the inhibition of cancer cell apoptosis, increased tumor cell proliferation, and promotion of angiogenesis [237]. Several recent reviews have focused on the therapeutic potential of targeting the activated cancer-associated stromal compartment [89, 106]. A number of studies have also implicated TAF in the capability to limit the impact on tumor growth and progression of cancer cell apoptosis [148]. Fibroblast activation protein-1α (FAP-1α), a cell surface protease with dipeptidyl peptidase and endopeptidase activity, is expressed by stromal cells in several different cancers [238] and has been used as a clinical therapeutic target by multiple immune-conjugate clinical studies in multiple cancer types [286]. Potential therapeutic targets of cancer-activated stromal signaling pathways that act as regulatory switches for tumor-promoting molecules have been identified, for example, IL-1β/IL-1R1, SDF-1α/CXCR4, GROα-1/CXCR-2, NF-κB p65, and Tn-C but detailed mechanisms remain to be fully elucidated.

Table 1 Tumor-associated fibroblast influence on tumor progression

Tumor source	Impact	References
Bone and soft tissue tumors	🟢	[71]
Breast cancer	🟢	[237, 306]
Breast cancer	🔴	[280]
Colorectal cancer	🔴	[22]
Invasive breast cancer	🟢	[290]
Lung adenocarcinoma	🟢	[131]
Melanoma	🟢	[18]
Oesophageal cancer	🟢	[93]
Oral squamous cell carcinoma	🟢	[153, 189]
Oral squamous cell carcinoma	🔴	[339]
Ovarian adenocarcinoma	🟢	[306]
Ovarian cancer	🟢	[37, 286]
Pancreatic adenocarcinoma	🟢	[137, 306]
Pleural mesothelioma	🟢	[186]
Prostate carcinoma	🟢	[145, 234]
Several solid tumors	🔴	[88]
Squamous skin carcinoma	🟢	[81]

Inhibition (🔴), Growth (🟢)

3 The Origin of Tumor-Associated Fibroblasts (TAF): From Resident Precursors to Circulating Progenitors

TAF are highly heterogeneous cellular elements and can be derived from different sources (see Fig. 3) [110]. Dissecting their origin may be important in a broader comprehension of cancer development. Major hypotheses indicate that TAF may be derived from resident elements, from a circulating pool, or both. All these possible activities are described below.

3.1 TAF from Resident Precursors

3.1.1 Resident Mesenchymal Fibroblasts

Emerging data indicate that TAF can be derived from pre-existing fibroblasts residing normally in tissues through signals from soluble products and ECM changes that are present at the tumor microenvironment, often originating from cancer cells themselves. A study of TAF, derived from liver colorectal cancer metastasis, suggested a local source of these cells as TAF were found to express markers similar to those of local resident fibroblasts [218]. After stimulation by members of the PDGF or TGF-β family, local fibroblasts or fibroblast precursors have generally been considered as the major source of TAF [148]. However, TAF

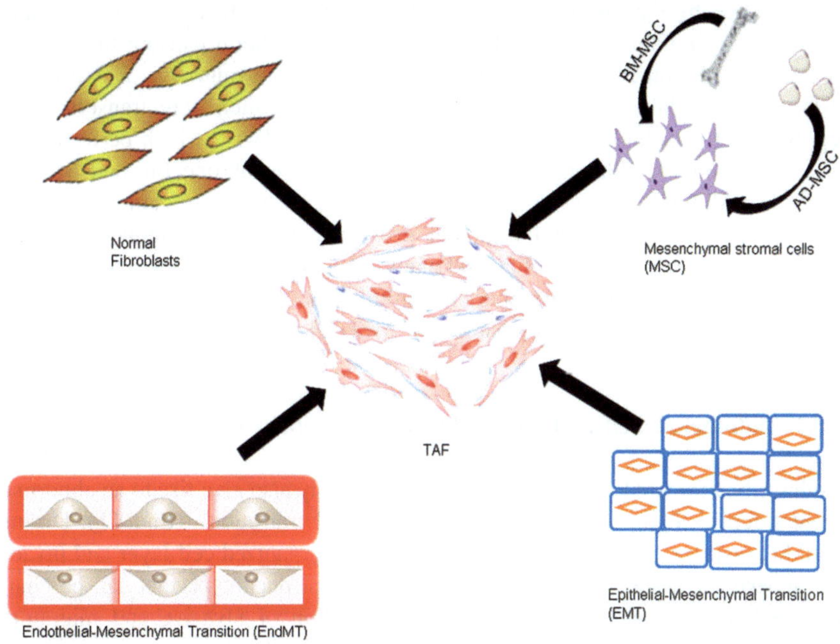

Fig. 3 *The origins of TAF*. TAF may be derived from different sources such as resident mesenchymal fibroblasts or from circulating mesenchymal progenitors derived from BM-MSC and AD-MSC. TAF may also originate from epithelial precursors occurring due to biological event epithelial to mesenchymal transition (EMT). Similarly, TAF may also result from a transition of endothelial cells to mesenchymal lineage (EndMT)

can also originate from pericytes, smooth muscle cells, and vascular cells [158]. Studies focusing on fibrosis leading to cancer development have identified activated tissue resident cells responsible for excessive ECM production, such as pancreatic stellate cells in pancreatitis that induce progression to pancreatic cancer [141] or peribronchiolar and perivascular lung fibroblasts that lead from lung fibrosis to lung cancer development [178].

3.1.2 Epithelial Cells: The Epithelial to Mesenchymal Transition

TAF may also originate from epithelial precursors accordingly to an abnormal biological event defined as "epithelial to mesenchymal transition" (EMT). EMT (first coined by Krug et al. in 1987) is a critical process in embryogenesis [316] and may occur in several epithelial cell types in post-natal life. It is a biological process relevant in development and tissue regeneration but has been linked to cancer progression as reported in breast, lung, prostate, pancreatic, and colorectal cancer [12]. EMT is defined as the switch from non-motile, polarized epithelial cells to motile, non-polarized mesenchymal cells, with the potential to migrate from a primary tumor site to distant organs, where they can "seed" and grow. It is characterized by the loss of E-cadherin, the major component of adherens junctions, and a simultaneous gain of mesenchymal N-cadherin causing the cells to lose affinity for other epithelial cells and become more migratory and invasive [108, 200]. Partially responsible for this "cadherin switch" are transcriptional repressors of E-cadherin: Snail 1, Snail 2 (Slug), and Twist, which perform pro-invasive functions, inducing $\alpha V\beta 3$-integrin expression, fibronectin, and MMP-9 [120]. In addition to transcriptional repressors, one of the most potent inducers of EMT is TGF-β1, a known pluripotent growth factor able to induce EMT in mammary, lung, pancreatic, colon, and many other cell types [84]. In cancer, EMT has been demonstrated during epithelial injury and can also occur in individual tumor cells as an important mechanism of invasion and metastasis.

3.1.3 Endothelial Cells: The Endothelial to Mesenchymal Transition

As described for EMT, endothelial cells can similarly also give rise to an abnormal event referred to as "endothelial to mesenchymal transition" (EndMT). During EndMT, resident endothelial cells detach from an organized cell layer and invade the underlying tissue. They acquire a mesenchymal phenotype that can be characterized by loss of cell–cell junctions, acquisition of invasive and migratory properties, loss of endothelial markers such as CD31, and gain of mesenchymal markers such as FSP1 or α-SMA [256]. Previous studies of EndMT focused largely on embryonic development of the heart, however, recent evidence suggests that EndMT can occur post-natally in a variety of pathological settings, including cancer, where EndMT can account for up to 40% TAF [351] and cardiac fibrosis [352]. The molecular mechanisms of EndMT in tumors are not fully understood but can be mediated by TGF-β1 or Notch ligands ex vivo and manipulations of the TGF-β1 or Notch pathways in vivo. However, it is still not clear whether Notch, TGF-β, or a combination of both pathways provides the initiating signal under physiological conditions and sheer stress may also have an important role [315]. It is also likely that other signaling pathways interact with TGF-β1 and Notch, including TGFβ2 and Rho signals to mediate EndMT [103, 208]. EndMT may be initiated by autocrine and paracrine inflammatory signals originating from the surrounding tissues. EndMT is often categorized as a specialized form of epithelial-to-mesenchymal transition (EMT).

3.2 TAF from Circulating Progenitors

3.2.1 Circulating Endothelial Cells

In addition to resident endothelial cells, bone marrow-derived endothelial progenitor cells (EPC) can be attracted to tumors differentiating into mature endothelial cells and generate capillaries [74, 197]. However, controversy exists on the relative contribution of the EPC to the tumor vasculature with estimates varying from less than 1% up to more than 50% [277]. Recently, it was shown that EPC egress the bone marrow and home to the tumor immediately after a certain type of chemotherapy, for example, paclitaxel. EPC are mobilized from the bone marrow and home to sites of tumor neovascularization in response to various cytokines, such as SDF-1α, MMP-9, VEGF, placental growth factor (PlGF), and granulocyte colony-stimulating factor (G-CSF). These findings have provided new insight into the mechanism of tumor regrowth, resistance to chemotherapy, early recurrence, and metastasis during or after chemotherapy [60, 276].

3.2.2 Circulating Mesenchymal Precursors

Similarly to endothelial progenitors, it has been reported that a circulating pool of cells deriving from bone marrow (BM) contributes to tumor stroma and, more specifically, to TAF. BM contains several progenitors having multipotent differentiation capacity [107]. In particular, bone marrow-derived mesenchymal elements have been called for this specific function.

Evidence for a BM source of TAF comes from studies using a gastric cancer mouse model, in which BM transplantation experiments indicated that subsets of gastric myofibroblasts were indeed derived from the BM [116]. Further studies based on BM transplants in animal models have estimated that BM-MSC may contribute as much as 20% to the TAF population [261]. The biological identity of these cells is still under investigation but mesenchymal stromal/stem cells (MSC) have been identified as possible precursors [68]. Similarly, white adipose tissue (WAT) provided a functional cellular contribution with soluble factors that migrated to tumors and promoted neovascularization [355]. Recently, it has also been shown that adipose tissue can release progenitors that migrate and contribute to tumor microenvironments [158].

4 Dissecting Biological Properties of MSC to Understand Their Tumor Stroma Contribution

4.1 What are MSC?

Because both marrow and adipose mesenchymal progenitors may contribute to tumor stroma, it is necessary to understand the origins and evolution of this complex scenario in light of the MSC qualities of homing and differentiation potential. Mesenchymal stem cells [aka mesenchymal stromal cells (MSC)] are multipotent progenitor cells, first identified by Friedenstein et al. who described a population of plastic adherent cells isolated from BM with fibroblast-like morphology [92]. MSC retain robust proliferation capacity thanks to a well-known clonogenic potential due to a self-renewal capacity [39].

In addition to a proliferative phenotype, MSC display differentiation capacities that allow the generation of adipocytic, osteogenic, chondrogenic, and myogenic lineages [254]. In addition, MSC can act by secreting bioactive molecules capable of influencing normal and pathological tissue homeostasis [129]. These functions are exerted either ex vivo or in vivo. In the latter case, a specific homing mechanism enhanced potential of therapeutic approaches aimed at regenerating several damaged organs [252].

Although MSC can be isolated from different tissues such as dental pulp [112], gingiva [354], cord blood [159, 279], Wharton's jelly [86, 87], and placenta [113] and are currently used for regenerative medicine, bone marrow and adipose tissues represent rich and more defined sources of MSC [27]. Curiously, these are the two tissues from where circulating TAF progenitors were reportedly derived.

These findings support the concept of MSC subpopulations homing into a wound-like microenvironment typical of a solid tumor participating with tumor stroma activities according to inherent phenotypic qualities.

4.2 Defining MSC Phenotype as TAF Precursors

The identification of a conversion from MSC to TAF should be based on comprehensive characterization of their phenotypes, whereas in vivo this still remains to be elucidated; in vitro findings are more established. Ex vivo MSC expanded are usually negative for the hematopoietic cell surface biomarkers such as CD34, CD45, CD14, CD11b, CD79α, or CD19 and HLA-DR expression [72] but are positive for CD105 (endoglin), CD73 (ecto-5'-nucleotidase), CD44, CD90, CD71 (transferrin receptor), the ganglioside GD2, CD271 (low-affinity nerve growth factor receptor), and STRO-1 [323]. Notably, ex vivo-isolated TAF display several similarities to MSC; in particular, TAF do express CD29, CD44, CD90, and CD73 [251]. Recently, our group reported that TAF isolated from primary lung tumor also expressed similar MSC immunophenotypic markers such as CD90, CD73, and CD105 [110]. In agreement, TAF isolated from several pediatric tumors, such as neuroblastoma and sarcomas expressed typical MSC markers such as CD90, CD73, and CD105 and were negative for hematopoietic antigens [144].

Cell surface biomarkers are also linked to functional activity: for example, HLA-DR serves as a putative antigen-presenting molecule. MSC do not habitually express HLA-DR but stimulation with IFN-γ or other cytokines such as bFGF increased their levels significantly, implying a role as immune regulators [181]. Similarly, it has been described that TAF may express low levels of HLA-DR with upregulation following IFN-γ exposure [225]. These few studies already indicate strong similarities between MSC and TAF. Analyzing the gene expression profile of these two populations is informative even if this may reflect a phenotypic equivalence rather than indicate cancer stroma is directly descended from MSC [357].

4.3 Driving MSC Homing Potential to Tumor Stroma

Systemically administered MSC have been observed to migrate to sites of injury by still unexplored mechanisms [43]. Tissue-specific engraftment is referred to as homing [16, 350], and this property of MSC-based therapy in diseases may be essential for a robust medicinal effect. Hoffman and his group have shown that ex vivo expanded allogenic and autologous MSC transduced with eGFP were distributed to a wide range of tissues in baboons, including lung, thymus, bone, skin, cerebellum, and gastrointestinal tract [64]. Moreover, MSC distributed widely to a variety of non-hematopoietic tissues following systemic infusion and may possess a proliferative capacity within these tissues. Intra-arterial and intravenous injections of MSC in rats led to early engraftment in the lung and later in the liver and other organs [99]. In osteogenesis imperfecta (OI) transgenic mice, infused with wild type MSC, homing involved many organs, including the lung, marrow, bone, skin, brain, and spleen. Human data further validate these findings within clinical trials for OI treatment [130].

Assuming that homing involves specific tropism beyond entrapment in the narrow vessels of filtering organs, the precise biological mechanisms remain to be fully understood. During homing the injured tissue generates an environment whereby soluble factors released by the ECM combine with mediators produced mainly by inflammatory cells. Notably, a high concentration of hyaluronic acid in a highly hydrated ECM favors diffusion of chemokines and growth factors stimulating cell migration that in turn favors the creation of chemoattractant gradients for cell attachment, migration, and tissue incorporation [102]. When engrafted at sites of tissue injury, MSC differentiated into connective tissue elements, supported vasculogenesis, and secreted cytokines and growth factors that facilitated healing. Chen et al. showed that BM-MSC-conditioned medium containing high levels of growth factors and chemokines enhanced wound healing in mice, implying a critical role of paracrine factors in MSC-mediated enhanced wound healing [44]. More specifically, BM-MSC were attracted by pancreatic islets ex vivo and in vivo, and the chemokine SDF-1 played a relevant role in this migration [304]. Similarly, MSC produced a wide array of cytokines including monocyte chemotactic protein-1 (MCP-1) and direct injection of MSC into an anatomical region forming collateral blood vessels improved perfusion and remodeling, lessened tissue damage, and enhanced limb function in a mouse model of hindlimb ischemia [161] These data suggest that it is the nature of BM-MSC to mobilize to specific tissues and induction of migration is a response to damaged tissue.

Similar to wound-homing, tumor-homing is a multistep process used by diverse cell types to travel from a distant site to a tumor [267]. MSC may "sense" cancer as a damaged tissue event with inflammation and home to tumors accordingly [180]. Houghton et al. observed MSC engraftment into gastric glands in a model of gastric cancer [132]. Fulfulling all the prerequisite steps, systemically administered hMSC homed to the tumor site, preferentially survived, proliferated in the presence of malignant cells, and became incorporated into the tumor architecture as stromal fibroblasts [222]. MSC tumor migration was motivated by many factors including tumor cell-specific receptors, ECM, and soluble tumor-derived factors such as SDF-1, TNF-α, and interleukins [98, 168]. Another factor likely to be involved in the chemoattraction and/or tissue engraftment of MSC is the chemokine, MCP-1 (CCL2). Secreted by breast tumor primary cultures and breast tumor explants, it promoted recruitment of MSC into the tumor [78] and enhanced mobilization during experimental brain ischemia [332].

In addition, a toll-like receptor ligand known as LL-37 has been linked to MSC tumor homing. Different tumor types, including ovarian, breast, and lung cancer, showed high expression of LL-37 which acted as a proliferative signal, pro-angiogeneic factor, and chemoattractant for various immune cells through activation of formyl peptide receptor like-1 (FPRL-1), a member of the toll-like receptor family [110]. LL-37, expressed on ovarian cancer cells activated MSC migration in a dose-dependent manner [51]. In vivo inhibition of MSC engraftment into tumor cells resulted in disorganization of the fibroblast–vascular network as well as a reduction of tumor growth.

Utilizing mouse xenografts, Lui et al. also showed that BM-MSC were recruited to sites of growing breast cancers by gradients of IL-6 [192], suggesting that IL-6 may mediate MSC homing and facilitate tumor growth. Similarly, AD-MSC secreted angiogenic factors that may further mediate tumor growth by stimulating tumor vascularization [320]. There is growing evidence for a specific mechanism by which MSC home to tumor microenvironments before incorporation as tumor stroma. MSC display many tumor-supporting roles including immune response suppression, inhibition of tumor apoptosis, and stimulation of EMT, angiogenesis, proliferation, extravasation, migration, and metastasis [267].

4.4 Driving MSC Differentiation into TAF

As stated, MSC are able to localize to the tumor together with other cells such as myofibroblasts, endothelial cells, pericytes, and inflammatory cells, to create a specific tumor micro-environment and modulate tumor growth and progression, however, the way by which MSC can directly contribute to the tumor microenvironment still requires investigation. One concept can be adopted from the paradigms of MSC differentiation involved in tissue regeneration.

Al-Khaldi et al. reported that MSC could support the tumor vasculature directly by differentiating into pericytes and perhaps endothelial cells [4], less directly by secreting vasculogenic growth factors [278]. In addition, pericytes isolated from the stromal–vascular compartment expressed MSC-like cell surface markers and a capacity to differentiate into tissues of mesenchymal lineage [55]. Furthermore, several studies have measured the contribution of bone marrow-derived cells to the tumor microenvironment within the context of transgenic mouse models and human bone marrow transplants [135]. The ability of MSC to travel to solid tumors after intravenous administration and the development of myofibroblast-like

Table 2 Tumor stroma markers expressed/secreted by MSC

Tumor stroma markers	MSC source		References
	BM	AD	
1. Cytokines			
HGF	+	–	[247, 314]
IFN-γ	+	+	[136, 242]
IL-1	+	+	[122, 136, 242]
IL-6	+	+	[15, 114, 122, 136, 139, 242, 247]
IL-8	+	–	[136, 242, 304]
IL-10	+	–	[136, 242, 247]
PDGF	+	–	[44]
SDF-1	+	–	[44, 69, 126, 136, 270, 304]
TGF-β	+	+	[48, 121, 175, 301]
TNF	+	+	[136, 242, 247]
VEGF	+	+	[20, 44, 121, 160, 175, 242]
2. Proteolytic enzymes			
MMPs	N/A	+	[48, 69, 205, 273, 301]
TIMPs	N/A	+	[48, 273, 329]
uPA	N/A	–	[226]
3. ECM proteins			
Collagens	+	+	[48, 69, 163, 174, 191, 216, 301]
Fibronectin	+	+	[48, 69, 163, 174, 216]
Integrins	+	–	[69, 163, 216]
Laminins	+	+	[48, 163, 216]
Tenascin-C	+	–	[191]

Positive (+), Negative (–); *BM* Bone Marrow, *AD* Adipose derived, N/A Not Applicable

characteristics under defined culture conditions [80] were consistent with the concept that MSC can be recruited into the developing tumor and assume a TAF-like phenotype under the influence of the tumor microenvironment. It has been demonstrated that hBM-MSC differentiated into myofibroblasts and expressed α-SMA, vimentin, FSP, SDF-1, and several other typical TAF markers [211, 312]. Similarly, Spaeth et al. reported that MSC-derived TAF were positive for FSP, FAP, and cell aggressive markers (Tn-C, Tsp-1, SL-1), tumor growth factors (HGF, EGF, IL-6), and angiogenic factors [306]. Similar findings described that transformed hBM-MSC can produce factors generally identifiable within the tumor microenvironment, including VEGF, EGF, bFGF, TGF-β, Tn-C, and IL-6 (Table 2).

Despite a close relationship, TAF and MSC display significant phenotypic differences, including soluble factors production and proliferation rate. This is likely to reflect dynamic specialization processes causing MSC to change within the tumor microenvironment in response to a selection pressure for MSC qualities that better "serve" the cancer cell [251]. MSC form a tumor–fibrovascular network differentiated into TAF and vascular pericytes [164] and hBM-MSC exposed to tumor-conditioned medium expressed TAF biomarkers and sustained expression of CXCL12 (SDF-1) which conferred an ability to promote tumor cell growth in an in vivo co-implantation model. In ovarian cancer, the tumor-conditioned media stimulated the differentiation of hAD-MSC to TAF, elevating the expression of SDF-1α through a TGF-β1–mediated autocrine stimulation of Smad2 [142]. Subsequently, Jotzu et al. also demonstrated that hAD-MSC exposed to conditioned media from human breast cancer cell lines differentiated into TAF-like myofibroblastic cells expressing α-SMA, Tn-C, SDF-1, and CCL5 with an ability to promote tumor cell invasion [146]. These observations support the notion that MSC give rise to a subset of "specialized" MSC referred to as TAF.

4.5 Secretory Potential of MSC into Tumor Environment

4.5.1 Secretory Potential of MSC from Regenerative Medicine to Neoplastic Settings

The secretory potential of MSC has been widely investigated and beyond the historically known hematopoietic supportive functions [24] and immunomodulatory factors capable of influencing B cells, T cells, and NK lymphocytes, a number of additional cytokines and secretory elements have been identified (Table 2). Regarding other tissues and organs, MSC were shown to produce and release factors directly involved in liver regeneration, such as HGF, TNF, and IL-6 [258] and they were sustained hepatocyte survival and differentiation in co-culture systems [139]. Similarly, in the heart and different paracrine mediators can influence the biology of adjacent parenchymal and stromal cells in normal and pathological conditions. Under hypoxic conditions, MSC express and secrete important paracrine factors including angiogenic cytokines and anti-apoptotic factors in a temporal and spatial manner that further enhances cell survival and activates endogenous repair and regeneration [329]. Therefore, in the case of tissue damage, MSC can be attracted to the damage site wherein they secrete bioactive factors that function to trophically assist the repair and regenerative process [40].

Cross-talk between tumors and MSC includes the secretion of growth factors such as EGF, VEGF, FGF-2, PDGF, IGF-1, G-CSF, granulocyte-macrophage colony stimulating factor (GM-CSF), and chemokines including CCL2, CCL5, and CXCL8 [23]. Within the tumor microenvironment, MSC themselves secrete the cytokines IL-6 and IL-8 [335]. MSC-secreted bioactive molecules can act both directly and/or indirectly: directly by inducing intracellular signaling or indirectly by causing another cell in the vicinity to secrete a further functionally active agent.

In addition to secreted soluble factors, intercellular communication also involves the release of microparticles or exosomes. Exosomes are lipid vesicles that are less than 1 mm in diameter that contain proteins or RNA molecules that can regulate intracellular signaling in adjacent cells [358]. MSC-secreted microparticles may also contain microRNAs in a precursor form. Purified exosomes secreted from MSC reduced infarct size for an in vivo mouse model of myocardial ischemia/reperfusion injury, highlighting a new exosome perspective for intercellular mediation of tissue injury and repair [177].

4.5.2 Secretory Potential of MSC in Immunity

Since tumors benefit from a reduced immune response from the host, the influence of MSC on tumors also involves a possible immunomodulatory role. Numerous studies indicate MSC reduce immune recognition or even inhibit ongoing immune reactions [224, 229]. Proposed mechanisms include activating apoptosis in T lymphocytes, induction of regulatory T lymphocytes, directing maturation of dendritic cells that attenuate regulatory T cell responses, and/or secretion of nitric oxide (NO) that inhibits T cell proliferation [284]. The interactions of MSC with T lymphocytes are well studied and many reports have shown that BM-MSC affect several properties of T cells, by efficiently suppressing the proliferation of activated CD4+ T helper cells and CD8+ cytotoxic T cells (CTL). In humans, most effects of BM-MSC on T cells are mediated through cell-contact independent processes, emphasizing the importance of secreted factors that include IL-1β, TGF-β1, HGF, prostaglandin E2 (PGE2), IDO, heme oxygenase-1 (HO-1), leukemia inhibitory factor (LIF), IGF, soluble human leukocyte antigen G5 (sHLA-G5), galectin-1, galectin-3, and Jagged-1 [45].

In contrast, MSC induce proliferation of B cells, T cells, and also Tregs, the latter demonstrated by the increase in the population of CD4+ CD25+ FoxP3+ cells in mixed lymphocyte cultures in the presence of MSC [198]. Indicating a role in autoimmunity, BM-derived MSC are able to promote proliferation and differentiation of transitional and naive B cells into immunoglobulin-secreting cells using cells derived from both healthy donors and pediatric patients with systemic lupus erythematosus [319].

Most of the inhibitory soluble factors are not constitutively secreted, but can be induced by the interaction between activated effector cells and MSC which resist T and NK mediated cell death. Accordingly, MSC in vitro inhibit proliferation, cytokine production, and cytotoxic activity of NK cells [3]. Also MSC reduce CD8+ T cell cytotoxicity towards allogenic cells, virally infected cells, and tumor cells [26]. MSC dampen B cell proliferation, differentiation into antibody secreting cells, and responsiveness to CXCL13 by downregulating the expression of the respective receptors CXCR4 and CXCR5. Finally, MSC promote the generation of Tregs through direct and indirect mechanisms [294].

MSC have been shown to protect breast cancer cells by expanding Tregs, with concomitant decrease of Th1 and increase of Th2 cytokines, an effect largely mediated by TGF-β1 [248]. PGE2 is a lipid intermediate implicated as a potential candidate responsible for T cell inhibition by MSC [343]. Although MSC may inhibit the immune system, from hundreds of patients treated with MSC for several therapeutic indications including co-existant neoplastic conditions, there is just one report stating that there was an increased risk of relapse after stem cell infusion [228]. The reasons for this discrepancy between ex vivo and in vivo preclinical data versus clinical outcome are under investigation.

4.5.3 Secretory Potential of MSC and Angiogenesis

Blood vessels and stromal components respond to pro- and anti-angiogenic factors that govern vascular remodeling during development and wound healing. The formation of new blood vessels is necessary to sustain the survival of newly formed granulation tissue. In two linked studies when wounds received implantation of BM-MSC they displayed enhanced angiogenesis with elevated levels of VEGF, SDF-1, and Angiopoietin-1 (Ang-1) along with increased numbers of cells positive for CD34, C-kit, or Flk-1 [326], indicating increased recruitment of endothelial cells and endothelial progenitor cells into the wound [47, 340]. Likewise, AD-MSC secreted multiple angiogenic and anti-apoptotic growth factors at bioactive levels that promoted angiogenesis when delivered to an ischemic hindlimb model [268].

As in tissue repair, neovascularization is a crucial process in tumor growth, progression, and metastasis [82]. Tumors require a blood supply to grow and to obtain this blood supply they can tilt the balance towards stimulatory angiogenic factors to drive vascular growth by attracting and activating cells from within the tumor microenvironment [336]. MSC within the tumor microenvironment support the tumor vasculature directly, by differentiating into pericytes and perhaps endothelial cells [5], by providing a supportive ECM [36] and by secreting several vasculogenic growth factors such as HGF, cyclooxygenase, IGF-1, PDGF, and TGF [278]. The presence of MSC during early tumor growth may facilitate the process of angiogenesis and co-injection of MSC with pancreatic cancer cells increased vessel density, which required MSC-derived VEGF expression [20]. Other studies also reported that MSC co-implanted with cancer cells in syngeneic animals accelerated tumor appearance, probably by favoring an angiogenic switch [11, 97]. Recently, Kidd et al. demonstrated that hAD-MSC gave rise to vascular and fibrovascular stroma and were capable of forming vessel structures within the tumor parenchyma [158].

4.6 Mesenchymal Stromal/Stem Cells Can Promote Tumor Growth

Given detailed interactions between MSC-TAF and tumor cells it is appreciated that the multiparametric mechanisms whereby MSC may enhance tumor growth are complex but generally dependent on cross-talk between the MSC and their target cells.

To a large extent, the function of MSC in tumors parallels their role in wound healing with MSC differentiating into pericytes and TAF [298]. Different reports have demonstrated that MSC can colonize metastatic tumors and in some models this favored metastasis [23]. MSC co-injected subcutaneously with breast carcinoma cells enhanced metastatic ability in a CCL5-RANTES-dependent manner, a chemokine secreted by the MSC upon induction by tumor cells [151]. Similar

Table 3 Mesenchymal stromal/stem cell influence on different tumor types

Tumor Source	Impact	References
Breast cancer	● (growth)	[97, 151, 217, 361]
Breast cancer	● (inhibition)	[260]
Colon carcinoma	● (inhibition)	[232]
Colon carcinoma	● (growth)	[296, 358, 359]
Gastric cancer	● (growth)	[356, 358]
Hepatoma and lymphoma	● (inhibition)	[196, 259]
Kaposi's sarcoma	● (inhibition)	[155]
Lewis lung melanoma	● (inhibition)	[199]
Lung or glioma	● (growth)	[346]
Melanoma	● (growth)	[70, 313]
Melanoma	● (inhibition)	[245]
Non-Hodgkin lymphoma	● (inhibition)	[293]
Pancreatic adenocarcinoma	● (growth)	[138]
Prostate cancer	● (growth)	[188, 257]

Inhibition (●), growth (●)

studies also reported an increase of metastatic osteosarcoma lesions mediated by MSC-derived CCL5 [312]. Reciprocally, MSC secreting large amounts of CXCL12 and CXCL13 may attract different circulating tumor cells, including breast and myeloma cells [215, 325]. Subsequent interactions between tumor cells and MSC lead to the production of soluble factors (i.e., PGE2 and Galectin-3 binding protein) and cytokines including IL-6, a potent osteoclast-activating factor that also promotes tumor cell growth, survival, and resistance to chemotherapy [32, 95, 302]. Studeny et al. showed that by using BrdUrd staining methods MSC proliferated in the presence of tumor cells in vivo, whereas MSC implanted without tumors did not proliferate. Hence, encouragement of tumor growth by MSC may involve a reciprocal encouragement of MSC growth by the tumor cells [309]. Both BM and adipose-derived endothelial and mesenchymal progenitor cells have been isolated, cultured, and injected back into mice to show that they possess both tumor tropism and tumor-promoting capacity [158] (see Table 3). BM-MSC has been shown to increase the in vivo growth of colon cancer, lymphoma, and melanoma cells [164]. Adult- and fetal-derived MSC were co-injected with colon cancer cells in a murine xenograft model [359], resulting in an increased incidence of tumors with enhanced vascularity and necrosis. Both adult and fetal MSC had similar growth-stimulating effects, but adult MSC appeared to promote tumor incidence more than fetal MSC. Reflecting an immunosuppressive function, MSC co-administered with B16 melanoma cells prevent the rejection of cancer cells in an allogenic animal model [70].

Tumor cells injected into nude mice in conjunction with MSC grew faster than those injected without MSC, however, this in vivo pro-tumorigenic effect was contrasted by ex vivo experiments, where MSC inhibited proliferation of tumor cells through induction of G1 phase arrest [263]. To explain this discrepancy Ramasamy et al. suggested that MSC may create an artificial niche in which tumor cells preserve their potential to proliferate. Accordingly, MSC niches conferred significant protection to leukemic lymphoblasts against asparaginase-induced cytotoxicity [140].

AD-MSC exhibited tumor tropism and were functionally similar to BM-MSC [73]. AD-MSC co-transplanted with mammary breast cancers in a syngeneic mouse model promoted the development of larger and more rapidly forming tumors [217]. Also when hAD-MSC were co-injected subcutaneously with lung cancer or glioma cells into nude mice, the number of viable tumor cells and relative tumor size increased [346].

4.7 Mesenchymal Stromal/Stem Cells Can Inhibit Tumor Growth

Whereas the above studies indicated MSC supported tumor growth, different studies described opposite effects, whereby MSC inhibited tumors [157]. Zipori et al. showed that BM-MSC inhibited sarcoma cells in vitro [362], observations later confirmed by demonstration that MSC potently antagonized Kaposi's sarcoma growth in vivo. The use of athymic nude mice in the latter experiments suggested that the inhibitory effects of MSC were not necessarily due to immunomodulatory effects [155]. Similar anti-tumor outcomes mediated by MSC were observed in an experimental model of pancreatic carcinoma [156] and in SCID mice with disseminated non-Hodgkin lymphoma (NHL). The extensive areas of necrosis within the tumor mass following injection of MSC into the NHL tumor probably reflected an anti-angiogenic effect, Since ex vivo experiments demonstrated MSC-induced endothelial cell apoptosis [293].

MSC inhibited the growth of rat colon carcinoma when co-injected with an equal number of MSC and tumor cells or with ten-fold more MSC [232]. Human fetal skin-derived MSC inhibited human liver cancer cell lines, with reduced proliferation, colony formation, and oncogene expression both in vitro and in vivo [260]. When these cell lines were co-injected with the same number of MSC, tumor development was delayed and tumor size decreased. The same fetal skin-derived MSC inhibited growth of breast cancer cells in vitro [260]. The researchers found that treatment with conditioned media resulted in downregulation of survival factors, such as β-catenin, c-Myc, and survivin. This effect was mediated by an inhibitor of β-catenin signaling, Dickkopf-related protein-1 (DKK-1), which was secreted by MSC. The DKK-1 effects were suppressed in MSC with the use of a neutralizing antibody and small interfering RNA, eliminating the growth inhibitory effects of MSC [260].

Cousin et al. demonstrated that AD-MSC suppressed pancreatic tumors by altering cell cycle progression. In vitro co-culture with AD-MSC increased rates of G1-phase arrest in pancreatic cancer cells and in vivo injection of AD-MSC into established pancreatic cancer xenografts further inhibited tumor growth [53]. In a similar approach, BM-MSC injected into established subcutaneous melanomas caused apoptosis and abrogation of tumor growth [245]. AD-MSC were also found to inhibit proliferation of primary leukemia cells and this effect was mediated by secreted DKK-1 regulated by the stem cell transcription factor NANOG [360].

4.8 Modified Mesenchymal Stromal/Stem Cells for Inhibiting Tumor Growth

Given evidence for both an inhibitory and a stimulatory influence of MSC on tumor growth and development one may strategically consider engineering an enhanced inhibitory potential and exploiting the MSC qualities of tumor homing and differentiation into TAF. Gene therapy is a promising novel therapeutic strategy for treatment of several heritable and nonheritable human diseases, including infections, degenerative disorders, and cancer [10]. MSC have been exploited as delivery vehicles to target anti-tumor agents to malignant cells. MSC possess several unique properties making them ideally suited for cellular therapies/ regenerative medicine and as vehicles for gene and drug delivery. These include: (1) relative ease of isolation; (2) the ability to differentiate into a wide variety of functional cell types of mesenchymal origin; (3) extensively expandable in culture before loss of differentiation potential; (4) hypoimmunogenic, they can induce immunosuppression upon transplantation; (5) they have pronounced anti-inflammatory properties; (6) following systemic administration they can home to damaged tissues, tumors, and metastasis [255]; (7) they are robust cells that can resist hypoxic stress and radiation therapy; (8) they can be readily genetically engineered ex vivo; (9) a cell-based therapy invoking reciprocal cellular interactions can respond more dynamically to tumor progression; and (10) in addition to influencing tumor cells, MSC may more broadly influence the tumor microenvironment helping to restore more normal vasculature and tissue homeostasis following tumor regression.

MSC can be readily transduced by the major clinically prevalent viral vector systems including those based upon adenovirus, the murine retroviruses, lentiviruses, and adeno-associated viruses (AAV) [255]. With the help of these viral vector systems MSC can efficiently express a wide range of cytoplasmic, membrane-bound, and secreted protein products. MSC can also be manipulated using physical (e.g., electroporation) and/or chemical agents (e.g., calcium phosphate or polycations) to enable gene transfer albeit with poorer efficiency than viral vectors [83, 292]. General ease of transduction coupled with the subsequent ability to select and expand ex vivo only the gene-modified cells to generate adequate numbers for clinical application make MSC one of the most promising stem cell populations for use in gene therapy studies and trials.

Use of MSC for therapeutic-gene delivery was originally proposed by Matthews and Keating [7, 202]. Initial experiments on virus-mediated transgene expression in MSC showed efficient cell transduction with retroviral vectors expressing lacZ or IL-3 genes, and no changes in differentiation potential of MSC after gene transfer.

Thereafter, different groups were able to detect systemic levels of human growth hormone or IL-3, produced by either canine- or murine-transduced MSC, up to several months after infusion [166]. MSC have also been used for tumor delivery of immunostimulatory cytokines and chemokines such as IL-2 [1, 223], IFN-β [222, 309, 310], INF-α [269], CX3CL1 [341], and tumor necrosis factor-related apoptosis-inducing ligand (TRAIL) [109, 194, 204, 283], suicide genes including thymidine kinase [209], cytosine deaminase [172], and carboxyesterase [58], growth factor antagonists (NK4) [149], and oncolytic viruses [167, 230], taking advantage of their tumor-homing capacities after systemic administration, or administered by intra-tumoral inoculation. The independent utility of many of these agents for cancer therapy was often limited by both their short half-life in vivo and their pronounced toxicity on normal non-malignant cells within the body. Using MSC to deliver these therapeutics may help minimize such problems, since MSC can selectively migrate to the tumor site and release their therapeutic effects locally, thus greatly increasing the agent's concentration within the tumor and significantly lowering its systemic toxicity.

Suitable modification of MSC with tailored viral vectors can cause engrafted MSC to release the therapeutic agent steadily, allowing a single administration to result in long-lasting effects. Elzaouk et al. stably transduced MSC with retroviral vector expressing cytokine IL-12 to inhibit the growth of melanoma [79]. Ren et al. stably transduced MSC with AAV vector expressing cytokine INF-α to inhibit the growth of melanoma [269]. An additional approach targeting melanoma, glioma, and breast carcinomas relies on the manipulation of MSC to express a pro-drug converting enzyme, such as cytosine deaminase. This enzyme converts 5-fluorocytosine (5-FC) to a soluble toxic molecule that kills both MSC and the neighboring cancer cells through a bystander effect [173]. Yong et al. showed that a delivery of the Δ24-RGD virus, a tumor-selective replication-competent adenovirus with specific cellular infectivity to tumors, produced long-term survival in an animal model of glioma [345]. MSC can be also modified to mimic plasma cells producing monoclonal antibodies such as the scFvEGFRv III, which specifically targets mutant EGF receptors on glioma cell surfaces [19]. These MSC-producing antibodies represent a novel strategy to deliver therapeutic molecules efficiently that would otherwise barely penetrate across the blood–brain barrier, thus introducing MSC as efficient therapeutic vehicles for malignant brain tumors.

As widely reported in the literature, TRAIL has a significantly higher therapeutic profile against cancer with virtually no toxicity towards normal tissues, but a short half-life in plasma limits its therapeutic potential. TRAIL signaling pathways are predominantly triggered by death receptors that selectively induce apoptosis in cancer cells without affecting normal cells [8]. Recently, our laboratory modified AD-MSC with members of the death ligand (DL) family which includes TNF-related apoptosis inducing ligands (TRAIL) as a powerful anti-cancer molecule against cervical carcinoma and pancreatic and colon cancer [109]. Grisendi et al. also investigated the feasibility of associating the AD-MSC TRAIL approach with other therapeutic agents, such as Bortezomib, a well-known proteasome inhibitor. TRAIL refractory tumors were treated in vitro in the attempt to sensitize them to our cell therapy approach. The results obtained demonstrated the synergic effect of AD-MSC TRAIL and Bortezomib against a breast cancer tumor cell line, known to be resistant to TRAIL. Using two different animal models, it was demonstrated that AD-MSC producing TRAIL were able to migrate to the tumor microenvironment and persist in tumors without a collateral toxic effect in the surrounding normal tissue. The selective anti-tumor effect was essentially due to cell-to-cell contact between AD-MSC expressing TRAIL and tumor cells, leading to caspase-8 activation within the tumor cells that rapidly activated and induced apoptotic cell death. Encouragingly, several investigators have shown that MSC expressing both transmembrane and secreted TRAIL are able to infiltrate and abrogate tumors [194, 204]. Kim et al. showed that human umbilical cord blood mesenchymal stromal cells (UCB-MSC) were suitable cellular vectors for TRAIL delivery, because these cells resisted TRAIL-mediated apoptosis and exhibited strong migratory ability and potent anti-tumoral activity towards glioma [159]. BM-MSC modified with a lentivirus expressing a secretable form of TRAIL (S-TRAIL) provided an effective drug delivery system for intra-cranial glioma [204]. Mohr et al. genetically modified BM-MSC using an adenoviral vector to express the full-length human TRAIL and apoptosis in a lung cancer cell line induced by cell-to-cell contact both in vitro and in vivo [214]. Further studies using murine BM-MSC successfully delivered S-TRAIL to induce human pancreatic cancer death without the need for cell-to-cell contact [213].

4.9 Conflicting Results of Using Mesenchymal Stromal/Stem Cells (MSC): The Janus Bifrons Nature of MSC

As described above, MSC appear to have a dual nature. This is true regarding their immunomodulatory potential but also with respect to cancer where they can promote tumor progression and metastasis in some studies, but seem to suppress tumor growth in others. These intriguing aspects recall the Roman myth of Janus Bifrons, a divinity with two faces, looking forwards and backwards in time: a god of transitions with a gatekeeper role. Just as Janus influenced the progress from one condition to another we might suppose that mesenchymal progenitors and tumor-related progeny could retain "Bifrons" properties reciprocally influencing microenvironmental cues specific for tumor progression. Reacting to discrete stimuli, MSC may change their phenotype to influence the transition from a pre-neoplastic condition to neoplasia or conversely, maintain a microenvironment that impedes tumor growth. Conceptually, MSC may thus have either a dynamic pro- or anti-tumorigenic effect according to different responses to specific tumor cell types or contact with their specific microenvironments.

Beyond speculation, the reasons for the discrepant actions of MSC on tumor growth are under investigation and may be attributable to differences in tumor models, the heterogeneity of MSC, the dose or timing of MSC injections, the animal host, or other factors that have not yet been appreciated. Zipori et al. initially demonstrated a bi-modal action of both murine and human BM-MSC against several tumor cell lines, describing that ex vivo MSC caused a dramatic increase in human lung and colon carcinoma cell line growth, but MSC inhibited the ex vivo cloning of both human and murine sarcoma cell lines. Such opposite effects seem to depend on the tumor type [362]. Nonetheless, in addition to experimental considerations even clinical studies on cancer patients provide some conflicting data about the effect mediated by MSC on cancer progression. Independent reports indicated that MSC do not affect the progression of breast or hematological malignancies [165, 179]. In contrast, a single cohort in patients suffering from hematological malignancies who were treated with chemotherapy and then co-transplanted with MSC and hematopoietic stem cells (HSC) showed greater incidence of disease relapse compared to those who received HSC alone [228]. Despite these data, in a larger cohort of more than 200 patients treated by MSC for different regenerative medicine applications [42] no increased risk of neoplasia was reported. Nonetheless, we cannot discount this possible adverse side effect of MSC treatment and further pre-clinical investigation is necessary regarding safe use of MSC as a novel tool of drug discovery. MSC are known to secrete exosomes or microparticles, with a growing appreciation that these may influence signaling within the tumor microenvironment [176, 358]. Introducing new paradigms, Roodhart et al. recently observed that treatment of an animal model of cancer with MSC could result in resistance to platinum chemotherapy [275]. They identified two distinct platinum-induced polyunsaturated fatty acids (PIFAs) released by MSC that conferred chemotherapeutic resistance. In addition, administration of PIFA alone sufficed to inhibit chemotherapeutic efficiency, leading the authors to conclude that PIFA antagonists may help generate beneficial effects in the context of novel chemotherapy combinations.

Therefore the "Bifrons" nature of MSC with regard to tumor interactions, though obscure, may be of utmost importance in experimental models aiming to clarify the complex relationship between a tumor and its milieu. In the meantime, the use of selected gene-modified MSC secreting pro-apoptotic inducing ligands may be introduced in the context of tumors where MSC may have a more neutral impact on cancer cell growth, as a novel approach to change the natural history of a still often fatal disease.

Acknowledgments This work was supported by: Ministero Italiano Istruzione Università e Ricerca-PRIN 2008 (MD, PP), Ministero Italiano della Salute-Bando Giovani Ricercatori (MD), Regione Emilia Romagna (PP, PC, MD), the Associazione ASEOP (PP) and Fondazione Cassa di Risparmio di Modena (MD).

References

1. Aboody KS, Najbauer J, Danks MK (2008) Stem and progenitor cell-mediated tumor selective gene therapy. Gene Ther 15:739–752
2. Adotevi O, Pere H, Ravel P, Haicheur N, Badoual C, Merillon N, Medioni J, Peyrard S, Roncelin S, Verkarre V, Mejean A, Fridman WH, Oudard S, Tartour E (2010) A decrease of regulatory T cells correlates with overall survival after sunitinib-based antiangiogenic therapy in metastatic renal cancer patients. J Immunother 33:991–998
3. Aggarwal S, Pittenger MF (2005) Human mesenchymal stem cells modulate allogeneic immune cell responses. Blood 105:1815–1822
4. Al-Khaldi A, Al-Sabti H, Galipeau J, Lachapelle K (2003) Therapeutic angiogenesis using autologous bone marrow stromal cells: improved blood flow in a chronic limb ischemia model. Ann Thorac Surg 75:204–209
5. Al-Khaldi A, Eliopoulos N, Martineau D, Lejeune L, Lachapelle K, Galipeau J (2003) Postnatal bone marrow stromal cells elicit a potent VEGF-dependent neoangiogenic response in vivo. Gene Ther 10:621–629
6. Allavena P, Mantovani A (2012) Immunology in the clinic review series; focus on cancer: tumour-associated macrophages: undisputed stars of the inflammatory tumour microenvironment. Clin Exp Immunol 167:195–205
7. Allay JA, Dennis JE, Haynesworth SE, Majumdar MK, Clapp DW, Shultz LD, Caplan AI, Gerson SL (1997) LacZ and interleukin-3 expression in vivo after retroviral transduction of marrow-derived human osteogenic mesenchymal progenitors. Hum Gene Ther 8:1417–1427
8. Almasan A, Ashkenazi A (2003) Apo2L/TRAIL: apoptosis signaling, biology, and potential for cancer therapy. Cytokine Growth Factor Rev 14:337–348
9. Amalinei C, Caruntu ID, Giusca SE, Balan RA (2010) Matrix metalloproteinases involvement in pathologic conditions. Rom J Morphol Embryol 51:215–228
10. Anderson WF (1992) Human gene therapy. Science 256:808–813
11. Annabi B, Naud E, Lee YT, Eliopoulos N, Galipeau J (2004) Vascular progenitors derived from murine bone marrow stromal cells are regulated by fibroblast growth factor and are avidly recruited by vascularizing tumors. J Cell Biochem 91:1146–1158
12. Ansieau S, Caron de Fromentel C, Bastid J, Morel AP, Puisieux A (2010) Role of the epithelial-mesenchymal transition during tumor progression. Bull Cancer 97:7–15
13. Apte RN, Voronov E (2008) Is interleukin-1 a good or bad 'guy' in tumor immunobiology and immunotherapy? Immunol Rev 222:222–241
14. Armstrong T, Packham G, Murphy LB, Bateman AC, Conti JA, Fine DR, Johnson CD, Benyon RC, Iredale JP (2004) Type I collagen promotes the malignant phenotype of pancreatic ductal adenocarcinoma. Clin Cancer Res 10:7427–7437
15. Arnulf B, Lecourt S, Soulier J, Ternaux B, Lacassagne MN, Crinquette A, Dessoly J, Sciaini AK, Benbunan M, Chomienne C, Fermand JP, Marolleau JP, Larghero J (2007) Phenotypic and functional characterization of bone marrow mesenchymal stem cells derived from patients with multiple myeloma. Leukemia 21:158–163
16. Askari AT, Unzek S, Popovic ZB, Goldman CK, Forudi F, Kiedrowski M, Rovner A, Ellis SG, Thomas JD, DiCorleto PE, Topol EJ, Penn MS (2003) Effect of stromal-cell-derived factor 1 on stem-cell homing and tissue regeneration in ischaemic cardiomyopathy. Lancet 362: 697–703
17. Ayala I, Baldassarre M, Caldieri G, Buccione R (2006) Invadopodia: a guided tour. Eur J Cell Biol 85:159–164
18. Balsamo M, Scordamaglia F, Pietra G, Manzini C, Cantoni C, Boitano M, Queirolo P, Vermi W, Facchetti F, Moretta A, Moretta L, Mingari MC, Vitale M (2009) Melanoma-associated fibroblasts modulate NK cell phenotype and antitumor cytotoxicity. Proc Natl Acad Sci U S A 106:20847–20852

19. Balyasnikova IV, Ferguson SD, Sengupta S, Han Y, Lesniak MS (2010) Mesenchymal stem cells modified with a single-chain antibody against EGFRvIII successfully inhibit the growth of human xenograft malignant glioma. PLoS One 5:e9750
20. Beckermann BM, Kallifatidis G, Groth A, Frommhold D, Apel A, Mattern J, Salnikov AV, Moldenhauer G, Wagner W, Diehlmann A, Saffrich R, Schubert M, Ho AD, Giese N, Buchler MW, Friess H, Buchler P, Herr I (2008) VEGF expression by mesenchymal stem cells contributes to angiogenesis in pancreatic carcinoma. Br J Cancer 99:622–631
21. Bensinger SJ, Christofk HR (2012) New aspects of the Warburg effect in cancer cell biology. Semin Cell Dev Biol 23:352–361
22. Berdiel-Acer M, Bohem ME, Lopez-Doriga A, Vidal A, Salazar R, Martinez-Iniesta M, Santos C, Sanjuan X, Villanueva A, Mollevi DG (2011) Hepatic carcinoma-associated fibroblasts promote an adaptative response in colorectal cancer cells that inhibit proliferation and apoptosis: nonresistant cells die by nonapoptotic cell death. Neoplasia 13:931–946
23. Bergfeld SA, DeClerck YA (2010) Bone marrow-derived mesenchymal stem cells and the tumor microenvironment. Cancer Metastasis Rev 29:249–261
24. Bernardo ME, Cometa AM, Locatelli F (2012) Mesenchymal stromal cells: a novel and effective strategy for facilitating engraftment and accelerating hematopoietic recovery after transplantation? Bone Marrow Transplant 47:323–329
25. Berndt A, Kollner R, Richter P, Franz M, Voigt A, Berndt A, Borsi L, Giavazzi R, Neri D, Kosmehl H (2010) A comparative analysis of oncofetal fibronectin and tenascin-C incorporation in tumour vessels using human recombinant SIP format antibodies. Histochem Cell Biol 133:467–475
26. Bianchi G, Borgonovo G, Pistoia V, Raffaghello L (2011) Immunosuppressive cells and tumour microenvironment: focus on mesenchymal stem cells and myeloid derived suppressor cells. Histol Histopathol 26:941–951
27. Bianco P, Robey PG, Saggio I, Riminucci M (2010) "Mesenchymal" stem cells in human bone marrow (skeletal stem cells): a critical discussion of their nature, identity, and significance in incurable skeletal disease. Hum Gene Ther 21:1057–1066
28. Binder DK, Berger MS (2002) Proteases and the biology of glioma invasion. J Neuro-Oncol 56:149–158
29. Blasi F, Sidenius N (2010) The urokinase receptor: focused cell surface proteolysis, cell adhesion and signaling. FEBS Lett 584:1923–1930
30. Bremnes RM, Donnem T, Al-Saad S, Al-Shibli K, Andersen S, Sirera R, Camps C, Marinez I, Busund LT (2011) The role of tumor stroma in cancer progression and prognosis: emphasis on carcinoma-associated fibroblasts and non-small cell lung cancer. J Thorac Oncol 6:209–217
31. Brentnall TA, Lai LA, Coleman J, Bronner MP, Pan S, Chen R (2012) Arousal of cancer-associated stroma: overexpression of palladin activates fibroblasts to promote tumor invasion. PLoS One 7:e30219
32. Brocke-Heidrich K, Kretzschmar AK, Pfeifer G, Henze C, Löffler D, Koczan D, Thiesen HJ, Burger R, Gramatzki M, Horn F (2004) Interleukin-6-dependent gene expression profiles in multiple myeloma INA-6 cells reveal a Bcl-2 family-independent survival pathway closely associated with Stat3 activation. Blood 103:242–251
33. Brooks PC, Silletti S, von Schalscha TL, Friedlander M, Cheresh DA (1998) Disruption of angiogenesis by PEX, a noncatalytic metalloproteinase fragment with integrin binding activity. Cell 92:391–400
34. Bui JD, Schreiber RD (2007) Cancer immunosurveillance, immunoediting and inflammation: independent or interdependent processes? Curr Opin Immunol 19:203–208
35. Burger JA, Kipps TJ (2006) CXCR4: a key receptor in the crosstalk between tumor cells and their microenvironment. Blood 107:1761–1767
36. Burns JS, Kristiansen M, Kristensen LP, Larsen KH, Nielsen MO, Christiansen H, Nehlin J, Andersen JS, Kassem M (2011) Decellularized matrix from tumorigenic human mesenchymal stem cells promotes neovascularization with galectin-1 dependent endothelial interaction. PLoS One 6:e21888

37. Cai J, Tang H, Xu L, Wang X, Yang C, Ruan S, Guo J, Hu S, Wang Z (2012) Fibroblasts in omentum activated by tumor cells promote ovarian cancer growth, adhesion and invasiveness. Carcinogenesis 33:20–29
38. Cammarota R, Bertolini V, Pennesi G, Bucci EO, Gottardi O, Garlanda C, Laghi L, Barberis MC, Sessa F, Noonan DM, Albini A (2010) The tumor microenvironment of colorectal cancer: stromal TLR-4 expression as a potential prognostic marker. J Transl Med 8:112
39. Caplan AI (1991) Mesenchymal stem cells. J Orthop Res 9:641–650
40. Caplan AI, Dennis JE (2006) Mesenchymal stem cells as trophic mediators. J Cell Biochem 98:1076–1084
41. Carmi Y, Rinott G, Dotan S, Elkabets M, Rider P, Voronov E, Apte RN (2011) Microenvironment-derived IL-1 and IL-17 interact in the control of lung metastasis. J Immunol 186:3462–3471
42. Centeno CJ, Schultz JR, Cheever M, Freeman M, Faulkner S, Robinson B, Hanson R (2011) Safety and complications reporting update on the re-implantation of culture-expanded mesenchymal stem cells using autologous platelet lysate technique. Curr Stem Cell Res Ther 6:368–378
43. Chen JR, Cheng GY, Sheu CC, Tseng GF, Wang TJ, Huang YS (2008) Transplanted bone marrow stromal cells migrate, differentiate and improve motor function in rats with experimentally induced cerebral stroke. J Anat 213:249–258
44. Chen L, Tredget EE, Wu PY, Wu Y (2008) Paracrine factors of mesenchymal stem cells recruit macrophages and endothelial lineage cells and enhance wound healing. PLoS One 3:e1886
45. Chen PM, Yen ML, Liu KJ, Sytwu HK, Yen BL (2011) Immunomodulatory properties of human adult and fetal multipotent mesenchymal stem cells. J Biomed Sci 18:49
46. Chen Y, Shi M, Yu GZ, Qin XR, Jin G, Chen P, Zhu MH (2012) Interleukin-8, a promising predictor for prognosis of pancreatic cancer. World J Gastroenterol 18:1123–1129
47. Chen ZZ, Jiang XD, Zhang LL, Shang JH, Du MX, Xu G, Xu RX (2008) Beneficial effect of autologous transplantation of bone marrow stromal cells and endothelial progenitor cells on cerebral ischemia in rabbits. Neurosci Lett 445:36–41
48. Chiellini C, Cochet O, Negroni L, Samson M, Poggi M, Ailhaud G, Alessi MC, Dani C, Amri EZ (2008) Characterization of human mesenchymal stem cell secretome at early steps of adipocyte and osteoblast differentiation. BMC Mol Biol 9:26
49. Cho HJ, Jung JI, Lim DY, Kwon GT, Her S, Park JH, Park JH (2012) Bone marrow-derived, alternatively-activated macrophages enhance solid tumor growth and lung metastasis of mammary carcinoma cells in a Balb/C mouse orthotopic model. Breast Cancer Res 14:R81
50. Clamp AR, Jayson GC (2005) The clinical potential of antiangiogenic fragments of extracellular matrix proteins. Br J Cancer 93:967–972
51. Coffelt SB, Marini FC, Watson K, Zwezdaryk KJ, Dembinski JL, LaMarca HL, Tomchuck SL, Honer zu Bentrup K, Danka ES, Henkle SL, Scandurro AB (2009) The pro-inflammatory peptide LL-37 promotes ovarian tumor progression through recruitment of multipotent mesenchymal stromal cells. Proc Natl Acad Sci U S A 106:3806–3811
52. Colognato H, Yurchenco PD (2000) Form and function: the laminin family of heterotrimers. Dev Dyn 218:213–234
53. Cousin B, Ravet E, Poglio S, De Toni F, Bertuzzi M, Lulka H, Touil I, Andre M, Grolleau JL, Peron JM, Chavoin JP, Bourin P, Penicaud L, Casteilla L, Buscail L, Cordelier P (2009) Adult stromal cells derived from human adipose tissue provoke pancreatic cancer cell death both in vitro and in vivo. PLoS One 4:e6278
54. Coussens LM, Werb Z (2002) Inflammation and cancer. Nature 420:860–867
55. Crisan M, Yap S, Casteilla L, Chen CW, Corselli M, Park TS, Andriolo G, Sun B, Zheng B, Zhang L, Norotte C, Teng PN, Traas J, Schugar R, Deasy BM, Badylak S, Buhring HJ, Giacobino JP, Lazzari L, Huard J, Peault B (2008) A perivascular origin for mesenchymal stem cells in multiple human organs. Cell Stem Cell 3:301–313

56. Cui W, Fowlis DJ, Bryson S, Duffie E, Ireland H, Balmain A, Akhurst RJ (1996) TGFbeta1 inhibits the formation of benign skin tumors, but enhances progression to invasive spindle carcinomas in transgenic mice. Cell 86:531–542
57. Czekay RP, Wilkins-Port CE, Higgins SP, Freytag J, Overstreet JM, Klein RM, Higgins CE, Samarakoon R, Higgins PJ (2011) PAI-1: An integrator of cell signaling and migration. Int J Cell Biol 2011:562481
58. Danks MK, Yoon KJ, Bush RA, Remack JS, Wierdl M, Tsurkan L, Kim SU, Garcia E, Metz MZ, Najbauer J, Potter PM, Aboody KS (2007) Tumor-targeted enzyme/prodrug therapy mediates long-term disease-free survival of mice bearing disseminated neuroblastoma. Cancer Res 67:22–25
59. Dano K, Behrendt N, Hoyer-Hansen G, Johnsen M, Lund LR, Ploug M, Romer J (2005) Plasminogen activation and cancer. Thromb Haemost 93:676–681
60. De Palma M, Nucera S (2012) Circulating endothelial progenitors and tumor resistance to vascular-targeting therapies. Cancer Discov 2:395–397
61. DeClerck YA, Bomann ET, Spengler BA, Biedler JL (1987) Differential collagen biosynthesis by human neuroblastoma cell variants. Cancer Res 47:6505–6510
62. Decock J, Thirkettle S, Wagstaff L, Edwards DR (2011) Matrix metalloproteinases: protective roles in cancer. J Cell Mol Med 15:1254–1265
63. Desgrosellier JS, Cheresh DA (2010) Integrins in cancer: biological implications and therapeutic opportunities. Nat Rev Cancer 10:9–22
64. Devine SM, Cobbs C, Jennings M, Bartholomew A, Hoffman R (2003) Mesenchymal stem cells distribute to a wide range of tissues following systemic infusion into nonhuman primates. Blood 101:2999–3001
65. Dieu-Nosjean MC, Antoine M, Danel C, Heudes D, Wislez M, Poulot V, Rabbe N, Laurans L, Tartour E, de Chaisemartin L, Lebecque S, Fridman WH, Cadranel J (2008) Long-term survival for patients with non-small-cell lung cancer with intratumoral lymphoid structures. J Clin Oncol 26:4410–4417
66. Dinarello CA (2006) The paradox of pro-inflammatory cytokines in cancer. Cancer Metastasis Rev 25:307–313
67. Ding Y, Song N, Luo Y. (2012). Role of bone marrow-derived cells in angiogenesis: focus on macrophages and pericytes. Cancer Microenviron, (in press). http://dx.doi.org/10.1007/s12307-012-0106-y
68. Direkze NC, Hodivala-Dilke K, Jeffery R, Hunt T, Poulsom R, Oukrif D, Alison MR, Wright NA (2004) Bone marrow contribution to tumor-associated myofibroblasts and fibroblasts. Cancer Res 64:8492–8495
69. Djouad F, Delorme B, Maurice M, Bony C, Apparailly F, Louis-Plence P, Canovas F, Charbord P, Noel D, Jorgensen C (2007) Microenvironmental changes during differentiation of mesenchymal stem cells towards chondrocytes. Arthritis Res Ther 9:R33
70. Djouad F, Plence P, Bony C, Tropel P, Apparailly F, Sany J, Noel D, Jorgensen C (2003) Immunosuppressive effect of mesenchymal stem cells favors tumor growth in allogeneic animals. Blood 102:3837–3844
71. Dohi O, Ohtani H, Hatori M, Sato E, Hosaka M, Nagura H, Itoi E, Kokubun S (2009) Histogenesis-specific expression of fibroblast activation protein and dipeptidylpeptidase-IV in human bone and soft tissue tumours. Histopathology 55:432–440
72. Dominici M, Le Blanc K, Mueller I, Slaper-Cortenbach I, Marini F, Krause D, Deans R, Keating A, Prockop D, Horwitz E (2006) Minimal criteria for defining multipotent mesenchymal stromal cells. The international society for cellular therapy position statement. Cytotherapy 8:315–317
73. Dubois SG, Floyd EZ, Zvonic S, Kilroy G, Wu X, Carling S, Halvorsen YD, Ravussin E, Gimble JM (2008) Isolation of human adipose-derived stem cells from biopsies and liposuction specimens. Methods Mol Biol 449:69–79
74. Duda DG, Cohen KS, Kozin SV, Perentes JY, Fukumura D, Scadden DT, Jain RK (2006) Evidence for incorporation of bone marrow-derived endothelial cells into perfused blood vessels in tumors. Blood 107:2774–2776

75. Duluc D, Delneste Y, Tan F, Moles MP, Grimaud L, Lenoir J, Preisser L, Anegon I, Catala L, Ifrah N, Descamps P, Gamelin E, Gascan H, Hebbar M, Jeannin P (2007) Tumor-associated leukemia inhibitory factor and IL-6 skew monocyte differentiation into tumor-associated macrophage-like cells. Blood 110:4319–4330
76. Dvorak HF (1986) Tumors: wounds that do not heal. Similarities between tumor stroma generation and wound healing. N Engl J Med 315:1650–1659
77. Dvorak HF, Weaver VM, Tlsty TD, Bergers G (2011) Tumor microenvironment and progression. J Surg Oncol 103:468–474
78. Dwyer RM, Potter-Beirne SM, Harrington KA, Lowery AJ, Hennessy E, Murphy JM, Barry FP, O'Brien T, Kerin MJ (2007) Monocyte chemotactic protein-1 secreted by primary breast tumors stimulates migration of mesenchymal stem cells. Clin Cancer Res 13:5020–5027
79. Elzaouk L, Moelling K, Pavlovic J (2006) Anti-tumor activity of mesenchymal stem cells producing IL-12 in a mouse melanoma model. Exp Dermatol 15:865–874
80. Emura M, Ochiai A, Horino M, Arndt W, Kamino K, Hirohashi S (2000) Development of myofibroblasts from human bone marrow mesenchymal stem cells cocultured with human colon carcinoma cells and TGF beta 1. In Vitro Cell Dev Biol Anim 36:77–80
81. Erez N, Truitt M, Olson P, Arron ST, Hanahan D (2010) Cancer-Associated Fibroblasts Are Activated in Incipient Neoplasia to Orchestrate Tumor-Promoting Inflammation in an NF-kappaB-dependent manner. Cancer Cell 17:135–147
82. Fidler IJ, Ellis LM (1994) The implications of angiogenesis for the biology and therapy of cancer metastasis. Cell 79:185–188
83. Fillat C, Carrio M, Cascante A, Sangro B (2003) Suicide gene therapy mediated by the Herpes Simplex virus thymidine kinase gene/Ganciclovir system: fifteen years of application. Curr Gene Ther 3:13–26
84. Finger EC, Giaccia AJ (2010) Hypoxia, inflammation, and the tumor microenvironment in metastatic disease. Cancer Metastasis Rev 29:285–293
85. Folkman J (1971) Tumor angiogenesis: therapeutic implications. N Engl J Med 285: 1182–1186
86. Fong CY, Gauthaman K, Cheyyatraivendran S, Lin HD, Biswas A, Bongso A (2012) Human umbilical cord Wharton's jelly stem cells and its conditioned medium support hematopoietic stem cell expansion ex vivo. J Cell Biochem 113:658–668
87. Fong CY, Richards M, Manasi N, Biswas A, Bongso A (2007) Comparative growth behaviour and characterization of stem cells from human Wharton's jelly. Reprod Biomed Online 15:708–718
88. Francia G, Emmenegger U, Kerbel RS (2009) Tumor-associated fibroblasts as "Trojan Horse" mediators of resistance to anti-VEGF therapy. Cancer Cell 15:3–5
89. Franco OE, Shaw AK, Strand DW, Hayward SW (2010) Cancer associated fibroblasts in cancer pathogenesis. Semin Cell Dev Biol 21:33–39
90. Franses JW, Baker AB, Chitalia VC, Edelman ER. (2011). Stromal endothelial cells directly influence cancer progression. Sci Transl Med 3:66ra5
91. Fredriksson L, Li H, Eriksson U (2004) The PDGF family: four gene products form five dimeric isoforms. Cytokine Growth Factor Rev 15:197–204
92. Friedenstein AJ, Chailakhyan RK, Latsinik NV, Panasyuk AF, Keiliss-Borok IV (1974) Stromal cells responsible for transferring the microenvironment of the hemopoietic tissues. Cloning in vitro and retransplantation in vivo. Transplantation 17:331–340
93. Fu L, Zhang C, Zhang LY, Dong SS, Lu LH, Chen J, Dai Y, Li Y, Kong KL, Kwong DL, Guan XY (2011) Wnt2 secreted by tumour fibroblasts promotes tumour progression in oesophageal cancer by activation of the Wnt/beta-catenin signalling pathway. Gut 60: 1635–1643
94. Fujita M, Khazenzon NM, Bose S, Sekiguchi K, Sasaki T, Carter WG, Ljubimov AV, Black KL, Ljubimova JY (2005) Overexpression of beta1-chain-containing laminins in capillary basement membranes of human breast cancer and its metastases. Breast Cancer Res 7:R411–R421

95. Fukaya Y, Shimada H, Wang LC, Zandi E, DeClerck YA (2008) Identification of galectin-3-binding protein as a factor secreted by tumour cells that stimulates interleukin-6 expression in the bone marrow stroma. J Biol Chem 283:18573–18581
96. Gabrilovich DI, Nagaraj S (2009) Myeloid-derived suppressor cells as regulators of the immune system. Nat Rev Immunol 9:162–174
97. Galie M, Konstantinidou G, Peroni D, Scambi I, Marchini C, Lisi V, Krampera M, Magnani P, Merigo F, Montani M, Boschi F, Marzola P, Orru R, Farace P, Sbarbati A, Amici A (2008) Mesenchymal stem cells share molecular signature with mesenchymal tumor cells and favor early tumor growth in syngeneic mice. Oncogene 27:2542–2551
98. Gao H, Priebe W, Glod J, Banerjee D (2009) Activation of signal transducers and activators of transcription 3 and focal adhesion kinase by stromal cell-derived factor 1 is required for migration of human mesenchymal stem cells in response to tumor cell-conditioned medium. Stem Cells 27:857–865
99. Gao J, Dennis JE, Muzic RF, Lundberg M, Caplan AI (2001) The dynamic in vivo distribution of bone marrow-derived mesenchymal stem cells after infusion. Cells Tissues Organs 169:12–20
100. Garcia-Tunon I, Ricote M, Ruiz AA, Fraile B, Paniagua R, Royuela M (2007) Influence of IFN-gamma and its receptors in human breast cancer. BMC Cancer 7:158
101. Garrafa E, Caprara V, Di Castro V, Rosano L, Bagnato A, Spinella F (2012) Endothelin-1 cooperates with hypoxia to induce vascular-like structures through vascular endothelial growth factor-C, -D and -A in lymphatic endothelial cells. Life Sci. doi:10.1016/j.lfs.2012.03.033
102. Gerdin B, Hallgren R (1997) Dynamic role of hyaluronan (HYA) in connective tissue activation and inflammation. J Intern Med 242:49–55
103. Ghosh AK, Nagpal V, Covington JW, Michaels MA, Vaughan DE (2012) Molecular basis of cardiac endothelial-to-mesenchymal transition (EndMT): differential expression of microRNAs during EndMT. Cell Signal 24:1031–1036
104. Giannelli G, Fransvea E, Bergamini C, Marinosci F, Antonaci S (2003) Laminin-5 chains are expressed differentially in metastatic and nonmetastatic hepatocellular carcinoma. Clin Cancer Res 9:3684–3691
105. Goel S, Duda DG, Xu L, Munn LL, Boucher Y, Fukumura D, Jain RK (2011) Normalization of the vasculature for treatment of cancer and other diseases. Physiol Rev 91:1071–1121
106. Gonda TA, Varro A, Wang TC, Tycko B (2010) Molecular biology of cancer-associated fibroblasts: can these cells be targeted in anti-cancer therapy? Semin Cell Dev Biol 21:2–10
107. Graf T (2002) Differentiation plasticity of hematopoietic cells. Blood 99:3089–3101
108. Gravdal K, Halvorsen OJ, Haukaas SA, Akslen LA (2007) A switch from E-cadherin to N-cadherin expression indicates epithelial to mesenchymal transition and is of strong and independent importance for the progress of prostate cancer. Clin Cancer Res 13:7003–7011
109. Grisendi G, Bussolari R, Cafarelli L, Petak I, Rasini V, Veronesi E, De Santis G, Spano C, Tagliazzucchi M, Barti-Juhasz H, Scarabelli L, Bambi F, Frassoldati A, Rossi G, Casali C, Morandi U, Horwitz EM, Paolucci P, Conte PF, Dominici M (2010) Adipose-derived mesenchymal stem cells as stable source of tumor necrosis factor-related apoptosis-inducing ligand delivery for cancer therapy. Cancer Res 70:3718–3729
110. Grisendi G, Bussolari R, Veronesi E, Piccinno S, Burns JS, De Santis G, Loschi P, Pignatti M, Di Benedetto F, Ballarin R, Di Gregorio C, Guarneri V, Piccinini L, Horwitz EM, Paolucci P, Conte PF, Dominici M (2011) Understanding tumor-stroma interplays for targeted therapies by armed mesenchymal stromal progenitors: the Mesenkillers. Am J Cancer Res 1:787–805
111. Grivennikov SI, Greten FR, Karin M (2010) Immunity, inflammation, and cancer. Cell 140:883–899
112. Gronthos S, Arthur A, Bartold PM, Shi S (2011) A method to isolate and culture expand human dental pulp stem cells. Methods Mol Biol 698:107–121
113. Gucciardo L, Lories R, Ochsenbein-Kolble N, Done' E, Zwijsen A, Deprest J (2009) Fetal mesenchymal stem cells: isolation, properties and potential use in perinatology and regenerative medicine. BJOG 116:166–172

114. Gunn WG, Conley A, Deininger L, Olson SD, Prockop DJ, Gregory CA (2006) A crosstalk between myeloma cells and marrow stromal cells stimulates production of DKK1 and interleukin-6: a potential role in the development of lytic bone disease and tumor progression in multiple myeloma. Stem Cells 24:986–991
115. Guo W, Giancotti FG (2004) Integrin signalling during tumour progression. Nat Rev Mol Cell Biol 5:816–826
116. Guo X, Oshima H, Kitmura T, Taketo MM, Oshima M (2008) Stromal fibroblasts activated by tumor cells promote angiogenesis in mouse gastric cancer. J Biol Chem 283:19864–19871
117. Guo Y, Xu F, Lu T, Duan Z, Zhang Z (2012) Interleukin-6 signaling pathway in targeted therapy for cancer. Cancer Treat Rev, (in press). http://www.sciencedirect.com/science/article/pii/S0305737212001053
118. Halama N, Michel S, Kloor M, Zoernig I, Benner A, Spille A, Pommerencke T, von Knebel DM, Folprecht G, Luber B, Feyen N, Martens UM, Beckhove P, Gnjatic S, Schirmacher P, Herpel E, Weitz J, Grabe N, Jaeger D (2011) Localization and density of immune cells in the invasive margin of human colorectal cancer liver metastases are prognostic for response to chemotherapy. Cancer Res 71:5670–5677
119. Han S, Khuri FR, Roman J (2006) Fibronectin stimulates non-small cell lung carcinoma cell growth through activation of Akt/mammalian target of rapamycin/S6 kinase and inactivation of LKB1/AMP-activated protein kinase signal pathways. Cancer Res 66:315–323
120. Haraguchi M, Okubo T, Miyashita Y, Miyamoto Y, Hayashi M, Crotti TN, McHugh KP, Ozawa M (2008) Snail regulates cell-matrix adhesion by regulation of the expression of integrins and basement membrane proteins. J Biol Chem 283:23514–23523
121. Hayashi Y, Tsuji S, Tsujii M, Nishida T, Ishii S, Iijima H, Nakamura T, Eguchi H, Miyoshi E, Hayashi N, Kawano S (2008) Topical transplantation of mesenchymal stem cells accelerates gastric ulcer healing in rats. Am J Physiol Gastrointest Liver Physiol 294:G778–G786
122. Haynesworth SE, Baber MA, Caplan AI (1996) Cytokine expression by human marrow-derived mesenchymal progenitor cells in vitro: effects of dexamethasone and IL-1 alpha. J Cell Physiol 166:585–592
123. Heusinkveld M, van der Burg SH (2011) Identification and manipulation of tumor associated macrophages in human cancers. J Transl Med 9:216
124. Hielscher AC, Qiu C, Gerecht S (2012) Breast cancer cell-derived matrix supports vascular morphogenesis. Am J Physiol Cell Physiol 302:C1243–C1256
125. Holler E (2005) Laminin isoform expression in breast tumors. Breast Cancer Res 7:166–167
126. Honczarenko M, Le Y, Swierkowski M, Ghiran I, Glodek AM, Silberstein LE (2006) Human bone marrow stromal cells express a distinct set of biologically functional chemokine receptors. Stem Cells 24:1030–1041
127. Honma K, Miyata T, Ochiya T (2007) Type I collagen gene suppresses tumor growth and invasion of malignant human glioma cells. Cancer Cell Int 7:12
128. Horimoto Y, Takahashi Y, Polanska U, Orimo A. (2012). Emerging roles of the tumor-associated stroma in promoting tumor metastasis. Cell Adh Migr 6(3):203–206 (in press)
129. Horwitz EM, Dominici M (2008) How do mesenchymal stromal cells exert their therapeutic benefit? Cytotherapy 10:771–774
130. Horwitz EM, Gordon PL, Koo WK, Marx JC, Neel MD, McNall RY, Muul L, Hofmann T (2002) Isolated allogeneic bone marrow-derived mesenchymal cells engraft and stimulate growth in children with osteogenesis imperfecta: Implications for cell therapy of bone. Proc Natl Acad Sci U S A 99:8932–8937
131. Hoshino A, Ishii G, Ito T, Aoyagi K, Ohtaki Y, Nagai K, Sasaki H, Ochiai A (2011) Podoplanin-positive fibroblasts enhance lung adenocarcinoma tumor formation: podoplanin in fibroblast functions for tumor progression. Cancer Res 71:4769–4779
132. Houghton J, Stoicov C, Nomura S, Rogers AB, Carlson J, Li H, Cai X, Fox JG, Goldenring JR, Wang TC (2004) Gastric cancer originating from bone marrow-derived cells. Science 306:1568–1571
133. Hsia HC, Schwarzbauer JE (2005) Meet the tenascins: multifunctional and mysterious. J Biol Chem 280:26641–26644

134. Huang P, Rani MR, Ahluwalia MS, Bae E, Prayson RA, Weil RJ, Nowacki AS, Hedayat H, Sloan AE, Lathia JD, Rich JN, Tipps R, Gladson CL (2012) Endothelial expression of TNF receptor-1 generates a proapoptotic signal inhibited by integrin alpha6beta1 in glioblastoma. Cancer Res 72:1428–1437
135. Hutchinson L, Stenstrom B, Chen D, Piperdi B, Levey S, Lyle S, Wang TC, Houghton J (2011) Human Barrett's adenocarcinoma of the esophagus, associated myofibroblasts, and endothelium can arise from bone marrow-derived cells after allogeneic stem cell transplant. Stem Cells Dev 20:11–17
136. Hwang JH, Shim SS, Seok OS, Lee HY, Woo SK, Kim BH, Song HR, Lee JK, Park YK (2009) Comparison of cytokine expression in mesenchymal stem cells from human placenta, cord blood, and bone marrow. J Korean Med Sci 24:547–554
137. Hwang RF, Moore T, Arumugam T, Ramachandran V, Amos KD, Rivera A, Ji B, Evans DB, Logsdon CD (2008) Cancer-associated stromal fibroblasts promote pancreatic tumor progression. Cancer Res 68:918–926
138. Ishii G, Sangai T, Oda T, Aoyagi Y, Hasebe T, Kanomata N, Endoh Y, Okumura C, Okuhara Y, Magae J, Emura M, Ochiya T, Ochiai A (2003) Bone-marrow-derived myofibroblasts contribute to the cancer-induced stromal reaction. Biochem Biophys Res Commun 309:232–240
139. Isoda K, Kojima M, Takeda M, Higashiyama S, Kawase M, Yagi K (2004) Maintenance of hepatocyte functions by coculture with bone marrow stromal cells. J Biosci Bioeng 97:343–346
140. Iwamoto S, Mihara K, Downing JR, Pui CH, Campana D (2007) Mesenchymal cells regulate the response of acute lymphoblastic leukemia cells to asparaginase. J Clin Invest 117:1049–1057
141. Jaster R (2004) Molecular regulation of pancreatic stellate cell function. Mol Cancer 3:26
142. Jeon ES, Moon HJ, Lee MJ, Song HY, Kim YM, Cho M, Suh DS, Yoon MS, Chang CL, Jung JS, Kim JH (2008) Cancer-derived lysophosphatidic acid stimulates differentiation of human mesenchymal stem cells to myofibroblast-like cells. Stem Cells 26:789–797
143. Ji J, Zhao L, Budhu A, Forgues M, Jia HL, Qin LX, Ye QH, Yu J, Shi X, Tang ZY, Wang XW (2010) Let-7 g targets collagen type I alpha2 and inhibits cell migration in hepatocellular carcinoma. J Hepatol 52:690–697
144. Johann PD, Vaegler M, Gieseke F, Mang P, Armeanu-Ebinger S, Kluba T, Handgretinger R, Muller I (2010) Tumour stromal cells derived from paediatric malignancies display MSC-like properties and impair NK cell cytotoxicity. BMC Cancer 10:501
145. Jones ML, Ewing CM, Isaacsa WB, Getzenberg RH (2012) Prostate cancer-derived angiogenin stimulates the invasion of prostate fibroblasts. J Cell Mol Med 16:193–201
146. Jotzu C, Alt E, Welte G, Li J, Hennessy BT, Devarajan E, Krishnappa S, Pinilla S, Droll L, Song YH (2010) Adipose tissue-derived stem cells differentiate into carcinoma-associated fibroblast-like cells under the influence of tumor-derived factors. Anal Cell Pathol (Amst) 33:61–79
147. Kalluri R (2003) Basement membranes: structure, assembly and role in tumour angiogenesis. Nat Rev Cancer 3:422–433
148. Kalluri R, Zeisberg M (2006) Fibroblasts in cancer. Nat Rev Cancer 6:392–401
149. Kanehira M, Xin H, Hoshino K, Maemondo M, Mizuguchi H, Hayakawa T, Matsumoto K, Nakamura T, Nukiwa T, Saijo Y (2007) Targeted delivery of NK4 to multiple lung tumors by bone marrow-derived mesenchymal stem cells. Cancer Gene Ther 14:894–903
150. Karin M (2006) Nuclear factor-kappaB in cancer development and progression. Nature 441:431–436
151. Karnoub AE, Dash AB, Vo AP, Sullivan A, Brooks MW, Bell GW, Richardson AL, Polyak K, Tubo R, Weinberg RA (2007) Mesenchymal stem cells within tumour stroma promote breast cancer metastasis. Nature 449:557–563
152. Kawamoto T, Ohga N, Akiyama K, Hirata N, Kitahara S, Maishi N, Osawa T, Yamamoto K, Kondoh M, Shindoh M, Hida Y, Hida K (2012) Tumor-Derived Microvesicles Induce Proangiogenic Phenotype in Endothelial Cells via Endocytosis. PLoS One 7:e34045

153. Kawashiri S, Tanaka A, Noguchi N, Hase T, Nakaya H, Ohara T, Kato K, Yamamoto E (2009) Significance of stromal desmoplasia and myofibroblast appearance at the invasive front in squamous cell carcinoma of the oral cavity. Head Neck 31:1346–1353
154. Kessenbrock K, Plaks V, Werb Z (2010) Matrix metalloproteinases: regulators of the tumor microenvironment. Cell 141:52–67
155. Khakoo AY, Pati S, Anderson SA, Reid W, Elshal MF, Rovira II, Nguyen AT, Malide D, Combs CA, Hall G, Zhang J, Raffeld M, Rogers TB, Stetler-Stevenson W, Frank JA, Reitz M, Finkel T (2006) Human mesenchymal stem cells exert potent antitumorigenic effects in a model of Kaposi's sarcoma. J Exp Med 203:1235–1247
156. Kidd S, Caldwell L, Dietrich M, Samudio I, Spaeth EL, Watson K, Shi Y, Abbruzzese J, Konopleva M, Andreeff M, Marini FC (2010) Mesenchymal stromal cells alone or expressing interferon-beta suppress pancreatic tumors in vivo, an effect countered by anti-inflammatory treatment. Cytotherapy 12:615–625
157. Kidd S, Spaeth E, Klopp A, Andreeff M, Hall B, Marini FC (2008) The (in) auspicious role of mesenchymal stromal cells in cancer: be it friend or foe. Cytotherapy 10:657–667
158. Kidd S, Spaeth E, Watson K, Burks J, Lu H, Klopp A, Andreeff M, Marini FC (2012) Origins of the tumor microenvironment: quantitative assessment of adipose-derived and bone marrow-derived stroma. PLoS One 7:e30563
159. Kim JS, Lee HK, Kim MR, Kim PK, Kim CW (2008) Differentially expressed proteins of mesenchymal stem cells derived from human cord blood (hUCB) during osteogenic differentiation. Biosci Biotechnol Biochem 72:2309–2317
160. Kinnaird T, Stabile E, Burnett MS, Lee CW, Barr S, Fuchs S, Epstein SE (2004) Marrow-derived stromal cells express genes encoding a broad spectrum of arteriogenic cytokines and promote in vitro and in vivo arteriogenesis through paracrine mechanisms. Circ Res 94:678–685
161. Kinnaird T, Stabile E, Burnett MS, Shou M, Lee CW, Barr S, Fuchs S, Epstein SE (2004) Local delivery of marrow-derived stromal cells augments collateral perfusion through paracrine mechanisms. Circulation 109:1543–1549
162. Kirschmann DA, Seftor EA, Hardy KM, Seftor RE, Hendrix MJ (2012) Molecular pathways: vasculogenic mimicry in tumor cells: diagnostic and therapeutic implications. Clin Cancer Res 18:2726–2732
163. Klees RF, Salasznyk RM, Kingsley K, Williams WA, Boskey A, Plopper GE (2005) Laminin-5 induces osteogenic gene expression in human mesenchymal stem cells through an ERK-dependent pathway. Mol Biol Cell 16:881–890
164. Klopp AH, Gupta A, Spaeth E, Andreeff M, Marini Fr (2011) Concise review: dissecting a discrepancy in the literature: do mesenchymal stem cells support or suppress tumor growth? Stem Cells 29:11–19
165. Koc ON, Gerson SL, Cooper BW, Dyhouse SM, Haynesworth SE, Caplan AI, Lazarus HM (2000) Rapid hematopoietic recovery after coinfusion of autologous-blood stem cells and culture-expanded marrow mesenchymal stem cells in advanced breast cancer patients receiving high-dose chemotherapy. J Clin Oncol 18:307–316
166. Koc ON, Lazarus HM (2001) Mesenchymal stem cells: heading into the clinic. Bone Marrow Transplant 27:235–239
167. Komarova S, Kawakami Y, Stoff-Khalili MA, Curiel DT, Pereboeva L (2006) Mesenchymal progenitor cells as cellular vehicles for delivery of oncolytic adenoviruses. Mol Cancer Ther 5:755–766
168. Komarova S, Roth J, Alvarez R, Curiel DT, Pereboeva L (2010) Targeting of mesenchymal stem cells to ovarian tumors via an artificial receptor. J Ovarian Res 3:12
169. Koppenol WH, Bounds PL, Dang CV (2011) Otto Warburg's contributions to current concepts of cancer metabolism. Nat Rev Cancer 11:325–337
170. Korkaya H, Liu S, Wicha MS (2011) Regulation of cancer stem cells by cytokine networks: attacking cancer's inflammatory roots. Clin Cancer Res 17:6125–6129

171. Kotyza J (2012) Interleukin-8 (CXCL8) in tumor associated non-vascular extracellular fluids: its diagnostic and prognostic values. A review. Int J Biol Markers. doi:10.5301/JBM.2012.9261
172. Kucerova L, Altanerova V, Matuskova M, Tyciakova S, Altaner C (2007) Adipose tissue-derived human mesenchymal stem cells mediated prodrug cancer gene therapy. Cancer Res 67:6304–6313
173. Kucerova L, Matuskova M, Pastorakova A, Tyciakova S, Jakubikova J, Bohovic R, Altanerova V, Altaner C (2008) Cytosine deaminase expressing human mesenchymal stem cells mediated tumour regression in melanoma bearing mice. J Gene Med 10:1071–1082
174. Kundu AK, Khatiwala CB, Putnam AJ (2009) Extracellular matrix remodeling, integrin expression, and downstream signaling pathways influence the osteogenic differentiation of mesenchymal stem cells on poly(lactide-co-glycolide) substrates. Tissue Eng Part A 15:273–283
175. Kunter U, Rong S, Djuric Z, Boor P, Muller-Newen G, Yu D, Floege J (2006) Transplanted mesenchymal stem cells accelerate glomerular healing in experimental glomerulonephritis. J Am Soc Nephrol 17:2202–2212
176. Lai RC, Arslan F, Lee MM, Sze NS, Choo A, Chen TS, Salto-Tellez M, Timmers L, Lee CN, El Oakley RM, Pasterkamp G, de Kleijn DP, Lim SK (2010) Exosome secreted by MSC reduces myocardial ischemia/reperfusion injury. Stem Cell Res 4:214–222
177. Lai RC, Chen TS, Lim SK (2011) Mesenchymal stem cell exosome: a novel stem cell-based therapy for cardiovascular disease. Regen Med 6:481–492
178. Lama VN, Phan SH (2006) The extrapulmonary origin of fibroblasts: stem/progenitor cells and beyond. Proc Am Thorac Soc 3:373–376
179. Lazarus HM, Koc ON, Devine SM, Curtin P, Maziarz RT, Holland HK, Shpall EJ, McCarthy P, Atkinson K, Cooper BW, Gerson SL, Laughlin MJ, Loberiza FRJ, Moseley AB, Bacigalupo A (2005) Cotransplantation of HLA-identical sibling culture-expanded mesenchymal stem cells and hematopoietic stem cells in hematologic malignancy patients. Biol Blood Marrow Transplant 11:389–398
180. Lazennec G, Jorgensen C (2008) Concise review: adult multipotent stromal cells and cancer: risk or benefit? Stem Cells 26:1387–1394
181. Le Blanc K (2003) Immunomodulatory effects of fetal and adult mesenchymal stem cells. Cytotherapy 5:485–489
182. Lemons JM, Feng XJ, Bennett BD, Legesse-Miller A, Johnson EL, Raitman I, Pollina EA, Rabitz HA, Rabinowitz JD, Coller HA (2010) Quiescent fibroblasts exhibit high metabolic activity. PLoS Biol 8:e1000514
183. Letterio JJ, Roberts AB (1998) Regulation of immune responses by TGF-beta. Annu Rev Immunol 16:137–161
184. Lewis AM, Varghese S, Xu H, Alexander HR (2006) Interleukin-1 and cancer progression: the emerging role of interleukin-1 receptor antagonist as a novel therapeutic agent in cancer treatment. J Transl Med 4:48
185. Li B, Pozzi A, Young PP (2011) TNFalpha accelerates monocyte to endothelial transdifferentiation in tumors by the induction of integrin alpha5 expression and adhesion to fibronectin. Mol Cancer Res 9:702–711
186. Li Q, Wang W, Yamada T, Matsumoto K, Sakai K, Bando Y, Uehara H, Nishioka Y, Sone S, Iwakiri S, Itoi K, Utsugi T, Yasumoto K, Yano S (2011) Pleural mesothelioma instigates tumor-associated fibroblasts to promote progression via a malignant cytokine network. Am J Pathol 179:1483–1493
187. Li XJ, Peng LX, Shao JY, Lu WH, Zhang JX, Chen S, Chen ZY, Xiang YQ, Bao YN, Zheng FJ, Zeng MS, Kang TB, Zeng YX, Teh BT, Qian CN (2012) As an independent unfavorable prognostic factor, IL-8 promotes metastasis of nasopharyngeal carcinoma through induction of epithelial-mesenchymal transition and activation of AKT signaling. Carcinogenesis, (in press) http://carcin.oxfordjournals.org/content/early/2012/05/17/carcin.bgs181.full.pdf+html

188. Lin G, Yang R, Banie L, Wang G, Ning H, Li LC, Lue TF, Lin CS (2010) Effects of transplantation of adipose tissue-derived stem cells on prostate tumor. Prostate 70: 1066–1073
189. Lin J, Liu C, Ge L, Gao Q, He X, Liu Y, Li S, Zhou M, Chen Q, Zhou H (2011) Carcinoma-associated fibroblasts promotes the proliferation of a lingual carcinoma cell line by secreting keratinocyte growth factor. Tumour Biol 32:597–602
190. Lin WW, Karin M (2007) A cytokine-mediated link between innate immunity, inflammation, and cancer. J Clin Invest 117:1175–1183
191. Liu H, Fan H, Wang Y, Toh SL, Goh JC (2008) The interaction between a combined knitted silk scaffold and microporous silk sponge with human mesenchymal stem cells for ligament tissue engineering. Biomaterials 29:662–674
192. Liu S, Ginestier C, Ou SJ, Clouthier SG, Patel SH, Monville F, Korkaya H, Heath A, Dutcher J, Kleer CG, Jung Y, Dontu G, Taichman R, Wicha MS (2011) Breast cancer stem cells are regulated by mesenchymal stem cells through cytokine networks. Cancer Res 71:614–624
193. Locksley RM, Killeen N, Lenardo MJ (2001) The TNF and TNF receptor superfamilies: integrating mammalian biology. Cell 104:487–501
194. Loebinger MR, Eddaoudi A, Davies D, Janes SM (2009) Mesenchymal stem cell delivery of TRAIL can eliminate metastatic cancer. Cancer Res 69:4134–4142
195. Lu P, Weaver VM, Werb Z (2012) The extracellular matrix: a dynamic niche in cancer progression. J Cell Biol 196:395–406
196. Lu YR, Yuan Y, Wang XJ, Wei LL, Chen YN, Cong C, Li SF, Long D, Tan WD, Mao YQ, Zhang J, Li YP, Cheng JQ (2008) The growth inhibitory effect of mesenchymal stem cells on tumor cells in vitro and in vivo. Cancer Biol Ther 7:245–251
197. Lyden D, Hattori K, Dias S, Costa C, Blaikie P, Butros L, Chadburn A, Heissig B, Marks W, Witte L, Wu Y, Hicklin D, Zhu Z, Hackett NR, Crystal RG, Moore MA, Hajjar KA, Manova K, Benezra R, Rafii S (2001) Impaired recruitment of bone-marrow-derived endothelial and hematopoietic precursor cells blocks tumor angiogenesis and growth. Nat Med 7:1194–1201
198. Maccario R, Podesta M, Moretta A, Cometa A, Comoli P, Montagna D, Daudt L, Ibatici A, Piaggio G, Pozzi S, Frassoni F, Locatelli F (2005) Interaction of human mesenchymal stem cells with cells involved in alloantigen-specific immune response favors the differentiation of CD4 + T-cell subsets expressing a regulatory/suppressive phenotype. Haematologica 90:516–525
199. Maestroni GJ, Hertens E, Galli P (1999) Factor(s) from nonmacrophage bone marrow stromal cells inhibit Lewis lung carcinoma and B16 melanoma growth in mice. Cell Mol Life Sci 55:663–667
200. Margulis A, Zhang W, Alt-Holland A, Pawagi S, Prabhu P, Cao J, Zucker S, Pfeiffer L, Garfield J, Fusenig NE, Garlick JA (2006) Loss of intercellular adhesion activates a transition from low- to high-grade human squamous cell carcinoma. Int J Cancer 118: 821–831
201. Massague J (2008) TGFbeta in Cancer. Cell 134:215–230
202. Matthews KE, Keating A (1996) Gene therapy with physical methods of gene transfer. Transfus Sci 17:29–34
203. McDonald DM, Munn L, Jain RK (2000) Vasculogenic mimicry: how convincing, how novel, and how significant? Am J Pathol 156:383–388
204. Menon LG, Kelly K, Yang HW, Kim SK, Black PM, Carroll RS (2009) Human bone marrow-derived mesenchymal stromal cells expressing S-TRAIL as a cellular delivery vehicle for human glioma therapy. Stem Cells 27:2320–2330
205. Mias C, Lairez O, Trouche E, Roncalli J, Calise D, Seguelas MH, Ordener C, Piercecchi-Marti MD, Auge N, Salvayre AN, Bourin P, Parini A, Cussac D (2009) Mesenchymal stem cells promote matrix metalloproteinase secretion by cardiac fibroblasts and reduce cardiac ventricular fibrosis after myocardial infarction. Stem Cells 27:2734–2743

206. Michieli P, Mazzone M, Basilico C, Cavassa S, Sottile A, Naldini L, Comoglio PM (2004) Targeting the tumor and its microenvironment by a dual-function decoy Met receptor. Cancer Cell 6:61–73
207. Midwood KS, Orend G (2009) The role of tenascin-C in tissue injury and tumorigenesis. J Cell Commun Signal 3:287–310
208. Mihira H, Suzuki HI, Akatsu Y, Yoshimatsu Y, Igarashi T, Miyazono K, Watabe T (2012) TGF-beta-induced mesenchymal transition of MS-1 endothelial cells requires Smad-dependent cooperative activation of Rho signals and MRTF-A. J Biochem 151:145–156
209. Miletic H, Fischer YH, Giroglou T, Rueger MA, Winkeler A, Li H, Himmelreich U, Stenzel W, Jacobs AH, von Laer D (2007) Normal brain cells contribute to the bystander effect in suicide gene therapy of malignant glioma. Clin Cancer Res 13:6761–6768
210. Mishra P, Banerjee D, Ben-Baruch A (2011) Chemokines at the crossroads of tumor-fibroblast interactions that promote malignancy. J Leukoc Biol 89:31–39
211. Mishra PJ, Merlino G (2008) A traitor in our midst: mesenchymal stem cells contribute to tumor progression and metastasis. Future Oncol 4:745–749
212. Mishra PJ, Mishra PJ, Humeniuk R, Medina DJ, Alexe G, Mesirov JP, Ganesan S, Glod JW, Banerjee D (2008) Carcinoma-associated fibroblast-like differentiation of human mesenchymal stem cells. Cancer Res 68:4331–4339
213. Mohr A, Albarenque SM, Deedigan L, Yu R, Reidy M, Fulda S, Zwacka RM (2010) Targeting of XIAP combined with systemic mesenchymal stem cell-mediated delivery of sTRAIL ligand inhibits metastatic growth of pancreatic carcinoma cells. Stem Cells 28:2109–2120
214. Mohr A, Lyons M, Deedigan L, Harte T, Shaw G, Howard L, Barry F, O'Brien T, Zwacka R (2008) Mesenchymal stem cells expressing TRAIL lead to tumour growth inhibition in an experimental lung cancer model. J Cell Mol Med 12:2628–2643
215. Molloy AP, Martin FT, Dwyer RM, Griffin TP, Murphy M, Barry FP, O'Brien T, Kerin MJ (2009) Mesenchymal stem cell secretion of chemokines during differentiation into osteoblasts, and their potential role in mediating interactions with breast cancer cells. Int J Cancer 124:326–332
216. Mruthyunjaya S, Manchanda R, Godbole R, Pujari R, Shiras A, Shastry P (2010) Laminin-1 induces neurite outgrowth in human mesenchymal stem cells in serum/differentiation factors-free conditions through activation of FAK-MEK/ERK signaling pathways. Biochem Biophys Res Commun 391:43–48
217. Muehlberg FL, Song YH, Krohn A, Pinilla SP, Droll LH, Leng X, Seidensticker M, Ricke J, Altman AM, Devarajan E, Liu W, Arlinghaus RB, Alt EU (2009) Tissue-resident stem cells promote breast cancer growth and metastasis. Carcinogenesis 30:589–597
218. Mueller L, Goumas FA, Affeldt M, Sandtner S, Gehling UM, Brilloff S, Walter J, Karnatz N, Lamszus K, Rogiers X, Broering DC (2007) Stromal fibroblasts in colorectal liver metastases originate from resident fibroblasts and generate an inflammatory microenvironment. Am J Pathol 171:1608–1618
219. Mueller L, von Seggern L, Schumacher J, Goumas F, Wilms C, Braun F, Broering DC (2010) TNF-alpha similarly induces IL-6 and MCP-1 in fibroblasts from colorectal liver metastases and normal liver fibroblasts. Biochem Biophys Res Commun 397:586–591
220. Mumm JB, Emmerich J, Zhang X, Chan I, Wu L, Mauze S, Blaisdell S, Basham B, Dai J, Grein J, Sheppard C, Hong K, Cutler C, Turner S, LaFace D, Kleinschek M, Judo M, Ayanoglu G, Langowski J, Gu D, Paporello B, Murphy E, Sriram V, Naravula S, Desai B, Medicherla S, Seghezzi W, McClanahan T, Cannon-Carlson S, Beebe AM, Oft M (2011) IL-10 elicits IFNgamma-dependent tumor immune surveillance. Cancer Cell 20:781–796
221. Nagy JA, Chang SH, Shih SC, Dvorak AM, Dvorak HF (2010) Heterogeneity of the tumor vasculature. Semin Thromb Hemost 36:321–331
222. Nakamizo A, Marini F, Amano T, Khan A, Studeny M, Gumin J, Chen J, Hentschel S, Vecil G, Dembinski J, Andreeff M, Lang FF (2005) Human bone marrow-derived mesenchymal stem cells in the treatment of gliomas. Cancer Res 65:3307–3318

223. Nakamura K, Ito Y, Kawano Y, Kurozumi K, Kobune M, Tsuda H, Bizen A, Honmou O, Niitsu Y, Hamada H (2004) Antitumor effect of genetically engineered mesenchymal stem cells in a rat glioma model. Gene Ther 11:1155–1164
224. Nauta AJ, Fibbe WE (2007) Immunomodulatory properties of mesenchymal stromal cells. Blood 110:3499–3506
225. Nazareth MR, Broderick L, Simpson-Abelson MR, Kelleher RJJ, Yokota SJ, Bankert RB (2007) Characterization of human lung tumor-associated fibroblasts and their ability to modulate the activation of tumor-associated T cells. J Immunol 178:5552–5562
226. Neuss S, Schneider RK, Tietze L, Knuchel R, Jahnen-Dechent W (2010) Secretion of fibrinolytic enzymes facilitates human mesenchymal stem cell invasion into fibrin clots. Cells Tissues Organs 191:36–46
227. Nguyen HT, Li C, Lin Z, Zhuang Y, Flemington EK, Burow ME, Lin Y, Shan B (2012) The microRNA expression associated with morphogenesis of breast cancer cells in three-dimensional organotypic culture. Oncol Rep 28:117–126
228. Ning H, Yang F, Jiang M, Hu L, Feng K, Zhang J, Yu Z, Li B, Xu C, Li Y, Wang J, Hu J, Lou X, Chen H (2008) The correlation between cotransplantation of mesenchymal stem cells and higher recurrence rate in hematologic malignancy patients: outcome of a pilot clinical study. Leukemia 22:593–599
229. Noel D, Djouad F, Bouffi C, Mrugala D, Jorgensen C (2007) Multipotent mesenchymal stromal cells and immune tolerance. Leuk Lymphoma 48:1283–1289
230. Numnum TM, Makhija S, Lu B, Wang M, Rivera A, Stoff-Khalili M, Alvarez RD, Zhu ZB, Curiel DT (2008) Improved anti-tumor therapy based upon infectivity-enhanced adenoviral delivery of RNA interference in ovarian carcinoma cell lines. Gynecol Oncol 108:34–41
231. Ohkouchi S, Block GJ, Katsha AM, Kanehira M, Ebina M, Kikuchi T, Saijo Y, Nukiwa T, Prockop DJ (2012) Mesenchymal stromal cells protect cancer cells from ROS-induced apoptosis and enhance the Warburg effect by secreting STC1. Mol Ther 20:417–423
232. Ohlsson LB, Varas L, Kjellman C, Edvardsen K, Lindvall M (2003) Mesenchymal progenitor cell-mediated inhibition of tumor growth in vivo and in vitro in gelatin matrix. Exp Mol Pathol 75:248–255
233. Oliveira ID, Petrilli AS, Tavela MH, Zago MA, de Toledo SR (2007) TNF-alpha, TNF-beta, IL-6, IL-10, PECAM-1 and the MPO inflammatory gene polymorphisms in osteosarcoma. J Pediatr Hematol Oncol 29:293–297
234. Olumi AF, Grossfeld GD, Hayward SW, Carroll PR, Tlsty TD, Cunha GR (1999) Carcinoma-associated fibroblasts direct tumor progression of initiated human prostatic epithelium. Cancer Res 59:5002–5011
235. Onimaru M, Yonemitsu Y (2011) Angiogenic and lymphangiogenic cascades in the tumor microenvironment. Front Biosci (Schol Ed) 3:216–225
236. Ooi LL, Dunstan CR (2009) CXCL12/CXCR4 axis in tissue targeting and bone destruction in cancer and multiple myeloma. J Bone Miner Res 24:1147–1149
237. Orimo A, Gupta PB, Sgroi DC, Arenzana-Seisdedos F, Delaunay T, Naeem R, Carey VJ, Richardson AL, Weinberg RA (2005) Stromal fibroblasts present in invasive human breast carcinomas promote tumor growth and angiogenesis through elevated SDF-1/CXCL12 secretion. Cell 121:335–348
238. Orimo A, Weinberg RA (2006) Stromal fibroblasts in cancer: a novel tumor-promoting cell type. Cell Cycle 5:1597–1601
239. Orimo A, Weinberg RA (2007) Heterogeneity of stromal fibroblasts in tumors. Cancer Biol Ther 6:618–619
240. Oskarsson T, Acharyya S, Zhang XH, Vanharanta S, Tavazoie SF, Morris PG, Downey RJ, Manova-Todorova K, Brogi E, Massague J (2011) Breast cancer cells produce tenascin C as a metastatic niche component to colonize the lungs. Nat Med 17:867–874
241. Oskarsson T, Massague J (2012) Extracellular matrix players in metastatic niches. EMBO J 31:254–256

242. Ostanin AA, Petrovskii YL, Shevela EY, Chernykh ER (2011) Multiplex analysis of cytokines, chemokines, growth factors, MMP-9 and TIMP-1 produced by human bone marrow, adipose tissue, and placental mesenchymal stromal cells. Bull Exp Biol Med 151:133–141
243. Ostman A, Augsten M (2009) Cancer-associated fibroblasts and tumor growth–bystanders turning into key players. Curr Opin Genet Dev 19:67–73
244. Ostrand-Rosenberg S, Sinha P, Beury DW, Clements VK (2012) Cross-talk between myeloid-derived suppressor cells (MDSC), macrophages, and dendritic cells enhances tumor-induced immune suppression. Semin Cancer Biol 22:275–281
245. Otsu K, Das S, Houser SD, Quadri SK, Bhattacharya S, Bhattacharya J (2009) Concentration-dependent inhibition of angiogenesis by mesenchymal stem cells. Blood 113:4197–4205
246. Pandya NM, Dhalla NS, Santani DD (2006) Angiogenesis–a new target for future therapy. Vascul Pharmacol 44:265–274
247. Parekkadan B, van Poll D, Megeed Z, Kobayashi N, Tilles AW, Berthiaume F, Yarmush ML (2007) Immunomodulation of activated hepatic stellate cells by mesenchymal stem cells. Biochem Biophys Res Commun 363:247–252
248. Patel SA, Meyer JR, Greco SJ, Corcoran KE, Bryan M, Rameshwar P (2010) Mesenchymal stem cells protect breast cancer cells through regulatory T cells: role of mesenchymal stem cell-derived TGF-beta. J Immunol 184:5885–5894
249. Patil AS, Sable RB, Kothari RM (2012) Occurrence, biochemical profile of vascular endothelial growth factor (VEGF) isoforms and their functions in endochondral ossification. J Cell Physiol 227:1298–1308
250. Patra D, Sandell LJ (2012) Antiangiogenic and anticancer molecules in cartilage. Expert Rev Mol Med 14:e10
251. Paunescu V, Bojin FM, Tatu CA, Gavriliuc OI, Rosca A, Gruia AT, Tanasie G, Bunu C, Crisnic D, Gherghiceanu M, Tatu FR, Tatu CS, Vermesan S (2011) Tumour-associated fibroblasts and mesenchymal stem cells: more similarities than differences. J Cell Mol Med 15:635–646
252. Pereira RF, Halford KW, O'Hara MD, Leeper DB, Sokolov BP, Pollard MD, Bagasra O, Prockop DJ (1995) Cultured adherent cells from marrow can serve as long-lasting precursor cells for bone, cartilage, and lung in irradiated mice. Proc Natl Acad Sci U S A 92: 4857–4861
253. Pietras K, Rubin K, Sjoblom T, Buchdunger E, Sjoquist M, Heldin CH, Ostman A (2002) Inhibition of PDGF receptor signaling in tumor stroma enhances antitumor effect of chemotherapy. Cancer Res 62:5476–5484
254. Pittenger MF, Mackay AM, Beck SC, Jaiswal RK, Douglas R, Mosca JD, Moorman MA, Simonetti DW, Craig S, Marshak DR (1999) Multilineage potential of adult human mesenchymal stem cells. Science 284:143–147
255. Porada CD, Almeida-Porada G (2010) Mesenchymal Stem Cells as Therapeutics and Vehicles for Gene and Drug Delivery. Adv Drug Deliv Rev 62(12):1156–1166
256. Potenta S, Zeisberg E, Kalluri R (2008) The role of endothelial-to-mesenchymal transition in cancer progression. Br J Cancer 99:1375–1379
257. Prantl L, Muehlberg F, Navone NM, Song YH, Vykoukal J, Logothetis CJ, Alt EU (2010) Adipose tissue-derived stem cells promote prostate tumor growth. Prostate 70:1709–1715
258. Puglisi MA, Tesori V, Lattanzi W, Piscaglia AC, Gasbarrini GB, D'Ugo DM, Gasbarrini A (2011) Therapeutic implications of mesenchymal stem cells in liver injury. J Biomed Biotechnol 2011:860578
259. Qiao L, Xu Z, Zhao T, Zhao Z, Shi M, Zhao RC, Ye L, Zhang X (2008) Suppression of tumorigenesis by human mesenchymal stem cells in a hepatoma model. Cell Res 18:500–507
260. Qiao L, Xu ZL, Zhao TJ, Ye LH, Zhang XD (2008) Dkk-1 secreted by mesenchymal stem cells inhibits growth of breast cancer cells via depression of Wnt signalling. Cancer Lett 269:67–77
261. Quante M, Tu SP, Tomita H, Gonda T, Wang SS, Takashi S, Baik GH, Shibata W, Diprete B, Betz KS, Friedman R, Varro A, Tycko B, Wang TC (2011) Bone marrow-derived

myofibroblasts contribute to the mesenchymal stem cell niche and promote tumor growth. Cancer Cell 19:257–272
262. Ramakrishnan R, Assudani D, Nagaraj S, Hunter T, Cho HI, Antonia S, Altiok S, Celis E, Gabrilovich DI (2010) Chemotherapy enhances tumor cell susceptibility to CTL-mediated killing during cancer immunotherapy in mice. J Clin Invest 120:1111–1124
263. Ramasamy R, Lam EW, Soeiro I, Tisato V, Bonnet D, Dazzi F (2007) Mesenchymal stem cells inhibit proliferation and apoptosis of tumor cells: impact on in vivo tumor growth. Leukemia 21:304–310
264. Rapisarda A, Melillo G. (2012) Overcoming disappointing results with antiangiogenic therapy by targeting hypoxia. Nat Rev Clin Oncol, (in press). http://dx.doi.org/10.1038/nrclinonc.2012.64
265. Rasanen K, Vaheri A (2010) Activation of fibroblasts in cancer stroma. Exp Cell Res 316:2713–2722
266. Rattigan YI, Patel BB, Ackerstaff E, Sukenick G, Koutcher JA, Glod JW, Banerjee D (2012) Lactate is a mediator of metabolic cooperation between stromal carcinoma associated fibroblasts and glycolytic tumor cells in the tumor microenvironment. Exp Cell Res 318: 326–335
267. Reagan MR, Kaplan DL (2011) Concise review: Mesenchymal stem cell tumor-homing: detection methods in disease model systems. Stem Cells 29:920–927
268. Rehman J, Traktuev D, Li J, Merfeld-Clauss S, Temm-Grove CJ, Bovenkerk JE, Pell CL, Johnstone BH, Considine RV, March KL (2004) Secretion of angiogenic and antiapoptotic factors by human adipose stromal cells. Circulation 109:1292–1298
269. Ren C, Kumar S, Chanda D, Chen J, Mountz JD, Ponnazhagan S (2008) Therapeutic potential of mesenchymal stem cells producing interferon alpha in a mouse melanoma lung metastasis model. Stem Cells 26:2332–2338
270. Rhodes LV, Antoon JW, Muir SE, Elliott S, Beckman BS, Burow ME (2010) Effects of human mesenchymal stem cells on ER-positive human breast carcinoma cells mediated through ER-SDF-1/CXCR4 crosstalk. Mol Cancer 9:295
271. Ricard-Blum S, Ballut L (2011) Matricryptins derived from collagens and proteoglycans. Front Biosci 16:674–697
272. Ridley AJ, Schwartz MA, Burridge K, Firtel RA, Ginsberg MH, Borisy G, Parsons JT, Horwitz AR (2003) Cell migration: integrating signals from front to back. Science 302:1704–1709
273. Ries C, Egea V, Karow M, Kolb H, Jochum M, Neth P (2007) MMP-2, MT1-MMP, and TIMP-2 are essential for the invasive capacity of human mesenchymal stem cells: differential regulation by inflammatory cytokines. Blood 109:4055–4063
274. Rintoul RC, Sethi T (2002) Extracellular matrix regulation of drug resistance in small-cell lung cancer. Clin Sci (Lond) 102:417–424
275. Roodhart JM, Daenen LG, Stigter EC, Prins HJ, Gerrits J, Houthuijzen JM, Gerritsen MG, Schipper HS, Backer MJ, van Amersfoort M, Vermaat JS, Moerer P, Ishihara K, Kalkhoven E, Beijnen JH, Derksen PW, Medema RH, Martens AC, Brenkman AB, Voest EE (2011) Mesenchymal stem cells induce resistance to chemotherapy through the release of platinum-induced fatty acids. Cancer Cell 20:370–383
276. Roodhart JM, Langenberg MH, Daenen LG, Voest EE (2009) Translating preclinical findings of (endothelial) progenitor cell mobilization into the clinic; from bedside to bench and back. Biochim Biophys Acta 1796:41–49
277. Roodhart JM, Langenberg MH, Vermaat JS, Lolkema MP, Baars A, Giles RH, Witteveen EO, Voest EE (2010) Late release of circulating endothelial cells and endothelial progenitor cells after chemotherapy predicts response and survival in cancer patients. Neoplasia 12:87–94
278. Roorda BD, ter Elst A, Kamps WA, de Bont ES (2009) Bone marrow-derived cells and tumor growth: contribution of bone marrow-derived cells to tumor micro-environments with special focus on mesenchymal stem cells. Crit Rev Oncol Hematol 69:187–198

279. Rosada C, Justesen J, Melsvik D, Ebbesen P, Kassem M (2003) The human umbilical cord blood: a potential source for osteoblast progenitor cells. Calcif Tissue Int 72:135–142
280. Sadlonova A, Novak Z, Johnson MR, Bowe DB, Gault SR, Page GP, Thottassery JV, Welch DR, Frost AR (2005) Breast fibroblasts modulate epithelial cell proliferation in three-dimensional in vitro co-culture. Breast Cancer Res 7:R46–R59
281. Sakai N, Yoshidome H, Shida T, Kimura F, Shimizu H, Ohtsuka M, Takeuchi D, Sakakibara M, Miyazaki M (2012) CXCR4/CXCL12 expression profile is associated with tumor microenvironment and clinical outcome of liver metastases of colorectal cancer. Clin Exp Metastasis 29:101–110
282. Sandler A, Gray R, Perry MC, Brahmer J, Schiller JH, Dowlati A, Lilenbaum R, Johnson DH (2006) Paclitaxel-carboplatin alone or with bevacizumab for non-small-cell lung cancer. N Engl J Med 355:2542–2550
283. Sasportas LS, Kasmieh R, Wakimoto H, Hingtgen S, van de Water JA, Mohapatra G, Figueiredo JL, Martuza RL, Weissleder R, Shah K (2009) Assessment of therapeutic efficacy and fate of engineered human mesenchymal stem cells for cancer therapy. Proc Natl Acad Sci U S A 106:4822–4827
284. Sato K, Ozaki K, Oh I, Meguro A, Hatanaka K, Nagai T, Muroi K, Ozawa K (2007) Nitric oxide plays a critical role in suppression of T-cell proliferation by mesenchymal stem cells. Blood 109:228–234
285. Sato T, Terai M, Tamura Y, Alexeev V, Mastrangelo MJ, Selvan SR (2011) Interleukin 10 in the tumor microenvironment: a target for anticancer immunotherapy. Immunol Res 51:170–182
286. Schauer IG, Sood AK, Mok S, Liu J (2011) Cancer-associated fibroblasts and their putative role in potentiating the initiation and development of epithelial ovarian cancer. Neoplasia 13:393–405
287. Scheller J, Chalaris A, Schmidt-Arras D, Rose-John S (2011) The pro- and anti-inflammatory properties of the cytokine interleukin-6. Biochim Biophys Acta 1813:878–888
288. Schmid MC, Varner JA (2010) Myeloid cells in the tumor microenvironment: modulation of tumor angiogenesis and tumor inflammation. J Oncol 2010:201026
289. Schmitt M, Sturmheit AS, Welk A, Schnelldorfer C, Harbeck N (2006) Procedures for the quantitative protein determination of urokinase and its inhibitor, PAI-1, in human breast cancer tissue extracts by ELISA. Methods Mol Med 120:245–265
290. Schoppmann SF, Berghoff A, Dinhof C, Jakesz R, Gnant M, Dubsky P, Jesch B, Heinzl H, Birner P (2012) Podoplanin-expressing cancer-associated fibroblasts are associated with poor prognosis in invasive breast cancer. Breast Cancer Res Treat 7:1669–1675
291. Schroder K, Hertzog PJ, Ravasi T, Hume DA (2004) Interferon-gamma: an overview of signals, mechanisms and functions. J Leukoc Biol 75:163–189
292. Secchiero P, Melloni E, Corallini F, Beltrami AP, Alviano F, Milani D, D'Aurizio F, di Iasio MG, Cesselli D, Bagnara GP, Zauli G (2008) Tumor necrosis factor-related apoptosis-inducing ligand promotes migration of human bone marrow multipotent stromal cells. Stem Cells 26:2955–2963
293. Secchiero P, Zorzet S, Tripodo C, Corallini F, Melloni E, Caruso L, Bosco R, Ingrao S, Zavan B, Zauli G (2010) Human bone marrow mesenchymal stem cells display anti-cancer activity in SCID mice bearing disseminated non-Hodgkin's lymphoma xenografts. PLoS One 5:e11140
294. Selmani Z, Naji A, Zidi I, Favier B, Gaiffe E, Obert L, Borg C, Saas P, Tiberghien P, Rouas-Freiss N, Carosella ED, Deschaseaux F (2008) Human leukocyte antigen-G5 secretion by human mesenchymal stem cells is required to suppress T lymphocyte and natural killer function and to induce CD4 + CD25highFOXP3 + regulatory T cells. Stem Cells 26: 212–222
295. Sharma M, Afrin F, Satija N, Tripathi RP, Gangenahalli GU (2011) Stromal-derived factor-1/CXCR4 signaling: indispensable role in homing and engraftment of hematopoietic stem cells in bone marrow. Stem Cells Dev 20(6):933–946

296. Shinagawa K, Kitadai Y, Tanaka M, Sumida T, Kodama M, Higashi Y, Tanaka S, Yasui W, Chayama K (2010) Mesenchymal stem cells enhance growth and metastasis of colon cancer. Int J Cancer 127:2323–2333
297. Shishodia S, Majumdar S, Banerjee S, Aggarwal BB (2003) Ursolic acid inhibits nuclear factor-kappaB activation induced by carcinogenic agents through suppression of IkappaBalpha kinase and p65 phosphorylation: correlation with down-regulation of cyclooxygenase 2, matrix metalloproteinase 9, and cyclin D1. Cancer Res 63:4375–4383
298. Short B, Brouard N, Occhiodoro-Scott T, Ramakrishnan A, Simmons PJ (2003) Mesenchymal stem cells. Arch Med Res 34:565–571
299. Sibony-Benyamini H, Gil-Henn H (2012) Invadopodia: the leading force. Eur J Cell Biol, in press. http://www.sciencedirect.com/science/article/pii/S0171933512000775
300. Silletti S, Kessler T, Goldberg J, Boger DL, Cheresh DA (2001) Disruption of matrix metalloproteinase 2 binding to integrin alpha vbeta 3 by an organic molecule inhibits angiogenesis and tumor growth in vivo. Proc Natl Acad Sci U S A 98:119–124
301. Silva WAJ, Covas DT, Panepucci RA, Proto-Siqueira R, Siufi JL, Zanette DL, Santos AR, Zago MA (2003) The profile of gene expression of human marrow mesenchymal stem cells. Stem Cells 21:661–669
302. Sohara Y, Shimada H, Minkin C, Erdreich-Epstein A, Nolta JA, DeClerck YA (2005) Bone marrow mesenchymal stem cells provide an alternate pathway of osteoclast activation and bone destruction by cancer cells. Cancer Res 65:1129–1135
303. Somasundaram R, Jacob L, Swoboda R, Caputo L, Song H, Basak S, Monos D, Peritt D, Marincola F, Cai D, Birebent B, Bloome E, Kim J, Berencsi K, Mastrangelo M, Herlyn D (2002) Inhibition of cytolytic T lymphocyte proliferation by autologous CD4 +/ CD25 + regulatory T cells in a colorectal carcinoma patient is mediated by transforming growth factor-beta. Cancer Res 62:5267–5272
304. Sordi V, Malosio ML, Marchesi F, Mercalli A, Melzi R, Giordano T, Belmonte N, Ferrari G, Leone BE, Bertuzzi F, Zerbini G, Allavena P, Bonifacio E, Piemonti L (2005) Bone marrow mesenchymal stem cells express a restricted set of functionally active chemokine receptors capable of promoting migration to pancreatic islets. Blood 106:419–427
305. Spaeth E, Klopp A, Dembinski J, Andreeff M, Marini F (2008) Inflammation and tumor microenvironments: defining the migratory itinerary of mesenchymal stem cells. Gene Ther 15:730–738
306. Spaeth EL, Dembinski JL, Sasser AK, Watson K, Klopp A, Hall B, Andreeff M, Marini F (2009) Mesenchymal stem cell transition to tumor-associated fibroblasts contributes to fibrovascular network expansion and tumor progression. PLoS One 4:e4992
307. St Croix B, Rago C, Velculescu V, Traverso G, Romans KE, Montgomery E, Lal A, Riggins GJ, Lengauer C, Vogelstein B, Kinzler KW (2000) Genes expressed in human tumor endothelium. Science 289:1197–1202
308. Stover DG, Bierie B, Moses HL (2007) A delicate balance: TGF-beta and the tumor microenvironment. J Cell Biochem 101:851–861
309. Studeny M, Marini FC, Champlin RE, Zompetta C, Fidler IJ, Andreeff M (2002) Bone marrow-derived mesenchymal stem cells as vehicles for interferon-beta delivery into tumors. Cancer Res 62:3603–3608
310. Studeny M, Marini FC, Dembinski JL, Zompetta C, Cabreira-Hansen M, Bekele BN, Champlin RE, Andreeff M (2004) Mesenchymal stem cells: potential precursors for tumor stroma and targeted-delivery vehicles for anticancer agents. J Natl Cancer Inst 96: 1593–1603
311. Subbaram S, Dipersio CM (2011) Integrin alpha3beta1 as a breast cancer target. Expert Opin Ther Targets 15:1197–1210
312. Subramanian A, Shu-Uin G, Kae-Siang N, Gauthaman K, Biswas A, Choolani M, Bongso A, Chui-Yee F (2012) Human umbilical cord Wharton's jelly mesenchymal stem cells do not transform to tumor-associated fibroblasts in the presence of breast and ovarian cancer cells unlike bone marrow mesenchymal stem cells. J Cell Biochem 113:1886–1895

313. Suzuki K, Sun R, Origuchi M, Kanehira M, Takahata T, Itoh J, Umezawa A, Kijima H, Fukuda S, Saijo Y (2011) Mesenchymal stromal cells promote tumor growth through the enhancement of neovascularization. Mol Med 17:579–587
314. Takai K, Hara J, Matsumoto K, Hosoi G, Osugi Y, Tawa A, Okada S, Nakamura T (1997) Hepatocyte growth factor is constitutively produced by human bone marrow stromal cells and indirectly promotes hematopoiesis. Blood 89:1560–1565
315. Ten Dijke P, Egorova AD, Goumans MJ, Poelmann RE, Hierck BP (2012) TGF-beta signaling in endothelial-to-mesenchymal transition: the role of shear stress and primary cilia. Sci Signal 5:pt2.
316. Thiery JP, Sleeman JP (2006) Complex networks orchestrate epithelial-mesenchymal transitions. Nat Rev Mol Cell Biol 7:131–142
317. Thyboll J, Kortesmaa J, Cao R, Soininen R, Wang L, Iivanainen A, Sorokin L, Risling M, Cao Y, Tryggvason K (2002) Deletion of the laminin alpha4 chain leads to impaired microvessel maturation. Mol Cell Biol 22:1194–1202
318. Tlsty TD (2001) Stromal cells can contribute oncogenic signals. Semin Cancer Biol 11: 97–104
319. Traggiai E, Volpi S, Schena F, Gattorno M, Ferlito F, Moretta L, Martini A (2008) Bone marrow-derived mesenchymal stem cells induce both polyclonal expansion and differentiation of B cells isolated from healthy donors and systemic lupus erythematosus patients. Stem Cells 26:562–569
320. Traktuev DO, Merfeld-Clauss S, Li J, Kolonin M, Arap W, Pasqualini R, Johnstone BH, March KL (2008) A population of multipotent CD34-positive adipose stromal cells share pericyte and mesenchymal surface markers, reside in a periendothelial location, and stabilize endothelial networks. Circ Res 102:77–85
321. Trevino V, Tadesse MG, Vannucci M, Al-Shahrour F, Antczak P, Durant S, Bikfalvi A, Dopazo J, Campbell MJ, Falciani F (2011) Analysis of normal-tumour tissue interaction in tumours: prediction of prostate cancer features from the molecular profile of adjacent normal cells. PLoS One 6:e16492
322. Tsuzuki Y, Fukumura D, Oosthuyse B, Koike C, Carmeliet P, Jain RK (2000) Vascular endothelial growth factor (VEGF) modulation by targeting hypoxia-inducible factor-1alpha– > hypoxia response element–> VEGF cascade differentially regulates vascular response and growth rate in tumors. Cancer Res 60:6248–6252
323. Uccelli A, Moretta L, Pistoia V (2008) Mesenchymal stem cells in health and disease. Nat Rev Immunol 8:726–736
324. Ugalde AP, Ordonez GR, Quiros PM, Puente XS, Lopez-Otin C (2010) Metalloproteases and the degradome. Methods Mol Biol 622:3–29
325. Urashima M, Chen BP, Chen S, Pinkus GS, Bronson RT, Dedera DA, Hoshi Y, Teoh G, Ogata A, Treon SP, Chauhan D, Anderson KC (1997) The development of a model for the homing of multiple myeloma cells to human bone marrow. Blood 90:754–765
326. Urbich C, Dimmeler S (2004) Endothelial progenitor cells: characterization and role in vascular biology. Circ Res 95:343–353
327. van der Schaft DW, Hillen F, Pauwels P, Kirschmann DA, Castermans K, Egbrink MG, Tran MG, Sciot R, Hauben E, Hogendoorn PC, Delattre O, Maxwell PH, Hendrix MJ, Griffioen AW (2005) Tumor cell plasticity in Ewing sarcoma, an alternative circulatory system stimulated by hypoxia. Cancer Res 65:11520–11528
328. Van Obberghen-Schilling E, Tucker RP, Saupe F, Gasser I, Cseh B, Orend G (2011) Fibronectin and tenascin-C: accomplices in vascular morphogenesis during development and tumor growth. Int J Dev Biol 55:511–525
329. van Poll D, Parekkadan B, Cho CH, Berthiaume F, Nahmias Y, Tilles AW, Yarmush ML (2008) Mesenchymal stem cell-derived molecules directly modulate hepatocellular death and regeneration in vitro and in vivo. Hepatology 47:1634–1643
330. Vander Heiden MG, Cantley LC, Thompson CB (2009) Understanding the Warburg effect: the metabolic requirements of cell proliferation. Science 324:1029–1033

331. Vermeulen L, De Sousa EMF, van der Heijden M, Cameron K, de Jong JH, Borovski T, Tuynman JB, Todaro M, Merz C, Rodermond H, Sprick MR, Kemper K, Richel DJ, Stassi G, Medema JP (2010) Wnt activity defines colon cancer stem cells and is regulated by the microenvironment. Nat Cell Biol 12:468–476
332. Wang L, Li Y, Chen X, Chen J, Gautam SC, Xu Y, Chopp M (2002) MCP-1, MIP-1, IL-8 and ischemic cerebral tissue enhance human bone marrow stromal cell migration in interface culture. Hematology 7:113–117
333. Wang R, Chadalavada K, Wilshire J, Kowalik U, Hovinga KE, Geber A, Fligelman B, Leversha M, Brennan C, Tabar V (2010) Glioblastoma stem-like cells give rise to tumour endothelium. Nature 468:829–833
334. Wang Y, McNiven MA (2012) Invasive matrix degradation at focal adhesions occurs via protease recruitment by a FAK-p130Cas complex. J Cell Biol 196:375–385
335. Waugh DJ, Wilson C (2008) The interleukin-8 pathway in cancer. Clin Cancer Res 14:6735–6741
336. Weis SM, Cheresh DA (2011) Tumor angiogenesis: molecular pathways and therapeutic targets. Nat Med 17:1359–1370
337. Wolf K, Friedl P (2009) Mapping proteolytic cancer cell-extracellular matrix interfaces. Clin Exp Metastasis 26:289–298
338. Woo EY, Chu CS, Goletz TJ, Schlienger K, Yeh H, Coukos G, Rubin SC, Kaiser LR, June CH (2001) Regulatory CD4(+)CD25(+) T cells in tumors from patients with early-stage non-small cell lung cancer and late-stage ovarian cancer. Cancer Res 61:4766–4772
339. Wu MH, Hong HC, Hong TM, Chiang WF, Jin YT, Chen YL (2011) Targeting galectin-1 in carcinoma-associated fibroblasts inhibits oral squamous cell carcinoma metastasis by downregulating MCP-1/CCL2 expression. Clin Cancer Res 17:1306–1316
340. Wu Y, Wang J, Scott PG, Tredget EE (2007) Bone marrow-derived stem cells in wound healing: a review. Wound Repair Regen 15(Suppl 1):S18–S26
341. Xin H, Kanehira M, Mizuguchi H, Hayakawa T, Kikuchi T, Nukiwa T, Saijo Y (2007) Targeted delivery of CX3CL1 to multiple lung tumors by mesenchymal stem cells. Stem Cells 25:1618–1626
342. Xu WT, Bian ZY, Fan QM, Li G, Tang TT (2009) Human mesenchymal stem cells (hMSCs) target osteosarcoma and promote its growth and pulmonary metastasis. Cancer Lett 281:32–41
343. Yagi H, Soto-Gutierrez A, Parekkadan B, Kitagawa Y, Tompkins RG, Kobayashi N, Yarmush ML (2010) Mesenchymal stem cells: mechanisms of immunomodulation and homing. Cell Transplant 19:667–679
344. Yilmaz M, Christofori G (2009) EMT, the cytoskeleton, and cancer cell invasion. Cancer Metastasis Rev 28:15–33
345. Yong RL, Shinojima N, Fueyo J, Gumin J, Vecil GG, Marini FC, Bogler O, Andreeff M, Lang FF (2009) Human bone marrow-derived mesenchymal stem cells for intravascular delivery of oncolytic adenovirus Delta24-RGD to human gliomas. Cancer Res 69:8932–8940
346. Yu JM, Jun ES, Bae YC, Jung JS (2008) Mesenchymal stem cells derived from human adipose tissues favor tumor cell growth in vivo. Stem Cells Dev 17:463–473
347. Yue J, Zhang K, Chen J (2012) Role of Integrins in Regulating Proteases to Mediate Extracellular Matrix Remodeling. Cancer Microenviron, in press. http://dx.doi.org/10.1007/s12307-012-0101-3
348. Zaidi MR, Merlino G (2011) The two faces of interferon-gamma in cancer. Clin Cancer Res 17:6118–6124
349. Zamarron BF, Chen W (2011) Dual roles of immune cells and their factors in cancer development and progression. Int J Biol Sci 7:651–658
350. Zanjani ED, Flake AW, Almeida-Porada G, Tran N, Papayannopoulou T (1999) Homing of human cells in the fetal sheep model: modulation by antibodies activating or inhibiting very late activation antigen-4-dependent function. Blood 94:2515–2522

351. Zeisberg EM, Potenta S, Xie L, Zeisberg M, Kalluri R (2007) Discovery of endothelial to mesenchymal transition as a source for carcinoma-associated fibroblasts. Cancer Res 67:10123–10128
352. Zeisberg EM, Tarnavski O, Zeisberg M, Dorfman AL, McMullen JR, Gustafsson E, Chandraker A, Yuan X, Pu WT, Roberts AB, Neilson EG, Sayegh MH, Izumo S, Kalluri R (2007) Endothelial-to-mesenchymal transition contributes to cardiac fibrosis. Nat Med 13:952–961
353. Zeng Z, Samudio IJ, Munsell M, An J, Huang Z, Estey E, Andreeff M, Konopleva M (2006) Inhibition of CXCR4 with the novel RCP168 peptide overcomes stroma-mediated chemoresistance in chronic and acute leukemias. Mol Cancer Ther 5:3113–3121
354. Zhang QZ, Su WR, Shi SH, Wilder-Smith P, Xiang AP, Wong A, Nguyen AL, Kwon CW, Le AD (2010) Human gingiva-derived mesenchymal stem cells elicit polarization of m2 macrophages and enhance cutaneous wound healing. Stem Cells 28:1856–1868
355. Zhang Y, Daquinag A, Traktuev DO, Amaya-Manzanares F, Simmons PJ, March KL, Pasqualini R, Arap W, Kolonin MG (2009) White adipose tissue cells are recruited by experimental tumors and promote cancer progression in mouse models. Cancer Res 69:5259–5266
356. Zhao BC, Zhao B, Han JG, Ma HC, Wang ZJ (2010) Adipose-derived stem cells promote gastric cancer cell growth, migration and invasion through SDF-1/CXCR4 axis. Hepatogastroenterology 57:1382–1389
357. Zhao H, Peehl DM (2009) Tumor-promoting phenotype of CD90hi prostate cancer-associated fibroblasts. Prostate 69:991–1000
358. Zhu W, Huang L, Li Y, Zhang X, Gu J, Yan Y, Xu X, Wang M, Qian H, Xu W (2012) Exosomes derived from human bone marrow mesenchymal stem cells promote tumor growth in vivo. Cancer Lett 315:28–37
359. Zhu W, Xu W, Jiang R, Qian H, Chen M, Hu J, Cao W, Han C, Chen Y (2006) Mesenchymal stem cells derived from bone marrow favor tumor cell growth in vivo. Exp Mol Pathol 80:267–274
360. Zhu Y, Sun Z, Han Q, Liao L, Wang J, Bian C, Li J, Yan X, Liu Y, Shao C, Zhao RC (2009) Human mesenchymal stem cells inhibit cancer cell proliferation by secreting DKK-1. Leukemia 23:925–933
361. Zimmerlin L, Donnenberg AD, Rubin JP, Basse P, Landreneau RJ, Donnenberg VS (2011) Regenerative therapy and cancer: in vitro and in vivo studies of the interaction between adipose-derived stem cells and breast cancer cells from clinical isolates. Tissue Eng Part A 17:93–106
362. Zipori D, Krupsky M, Resnitzky P (1987) Stromal cell effects on clonal growth of tumors. Cancer 60:1757–1762
363. Zitvogel L, Kepp O, Galluzzi L, Kroemer G (2012) Inflammasomes in carcinogenesis and anticancer immune responses. Nat Immunol 13:343–351

Adv Biochem Eng Biotechnol (2013) 130: 267–286
DOI: 10.1007/10_2012_161

Published Online: 2 November 2012

Sources of Mesenchymal Stem Cells: Current and Future Clinical Use

Michela Pozzobon, Martina Piccoli and Paolo De Coppi

Abstract Despite the lack of international criteria defining the biological properties and surface markers that must possess mesenchymal stem cells (MSC), it become of paramount importance to know the different sources and the clinical applications of these promising stem cells. In this chapter we overview the most important sources of MSC from the inner cell mass of the blastocyst to the adult source and landing to the induced pluripotent stem cells (iPS). Following the criteria defining MSC properties so far observed, we drew the attention on the role of MSC as tool for regenerative medicine and therapeutic purposes.

Keywords Adult MSC · Amniotic fluid · Adipose tissue · Cord blood · iPS · Therapeutic applications of MSC

Contents

M. Pozzobon · M. Piccoli
Department of Woman and Child Health, University of Padova,
35128 Padua, Italy

M. Pozzobon · M. Piccoli
Città della Speranza Foundation, 36030 Monte di Malo, Vicenza, Italy

P. De Coppi (✉)
UCL Institute of Child Health, Surgery Unit, Great Ormond Street Hospital,
University College, 30 Guilford Street, London WC1N 1EH, UK
e-mail: p.decoppi@ich.ucl.ac.uk

P. De Coppi
Department of Pediatrics and Pediatric Surgery, University of Padova,
35128 Padua, Italy

1 Introduction

Stem cells are primal cells found in all multi-cellular organisms and are defined by two characteristics. They are unspecialized cells that renew for a lifetime by cell division to maintain the stem cell pool, and they can differentiate into cells with special functions under particular physiologic or experimental conditions. To fulfil this dual function, they undergo symmetric and asymmetric divisions during development. Mesenchymal stem cells (MSC), which originate in the embryo from the mesodermal layer, can be found during all phases of the development of mammalians. In this chapter we will focus on defining the characteristics of MSC derived from the three different stages of development with particular emphasis on MSC-ES- or iPS-derived, foetal MSC (amniotic fluid, AF), neonatal [placenta and cord blood (CB)], and adult MSC (bone marrow and adipose tissue) (Fig. 1).

Since the discovery of MSC in bone marrow (BM) [26], they have been isolated from a wide range of adult tissues such as trabecular bone, periosteum [32], neural system [76], skeletal muscle, skin, pericytes, peripheral blood, deciduous teeth, periodontal ligament, placenta, cord blood, and adipose tissues [11, 22, 23, 45, 51, 115, 117].

The definition "mesenchymal stem cells" has been considered unclear through the years; "mesenchymal" was based on the hypothesis that multiple tissues beyond skeletal lineages, such as skeletal muscle, myocardium, and smooth muscle could be generated by MSC and secondly during embryonic organogenesis. However, post-natal MSC-related tissues are generated by a system of distinct progenitors, rather than from a common precursor. In order to deal with nomenclature problems, three major criteria were introduced to define MSC by the International Society for Cell Therapy [41]. First, cells must be plastic-adherent when maintained under standard culture conditions. When measured by flow cytometry, 95 % of the cell population must express CD73 (50-nucleotidase ecto, NT5E), CD90 (Thy-1) and CD105 (SH2, MCAM, or endoglin), CD271 (Low affinity Nerve Growth Factor Receptor, LNGFR), CD166 (ALCAM adhesion protein), CD146 (P1H12), CD29, CD106 (vascular adhesion molecule-1, VCAM-1), and 98 % of the cells should be negative for the following haematopoietic cell surface antigens: CD45, a pan-leukocyte marker; CD34, a marker of primitive haematopoietic progenitors and endothelial cells; either CD11b or CD14, markers for monocytes; CD19 or CD79a,

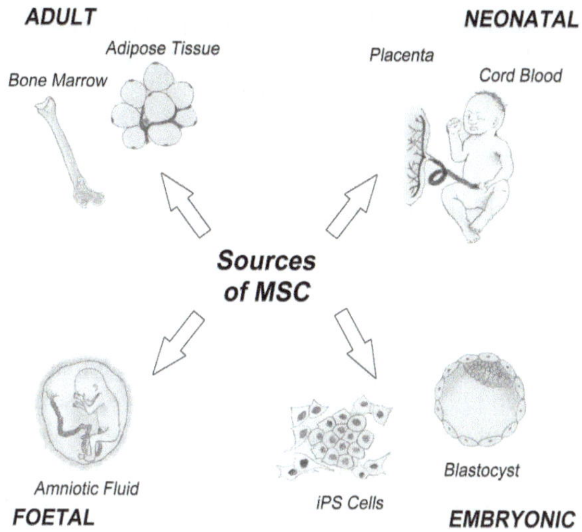

Fig. 1 Cartoon representing the most important sources of MSC from the different phases of life

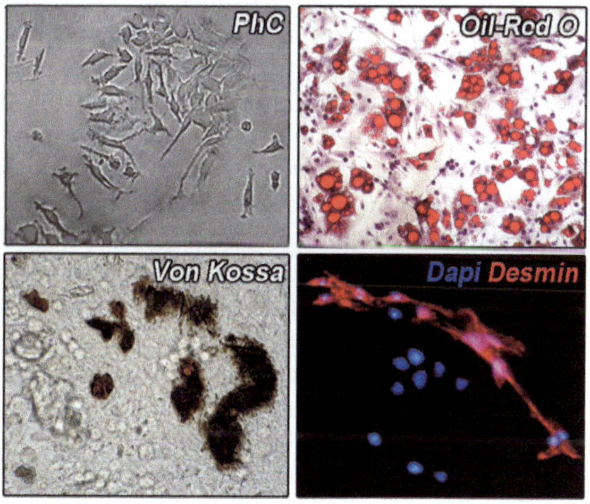

Fig. 2 *MSC and differentiations.* Phase contrast of BM MSC on the left inside. Three pictures of MSC under differentation: adipogenic (Oil Red O staining. Lipidic vacuoli detection), osteogenic (Von Kossa staining. Calcium deposition detection.) and miogenic (Immunofluorescence for Desmin, protein expressed by muscle committed cells) differentation assays

B cell markers, and Human Leukocyte Antigen II (HLA Class 2). Finally, to be defined as MSC, cells should be capable of differentiating into osteoblasts, chondroblasts, and adipocytes when placed into an appropriate induction/differentiation medium (Fig. 2). Among the MSC collected from different tissues, there is no clear evidence of phenotypic or functional differences in surface antigen expression. However, the success rate of MSC isolation varies among tissues: MSC can be isolated from only 63 % of cord blood samples, while they can be easily derived from 100 % of both processed bone marrow and adipose tissue [51].

Starting with MSC derived from pluripotent tissues in the following paragraphs we have attempted to summarize the main characteristics of MSC derived at different stages of development.

2 Pluripotent-Derived MSC

2.1 MSC Derived from ES Cells

Embryonic stem (ES) cells derive from the inner cell mass of a blastocyst-stage embryo [79]. They are pluripotent and give rise during development to all derivatives of the three primary germ layers: ectoderm, endoderm, and mesoderm. Hence, they possess the potential to develop into most of the cell types within the body [50, 90, 109]. ES cells, being pluripotent, require specific signals for correct differentiation and if injected in vivo prior to commitment, they will give rise to many different types of cells, causing teratomas [37, 104]. Before any human therapeutic applications can be achieved, there must be reproducible, efficient, and safe methodologies for directed differentiation of human ES (hES) cells into desired cell types, either in vitro or in vivo. The derivation of primitive or differentiated cells, originally from murine ES cells and, more recently, from hES cells, has been the subject of intensive research. Differentiating hES cells into MSC before undergoing lineage-specific differentiation provides the advantage of producing a large source of multipotent progenitor cells that can be expanded and differentiated into specified lineages such as bone, cartilage, or fat [25, 35, 36]. To date, the differentiation conditions for deriving MSC from hES cells have required long culture periods [81], were dependent on a feeder layer, and demonstrated low yields of MSC [4, 112]. Generating MSC in serum-free conditions supplemented with PDGF AB and FGF2 has also been reported [63]. To satisfy the likely demand for high numbers of progenitor cells to regenerate skeletal defects via a tissue engineering approach, cell culture conditions must be improved to assure appropriate and consistent differentiation of hES cells into MSC on a large scale. Interestingly, Arpornmaeklong and colleagues suggested a new method for an osteogenic cell enrichment strategy that could provide large numbers of osteoprogenitor cells for analysis and cell transplantation [3]. Another crucial point in obtaining MSC from ES cells is whether they are comparable to the "classical" MSC in terms of immunogenicity and immunosuppression. Their immunotolerance properties in vitro have been partially reported [113], but no information is available about their potential immunotolerance and anti-inflammatory properties in vivo. The transforming growth factor (TGF-b) signalling through SMAD-2/3 downstream effectors is of increasing interest for regenerative medicine and lineage specification during human embryonic development [66, 91]. In contrast to specific bone morphogenic proteins (BMPs), which are known to play an important role in directing cell fate decisions toward mesoderm and further differentiation into MSC, the mesoderm effects of TGF-b signalling in human embryonic development is controversial [66, 91]. Sanchez et al. studied the role of SMAD-2/3 inhibition in the potential enrichment of functional and multipotent MSC from hES cells and reported a robust and efficient enrichment of MSC from hES cells through specific inhibition of the SMAD-2/3 pathway. These hES-derived MSC display, in fact, multi-lineage differentiation potential and exhibited potent immunosuppressive and anti-inflammatory properties in vitro and in vivo, which are capable of protecting against experimental inflammatory bowel disease [95].

2.2 MSC Derived from iPS Cells

A new avenue to overcome the ethical problems associated with ES and allow the creation of patient-specific embryonic-like stem cells (ESCs) is represented by the relatively recent demonstration that it is possible to generate induced pluripotent stem (iPS) cells [105]. The production of iPS cells with almost identical genetic and functional properties offers the possibility to bypass both moral conflicts and different genetic background inherent to ES cells. Moreover, the source will be kept autologous leading to the absence of immunological reactions. An iPS cell is a pluripotent stem cell developed from a non-pluripotent cell, usually an adult somatic cell, by causing a forced expression of defined genetic sequences and they were first described in 2006 by Takahashi and Yamanaka using mouse somatic cells [105]. The key genes Oct3/4 (POU5F1), the transcription factor Sox2, c-Myc proto-oncogene protein and Klf4 (Krueppel-like factor 4) were sufficient to reprogram mouse fibroblasts to cells closely resembling mouse ES cells. Since 2006 several advancements have been made in iPS technology: iPS can now be generated using (i) transfection systems with a lower risk of tumour formation [71, 80], (ii) with a lower number of transfected reprogramming factors [100], (iii) without virus [119], (iv) with microRNAs [2, 73], (v) using only proteins [14, 52], (vi) piggyBac [121], or (vii) episome [126]. All these techniques were addressed to turn iPS creation into a more suitable method for possible clinical application, indeed today iPS can be derived from many different tissues or body fluids, including urine [130], cord blood [34], keratinocytes [1] and amniotic fluid [75].

Similarly to ES cells, iPS could also be differentiated into MSC, which were able to ameliorate vascularization in a classical model of limb ischemia [64]. Similarly iPS could be differentiated into osteoblasts in vitro using soluble factors in conditioned media [62] or, more recently by means of a biomaterial matrix such as collagen-coated dishes [65]. iPS from dermal fibroblasts were also used and differentiated into adipocytes, bone, cartilage in vitro and, in vivo, subcutaneously implanted iPS-derived osteoblasts seeded on matrix, highlighting the ability of originating functional osteoblasts from the differentiated iPS [5].

3 Foetal-Derived MSC

Although MSC have become a very common cellular source for research, in the human adult their quantity is very low (about 0.001–0.01 %); self-renewal, proliferative and expansion potencies are limited, and their multipotency differentiation capacity decreases with age [24, 86]. Therefore, foetal tissues have attracted attention as alternative sources of stem cells. They represent a relatively new entrant into the stem cell field, exhibiting unique and fascinating peculiarity. Relevant studies have demonstrated that several foetal tissues are abundant sources of MSC, including foetal bone marrow, blood, lung, liver and spleen [28, 45], but during pregnancy one of the most available sources of MSC is the amniotic fluid.

3.1 Amniotic Fluid Stem Cells

Amniotic fluid (AF) contains a heterogeneous population of cells displaying a range of morphologies. Most of these cells are epithelial in nature and have a limited capacity to proliferate in culture. AF has been used for decades as a tool for pre-natal diagnosis, but recent studies have provided important evidences about the potential of AF as an alternative source of stem cells [12]. Many works have characterized putative stem cell populations isolated from AF. Prusa et al. demonstrated the expression of Oct4 within a subset of AF cells [87]. This is important, as Oct4 expression is associated with pluripotent cells such as embryonic germ and embryonic stem cells. Demonstration of proliferation within this population further suggests that pluripotent stem cells can be both isolated and propagated from the AF of humans. The AF-MSC exhibit typical mesenchymal markers, such as CD29, CD90, CD166, CD73, CD105, CD49e and CD44, while they are negative for the lineage-committed markers such as the haematopoietic CD45, CD34 and CD14 [53, 87, 88, 114, 115, 128] or myogenic antigen [6]. Undifferentiated AF-MSC expand efficiently with a doubling time of 36 h. Comparison of growth kinetics between the AF-MSC and MSC either from BM or cord blood (CB) [93] revealed a four to eightfold difference in expansion capacity for the AF-MSC with an average doubling time of about 18 h. AF-MSC have been successfully cultured for a long time (over 8 months), displaying a high proliferation rate and stable karyotype [53]. The in vivo application of AF-MSC is still at an early phase, but when used in a model of disease they are capable of acting through paracrine pathways that are similar to their adult counterpart [18] On the other hand, they are unable to display a pluripotency status and to differentiate into non-mesenchymal lineages. AF-MSC can therefore be used to ameliorate ischemic heart disease through neovascularization [96], can be engineered to generate autologous human heart valves [98], or to regenerate injured sciatic nerves [82]. They can differentiate in vivo into smooth muscle cells and promote the regeneration of an injured bladder [18].

In contrast, employing cell selection through the use of a specific antigen, AF cells expressing the cell surface antigen CD117 (c-Kit) were purified from primary amniocentesis cultures [17]. The selected cells, defined as Amniotic Fluid Stem (AFS) cells, grew rapidly in culture and were capable of more than 250 population doublings. Importantly, AFS cells display a normal karyotype and maintain telomere length during long-term culture. This latter attribute facilitated the establishment of clonal lines from AFS cells, necessary to set the 'stemness' of a population. This possibility makes the stem cells from AF of greater interest because it is possible to obtain a very homogeneous population that is able to respond to environmental stimuli in a synchronous manner and in a single direction. Clonal AFS cell lines can differentiate in vitro to putative adipocytes, endothelial cells, hepatocytes, osteocytes, myocytes and neurons, derivatives of all germ layers. These cells still retain a strong mesenchymal potential, as a report regarding their chondrogenic differentiation potential has highlighted [56]. However, freshly isolated cells possess both haematopoietic [21] and myogenic [85] potentials which have never been demonstrated for MSC.

While the employment of AF-MSC are limited, they may be significant in the paediatric field, where they could play an important role in pre-natally diagnosed structural defects. In this particular scenario, there is the possibility of obtaining homologous cells at the time of invasive sampling; foetal cells could be harvested, cultured, and manipulated in vitro, during the remainder of pregnancy and later used for tissue engineering of graft material that can find clinical applications pre-natal/post-natal reconstruction. They could also be stored for future use.

4 Neonatal-Derived MSC

Because of the difficulty in obtaining BM samples that are accessible for transplantation, in recent years much attention has been given to the collection of discarded samples that are able to provide MSC with properties similar to or better than those of BM and therefore are suitable for clinical transplantation. Foetal annexes are embryonic tissues that no longer serve the foetus after birth, and are routinely discarded. For this reason, they are considered an easy and constant source of MSC.

4.1 Placenta Stem Cells

The foetal annexes, which are normally discarded after childbirth, have been considered to obtain more primitive stem cells than adult ones.

In particular, the unique transitory foetomaternal organ, the placenta, besides the presence of haematopoietic stem cells (HSC), contains placenta mesenchymal stem cells (P-MSC), which have been reported to have a higher expansion and engraftment ability than BM-MSC. Placental tissues can be foetal or maternal in origin requiring the two types of tissue to be identified with respect to MSC function. According to the first workshop on placenta-derived stem cells [83], four regions of foetal placenta can be featured (amniotic epithelial, amniotic mesenchymal, chorionic mesenchymal and chorionic trophoblastic tissue) and the stem cells from the foetal part possess a shorter life span than the maternal ones. Stem cell populations derived from human placenta tissues are the chorionic mesenchymal stromal cells and the chorionic trophoblastic cells [83], which demonstrated variable plasticity. The expression of the peculiar marker frizzled 9 (FZD9 or CD349) by the placenta MSC enables a specific method for the selective isolation of the cells. Moreover, the cells must be positive for CD90, CD73 and CD105 and negative for CD45, CD34, CD14, and HLA-DR. P-MSC are more efficient than BM-MSC when used as a supportive feeder layer [54] for human embryonic stem cell propagation and exhibit superior engraftment due to a more efficient utilization of VL4-mediating binding [7]. P-MSC possess immunomodulatory properties and the expression of some pluripotency markers can be induced such as SSEA-4, Nanog, Oct4 and Rex-1 [44]. Several reports demonstrated the ability of P-MSC to differentiate in vitro into different cell types, such as hepatocytes, vascular-endothelial cells, pancreatic (also in vivo) and neural-like cells [13, 125]. Moreover, human P-MSC were able to undergo hepatic differentiation when injected into immunocompromised mice, and when pre-treated with a hyaluronan mixed ester of butyric and retinoic acid they displayed reparative potential in rat and pig infarction [49]. It is worth noting that the pluripotency marker OCT-4 was found to be down-regulated epigenetically after methylation in the placenta and this process may be relevant to the study of the pathogenesis of gestational trophoblastic disease [127]. An additional multipotent placenta stem cell population isolated from the maternal part that evenly exhibited pluripotency markers (SSEA-4, Oct-4, Stro-1 and Tra1-80) along with mesenchymal and haematopoietic markers has also been described [101]. This population could be a promising source of broadly multipotent stem cells.

In conclusion, MSC from this post-natal derivative have the advantage of being easily cultured and expanded, without ethical concerns, with high differentiation potential and no teratoma formation. For all these reasons P-MSC may be relevant for future clinical applications.

4.2 Cord Blood Mesenchymal Stem Cells

During the last decades, umbilical cord blood has been exploited as a source of haematopoietic and progenitor stem cells to establish therapeutically effective transplantation, both for malignant and non-malignant disorders [9].

When considering the total CB cells about 1 % of the mononuclear fraction expresses the CD34 antigen, which represents the most important marker for HSC. The CD34+ cells isolated from CB are able to self-renew and differentiate into several cell lineages not only in vitro, but most importantly in vivo, repopulating the BM of severe combined immunodeficient mice [8]. Significant progress has been made regarding the important role of CB-MSC in supporting the in vivo expansion of HSC [118] and functioning as a supportive cell population for engraftment [46]. It was outstanding news when it became clear that MSC can modulate the immune system in vitro. They could inhibit T cell proliferation. CD3-ligation, mitogens, and alloantigen are responsible for inhibition of T cell proliferation. It seems that other cells such as antigen-presenting cells and NK cells are also affected by MSC. These aspects can be clinically used to support HSC transplantation and to reduce the possibility of a graft versus host disease (GvDH). In fact, similarly to BM-MSC, stromal cells from CB express HLA class I (MHCI) at low levels and are negative for MHC class II (MHCII) and are therefore particularly attractive to manipulate or modify GvHD [110]. In addition, it was demonstrated that MHC-mismatched inactivated CB cells did not induce a detectable immune response in an animal model [15]. However, it was shown that an IFNγ-stimulation of CB cells increases MHCII expression [19, 116]. Therefore, differences in the degree of immunogenic response can be postulated dependent on the local cytokine profile of the anatomic transplantation site of the host. In addition, some data from the literature also indicates that activated CB-MSC have a cytotoxic effect against tumour cells, such as malignant glioma cells [47].

The current data so far suggests that CB-MSC may harbour a subpopulation of stem cells leading to improvement of several deficits [29, 57, 74, 124] via mechanisms involving either the release of growth factors or cytokines, or by neovascularization at the ischemic zones, or by enhancing the regeneration processes without a detectable long-term engraftment [57].

CB-MSC, representing the second major cell population of CB, possess a phenotype that more closely resembles that of ES. Besides the typical markers for MSC, CB-MSC also express stem cell markers such as Oct4, SSEA3, SSEA4, Tra-1-60, Tra-1-80 and Nanog [67, 69, 103, 129]. Moreover, their multipotency and differentiation capacity to cell lineages derived from all three germ layers has been documented [61].

Bone marrow and CB have been traditionally considered as the two classical and main sources of MSC and have eventually become a major cellular population of precursor cells suitable for functional studies and therapeutic applications.

5 Adult MSC

Adult stem cells have been thoroughly investigated and this cell source has recently been adopted for clinical applications.

5.1 Bone Marrow Mesenchymal Stem Cells

Bone marrow is known to have two populations of stem cells: HSC and MSC. HSC are generally accepted to give rise to the different classes of blood cells (myeloid, erythroid, lymphoid, platelets, and mast cells), while MSC are precursors of the stromal parts and give rise to the structural elements of the skeleton, such as bone, cartilage, and marrow fat.

Mesenchymal stromal cells were first identified in the bone marrow by Friedenstein in 1976 [27] who described a plastic adherent fibroblast-like population able to differentiate into bone that he referred to as osteogenic precursor cells. Later, it was discovered that the mesenchymal stroma of nearly all tissues harbours this important population of cells that possess stem cell-like characteristics including self-renewal, differentiation capacities and are mostly located in perivascular niches.

Many scientific reports indicate that MSC possess immunomodulatory properties and may play specific roles as immunomodulators in transplantation tolerance, autoimmunity, as well as foetal-maternal tolerance [59]. MSC suppress T cell proliferation, but express different ligands that are recognized by activating natural killer (NK) receptors that trigger NK alloreactivity. Treatment of MSC with IFN-gamma up-regulate expression of HLA class I molecules and decrease NK activity [48]. Recently, it has been supposed that MSC may exert a more significant role through the release of different factors via paracrine action, rather than adopting a particular differentiated state after engraftment in target tissue [16]. In contrast with the aforementioned cells, MSC also have a limited life span and become senescent when cultured in vitro. Several mechanisms were involved to explain the acquisition of this phenotype such as loss of telomeres, and various experimental strategies have been adopted to extend MSC life span [58, 106, 108]. For istance, proliferation capacity of MSC can be significantly increased by the presence of oncogenes (E6–E7) from HPV. Unexpectedly, transfected MSC showed no signs of neoplastic transformation [107]. Nevertheless, the acquisition of neoplastic features in these engineered cells could not be totally excluded and might occur.

Regardless of the isolation procedure, quantities of MSC obtained from primary tissues are not sufficient for any application in clinical settings. In vitro expansion can affect biological properties of the cells; in fact MSC go through very significant changes in phenotype and gene expression as a result of cell culture adaptation. Although considered a safer source, if compared to embryonic stem cells, the prospective clinical applications of MSC require meticulous examination. Some approaches aimed at improving safety have been established to evaluate the possibility of eliminating xenoproteins or xenoproducts like foetal calf serum in the feeding medium, to reduce the risk of potential viral-transmission-like unidentified zoonosis or prions and reduce immunogenicity related to serum component absorption [84].

Table 1 Selected ongoing clinical trials of the different categories of mesenchymal stem cells www.clinicaltrials.gov

	BM	ADS	Placenta	CB	AFS	ES/iPS derived
Ankilosing spondylitis			+	+		
Aplastic anemia	+		+	+		
Colitis				+		
Chronic GvHD	+	+		+		
Diabetes mellitus	+			+		
Diabetes type II			+	+		
Ischemic stroke			+			
Limb ischemia in diabetics		+				
Multiple sclerosis		+				
Myelodysplastic syndromes			+	+		
Osteoarthritis	+					
Parkinson	+					
Spinal cord injury	+					
Pulmonary Sarcoidosis			+			

The ability of MSC to give rise to different lineages has been a matter of intense study and plasticity and mechanisms of action have been studied in models of small and large animals. MSC can differentiate beyond their traditional mesodermal lineage, at least in vitro, into cells of both ectodermal (neurons) and endodermal (hepatocytes) nature [39, 94, 102]. However, the broad abilities of MSC are questionable and in several publications it has been demonstrated that MSC do not undergo a proper trans-differentiation (irreversible switch of one differentiated cell into another), but rather fuse with specialized differentiated cells; thus, more studies are required to achieve a better understanding of this issue [20].

To date, MSC have been tested on patients for several clinical indications, such as inborn error of metabolism (Metachromatic leukodystrophy, Hurler syndrome, Infantile hypophosphatasemia), osteogenesis imperfecta, and GVHD [40, 42, 55, 60, 120]. Preliminary studies have been assessed in patients with amyotrophic lateral sclerosis and autologous MSC transplantation has also been evaluated in patients after acute myocardial infarction [10, 68]. For the latest clinical trials see Table 1.

5.2 *Adipose Tissue Mesenchymal Stem Cells*

Adipose tissue, like bone marrow, is derived from the mesenchymal embryonic layer and contains a stroma that is easily isolated. This tissue is highly complex consisting of mature adipocytes, which constitute more than 90 % of the tissue, and a stromal vascular fraction (SVF), which includes pre-adipocytes, fibroblasts, vascular smooth muscle cells, endothelial cells, resident monocytes/macrophages, lymphocytes and stem cells [122, 123]. Numerous acronyms are used to identify the stem cell fraction in adipose tissue: adipose derived stem/stromal cells (ASCs), adipose-derived adult stem (ADAS) cells, adipose-derived adult stromal cells, adipose-derived stromal cells (ADSCs), adipose stromal cells (ASCs), adipose mesenchymal stem cells (AdMSCs), lipoblast, pericyte, pre-adipocyte, and processed lipoaspirate (PLA) cells [131, 132]. To address the issue of adopting a uniform name for this stem cell category, the International Fat Applied Technology Society reached a consensus to adopt the term "adipose-derived stem cells" (ASC) to identify the isolated, plastic-adherent, multipotent cell population.

At least five different types of adipose tissue exist: bone marrow, brown, mammary, mechanical, and white. Briefly, bone marrow adipose tissue space is no longer required for haematopoiesis and serves as an energy reservoir and cytokine source for osteogenic and haematopoietic processes. Brown adipose tissue is thermogenic, generating heat through the expression of a unique uncoupling protein that short-circuits the mitochondrial pH gradient. Whereas brown adipose tissue is found around the major organs (heart, kidney, aorta, and gonads) in the newborn infant, it disappears as humans mature. Mammary adipose tissue provides nutrients and energy during lactation and is regulated, in part, by pregnancy-associated hormones. Mechanical adipose depots, such as the retro orbital and palmar fat pads, provide support to the eye, hand, and other critical structures. Finally, white adipose tissue serves to store energy and provide insulation.

Whereas multipotent stem cells are abundant within murine white adipose tissue, their numbers and differentiation potential are reduced in brown adipose tissue. In humans, differences in stem cell recovery have been noted between subcutaneous white adipose tissue depots, with the greatest numbers recovered from the arm when compared to the thigh, abdomen, and breast [97].

Initial enzymatic digestion of adipose tissue yields a mixture of stromal and vascular cells referred to as the SVF [111]. SFV is a heterogeneous cell population including circulating blood cells, fibroblasts, pericytes and endothelial cells as well as "preadipocytes" or adipocyte progenitors. The final isolation step focuses on enrichment by adhesion of the pre-adipocyte fraction.

Identification of the ASC surface immunophenotype has provided a mechanism to enrich or purify the stem cell population directly from the heterogeneous SVF cells [43, 72, 99]. Investigators have used immunomagnetic beads or flow cytometry to both positively and negatively select for a subpopulation of cells within the SVF. For example, endothelial progenitors can be removed by negatively selecting for cells expressing CD31 or platelet endothelial cell adhesion molecule-1 [43, 72, 99]. Likewise, positive selection has been performed using CD34. Nevertheless, it has been reported there are still major differences in markers chosen for ASC selection and characterisation, for instance the marker Stro-1, a classic BM-associated marker, has been reported to be present [131] or absent [33] on human ASC.

Freshly isolated SVF cells are a heterogeneous cell population that includes ASC (CD31-, CD34±, CD45-, CD90+, CD105-, CD146-), endothelial progenitor cells (CD31+, CD34+, CD45-, CD90+, CD105-, CD146+), vascular smooth muscle cells or pericytes (CD31-, CD34±, CD45-, CD90+, CD105-, CD146+), and haematopoietic cells (CD45+) in uncultured conditions [38]. It remains, in

Table 2 Summary of the most important markers characterizing the different described human mesenchymal stem cells

Antigen category surface		BM	ADS	Placenta	CB	AF	ES/iPS derived
Adhesion molecules	CD9 (tetraspan)	+	+			-	
	CD29 (β1 integrin)	+	+			+	
	CD49 days (4 integrin)	+	+			+	
	CD54 (ICAM-1)	+	+			-	
	CD105 (endoglin)	+	+			+	
	CD166 (ALCAM)	+	+	+		+	
	CD11b (α integrin)	+	+			-	
	CD56 (NCAM)	-	-			-	
Hematopoietic	CD14	-	-	-	-	-	-
	CD34	-	-	-	-	-	-
	CD45	-	-	-	-	-	-
Extracellular matrix molecules	CD90 (Thy1)	+		+		+	
	CD146 (Muc18)	+				+	
Stromal factors	CD29	+				+	
	CD44	+				+	
	CD73	+		+		+	
	CD166	+				+	
Pluripotency markers	POU5F1 (OCT-4)			+	+	+	+
	Sox-2					+	+
	SSEA-4	+	+	+	+	+	+
	Tra 1-60			+	+	+	+
	Tra 1-81				+	+	+
Histocompatibility antigen	HLA-ABC	+		+	+	+	
	HLA-DR	-		-	-	-	

fact, to be proven whether the origin of the cells correlates with the endothelial, pericyte or stromal compartments. Additionally, compared with ASC from later passages, freshly isolated SVF cells and early passage ASC express higher levels of CD117 (c-kit), HLA-ABC, and stem cell-associated markers such as CD34, along with lower levels of stromal cell markers such as CD13, CD29, CD44, CD63, CD73, CD90, CD105 and CD166 [92] (Table 2).

Since human adipose tissue is ubiquitous and easily obtained in large quantities with little donor site morbidity or patient discomfort, the use of autologous ASC as both research tools and as cellular therapeutics is feasible, and has been shown to be both safe and efficacious in pre-clinical and clinical studies of injury and disease. As for BM-MSC, it has been proposed that transplantation into an injured or diseased tissue may cause secretion of cytokines and growth factors that stimulate recovery in a paracrine manner. It is important to consider the potential use of both autologous and allogeneic ASC. Autologous ASC offer advantages from regulatory and histocompatibility perspectives. Passaged human ASC, as opposed to freshly isolated SVF cells, reduce their expression of surface histocompatibility antigens and no longer stimulate a mixed lymphocyte reaction when co-cultured with allogeneic peripheral blood monocytes [70, 89].

It has been reported that ASC can differentiate into all the classical mesenchymal lineages such as adipogenic, osteogenic, chondrogenic, myogenic, cardiomyogenic, angiogenic, tenogenic and periodontogenic lineages, and tissue regeneration studies with suitable scaffolds and growth factors in appropriate external environments have been carried out [132]. Recent breakthroughs in understanding the roles played by ASC in wound healing and tissue regeneration have provided new options for treating wounds [78]. Pre-clinical studies in a murine model have shown that the topical administration of autologous ASC, together with a type I collagen sponge matrix, into a diabetic animal accelerated the healing of diabetic ulcers [77]. In addition, ASC have been used for Crohn's and non-Crohn's disease and shown to be effective following direct injection of ASC into the tract wall together with a fibrin glue sealant, and no adverse effects were observed [30, 31].

Despite the limitations upon the differentiation ability due to the body site from which ASC are taken, ASC have practical advantages in clinical medicine and their use has become more realistic because adipose tissue, the primary source of ASC, is abundant and easy to obtain with less donor site morbidity.

6 Conclusion

MSC can be derived from all stages of development and represent a promising tool for therapeutic applications. Adult MSC have a limited proliferation and differentiation capability. In contrast, MSC derived from pluripotent stem cells may overcome these limitations but could be problematic because of allogenic rejection (in the case of ES cells) or teratogenesis (for both ES and iPS cells). Foetal- or neonatal-derived MSC may have better potential in the long term because of their expansion capability and lack of teratoma formation.

References

1. Aasen T, Izpisua Belmonte JC (2010) Isolation and cultivation of human keratinocytes from skin or plucked hair for the generation of induced pluripotent stem cells. Nat Protoc 5(2):371–382
2. Anokye-Danso F, Trivedi CM et al (2011) Highly efficient miRNA-mediated reprogramming of mouse and human somatic cells to pluripotency. Cell Stem Cell 8(4):376–388
3. Arpornmaeklong P, Brown SE et al (2009) Phenotypic characterization, osteoblastic differentiation, and bone regeneration capacity of human embryonic stem cell-derived mesenchymal stem cells. Stem Cells Dev 18(7):955–968
4. Barberi T, Willis LM et al (2005) Derivation of multipotent mesenchymal precursors from human embryonic stem cells. PLoS Med 2(6):e161
5. Bilousova G, Jun du H et al (2011) Osteoblasts derived from induced pluripotent stem cells form calcified structures in scaffolds both in vitro and in vivo. Stem Cells 29(2):206–216
6. Bossolasco P, Montemurro T et al (2006) Molecular and phenotypic characterization of human amniotic fluid cells and their differentiation potential. Cell Res 16(4):329–336
7. Brooke G, Tong H et al (2008) Molecular trafficking mechanisms of multipotent mesenchymal stem cells derived from human bone marrow and placenta. Stem Cells Dev 17(5):929–940
8. Broxmeyer HE, Cooper S et al (2005) Identification of a massive reserve of hematopoietic progenitors in mice. Stem Cells Dev 14(2):105–110
9. Broxmeyer HE, Srour E et al (2006) Cord blood stem and progenitor cells. Methods Enzymol 419:439–473
10. Burt RK, Loh Y et al (2008) Clinical applications of blood-derived and marrow-derived stem cells for nonmalignant diseases. JAMA 299(8):925–936
11. Campagnoli C, Roberts IA et al (2001) Identification of mesenchymal stem/progenitor cells in human first-trimester fetal blood, liver, and bone marrow. Blood 98(8):2396–2402
12. Cananzi M, Atala A et al (2009) Stem cells derived from amniotic fluid: new potentials in regenerative medicine. Reprod Biomed Online 18(Suppl 1):17–27
13. Chang CM, Kao CL et al (2007) Placenta-derived multipotent stem cells induced to differentiate into insulin-positive cells. Biochem Biophys Res Commun 357(2):414–420
14. Cho HJ, Lee CS et al (2010) Induction of pluripotent stem cells from adult somatic cells by protein-based reprogramming without genetic manipulation. Blood 116(3):386–395
15. Cho PS, Messina DJ et al (2008) Immunogenicity of umbilical cord tissue derived cells. Blood 111(1):430–438
16. Crisostomo PR, Markel TA et al (2008) Surgically relevant aspects of stem cell paracrine effects. Surgery 143(5):577–581
17. De Coppi P, Bartsch G Jr et al (2007) Isolation of amniotic stem cell lines with potential for therapy. Nat Biotechnol 25(1):100–106
18. De Coppi P, Callegari A et al (2007) Amniotic fluid and bone marrow derived mesenchymal stem cells can be converted to smooth muscle cells in the cryo-injured rat bladder and prevent compensatory hypertrophy of surviving smooth muscle cells. J Urol 177(1):369–376
19. Derzic S, Slone V et al (2005) IL-2-activated cord blood mononuclear cells. Cytotherapy 7(5):408–416
20. Dimmeler S, Zeiher AM (2009) Cell therapy of acute myocardial infarction: open questions. Cardiology 113(3):155–160

21. Ditadi A, de Coppi P et al (2009) Human and murine amniotic fluid c-Kit + Lin- cells display hematopoietic activity. Blood 113(17):3953–3960
22. Eyckmans J, Luyten FP (2006) Species specificity of ectopic bone formation using periosteum-derived mesenchymal progenitor cells. Tissue Eng 12(8):2203–2213
23. Fan CG, Tang FW et al (2005) Characterization and neural differentiation of fetal lung mesenchymal stem cells. Cell Transplant 14(5):311–321
24. Fehrer C, Lepperdinger G (2005) Mesenchymal stem cell aging. Exp Gerontol 40(12):926–930
25. Fenno LE, Ptaszek LM et al (2008) Human embryonic stem cells: emerging technologies and practical applications. Curr Opin Genet Dev 18(4):324–329
26. Friedenstein AJ, Chailakhyan RK et al (1974) Stromal cells responsible for transferring the microenvironment of the hemopoietic tissues. Cloning in vitro and retransplantation in vivo. Transplantation 17(4):331–340
27. Friedenstein AJ, Gorskaja JF et al (1976) Fibroblast precursors in normal and irradiated mouse hematopoietic organs. Exp Hematol 4(5):267–274
28. Fukuchi Y, Nakajima H et al (2004) Human placenta-derived cells have mesenchymal stem/progenitor cell potential. Stem Cells 22(5):649–658
29. Gang EJ, Jeong JA et al (2004) Skeletal myogenic differentiation of mesenchymal stem cells isolated from human umbilical cord blood. Stem Cells 22(4):617–624
30. Garcia-Olmo D, Herreros D et al (2009) Expanded adipose-derived stem cells for the treatment of complex perianal fistula: a phase II clinical trial. Dis Colon Rectum 52(1):79–86
31. Garcia-Olmo D, Herreros D et al (2009) Treatment of enterocutaneous fistula in Crohn's Disease with adipose-derived stem cells: a comparison of protocols with and without cell expansion. Int J Colorectal Dis 24(1):27–30
32. Granero-Molto F, Weis JA et al (2008) Role of mesenchymal stem cells in regenerative medicine: application to bone and cartilage repair. Expert Opin Biol Ther 8(3):255–268
33. Gronthos S, Franklin DM et al (2001) Surface protein characterization of human adipose tissue-derived stromal cells. J Cell Physiol 189(1):54–63
34. Haase A, Olmer R et al (2009) Generation of induced pluripotent stem cells from human cord blood. Cell Stem Cell 5(4):434–441
35. Heng BC, Cao T et al (2004) Strategies for directing the differentiation of stem cells into the osteogenic lineage in vitro. J Bone Miner Res 19(9):1379–1394
36. Heng BC, Haider H et al (2004) Strategies for directing the differentiation of stem cells into the cardiomyogenic lineage in vitro. Cardiovasc Res 62(1):34–42
37. Hilberg F, Wagner EF (1992) Embryonic stem (ES) cells lacking functional c-jun: consequences for growth and differentiation, AP-1 activity and tumorigenicity. Oncogene 7(12):2371–2380
38. Hipp J, Atala A (2008) Sources of stem cells for regenerative medicine. Stem Cell Rev 4(1):3–11
39. Hong SH, Gang EJ et al (2005) In vitro differentiation of human umbilical cord blood-derived mesenchymal stem cells into hepatocyte-like cells. Biochem Biophys Res Commun 330(4):1153–1161
40. Horwitz EM, Gordon PL et al (2002) Isolated allogeneic bone marrow-derived mesenchymal cells engraft and stimulate growth in children with osteogenesis imperfecta: Implications for cell therapy of bone. Proc Natl Acad Sci U S A 99(13):8932–8937
41. Horwitz EM, Le Blanc K et al (2005) Clarification of the nomenclature for MSC: the international society for cellular therapy position statement. Cytotherapy 7(5):393–395
42. Horwitz EM, Prockop DJ et al (1999) Transplantability and therapeutic effects of bone marrow-derived mesenchymal cells in children with osteogenesis imperfecta. Nat Med 5(3):309–313
43. Hutley LJ, Herington AC et al (2001) Human adipose tissue endothelial cells promote preadipocyte proliferation. Am J Physiol Endocrinol Metab 281(5):E1037–E1044

44. In't Anker PS, Scherjon SA et al (2004) Isolation of mesenchymal stem cells of fetal or maternal origin from human placenta. Stem Cells 22(7):1338–1345
45. In't Anker PS, Scherjon SA et al (2003) Amniotic fluid as a novel source of mesenchymal stem cells for therapeutic transplantation. Blood 102(4):1548–1549
46. Javazon EH, Beggs KJ et al (2004) Mesenchymal stem cells: paradoxes of passaging. Exp Hematol 32(5):414–425
47. Kang SG, Jeun SS et al (2008) Cytotoxicity of human umbilical cord blood-derived mesenchymal stem cells against human malignant glioma cells. Childs Nerv Syst 24(3):293–302
48. Karlsson H, Samarasinghe S et al (2008) Mesenchymal stem cells exert differential effects on alloantigen and virus-specific T-cell responses. Blood 112(3):532–541
49. Kaviani A, Perry TE et al (2002) The placenta as a cell source in fetal tissue engineering. J Pediatr Surg 37(7):995–999 (discussion 995–999)
50. Kehat I, Kenyagin-Karsenti D et al (2001) Human embryonic stem cells can differentiate into myocytes with structural and functional properties of cardiomyocytes. J Clin Invest 108(3):407–414
51. Kern S, Eichler H et al (2006) Comparative analysis of mesenchymal stem cells from bone marrow, umbilical cord blood, or adipose tissue. Stem Cells 24(5):1294–1301
52. Kim D, Kim CH et al (2009) Generation of human induced pluripotent stem cells by direct delivery of reprogramming proteins. Cell Stem Cell 4(6):472–476
53. Kim J, Lee Y et al (2007) Human amniotic fluid-derived stem cells have characteristics of multipotent stem cells. Cell Prolif 40(1):75–90
54. Kim SJ, Song CH et al (2007) Human placenta-derived feeders support prolonged undifferentiated propagation of a human embryonic stem cell line, SNUhES3: comparison with human bone marrow-derived feeders. Stem Cells Dev 16(3):421–428
55. Koc ON, Day J et al (2002) Allogeneic mesenchymal stem cell infusion for treatment of metachromatic leukodystrophy (MLD) and Hurler syndrome (MPS-IH). Bone Marrow Transplant 30(4):215–222
56. Kolambkar YM, Peister A et al (2007) Chondrogenic differentiation of amniotic fluid-derived stem cells. J Mol Histol 38(5):405–413
57. Koponen JK, Kekarainen T et al (2007) Umbilical cord blood-derived progenitor cells enhance muscle regeneration in mouse hindlimb ischemia model. Mol Ther 15(12):2172–2177
58. Ksiazek K (2009) A comprehensive review on mesenchymal stem cell growth and senescence. Rejuvenation Res 12(2):105–116
59. Le Blanc K (2003) Immunomodulatory effects of fetal and adult mesenchymal stem cells. Cytotherapy 5(6):485–489
60. Le Blanc K, Gotherstrom C et al (2005) Fetal mesenchymal stem-cell engraftment in bone after in utero transplantation in a patient with severe osteogenesis imperfecta. Transplantation 79(11):1607–1614
61. Lee OK, Kuo TK et al (2004) Isolation of multipotent mesenchymal stem cells from umbilical cord blood. Blood 103(5):1669–1675
62. Li F, Bronson S et al (2010) Derivation of murine induced pluripotent stem cells (iPS) and assessment of their differentiation toward osteogenic lineage. J Cell Biochem 109(4):643–652
63. Lian Q, Lye E et al (2007) Derivation of clinically compliant MSCs from CD105 + , CD24- differentiated human ESCs. Stem Cells 25(2):425–436
64. Lian Q, Zhang Y et al (2010) Functional mesenchymal stem cells derived from human induced pluripotent stem cells attenuate limb ischemia in mice. Circulation 121(9):1113–1123
65. Liu Y, Goldberg AJ et al (2012) One-step derivation of mesenchymal stem cell (MSC)-like cells from human pluripotent stem cells on a fibrillar collagen coating. PLoS ONE 7(3):e33225
66. Lorda-Diez CI, Montero JA et al (2009) Transforming growth factors beta coordinate cartilage and tendon differentiation in the developing limb mesenchyme. J Biol Chem 284(43):29988–29996

67. Markov V, Kusumi K et al (2007) Identification of cord blood-derived mesenchymal stem/stromal cell populations with distinct growth kinetics, differentiation potentials, and gene expression profiles. Stem Cells Dev 16(1):53–73
68. Mazzini L, Mareschi K et al (2008) Stem cell treatment in amyotrophic lateral sclerosis. J Neurol Sci 265(1–2):78–83
69. McGuckin CP, Forraz N et al (2005) Production of stem cells with embryonic characteristics from human umbilical cord blood. Cell Prolif 38(4):245–255
70. McIntosh K, Zvonic S et al (2006) The immunogenicity of human adipose-derived cells: temporal changes in vitro. Stem Cells 24(5):1246–1253
71. Meissner A, Wernig M et al (2007) Direct reprogramming of genetically unmodified fibroblasts into pluripotent stem cells. Nat Biotechnol 25(10):1177–1181
72. Miranville A, Heeschen C et al (2004) Improvement of postnatal neovascularization by human adipose tissue-derived stem cells. Circulation 110(3):349–355
73. Miyoshi N, Ishii H et al (2011) Reprogramming of mouse and human cells to pluripotency using mature microRNAs. Cell Stem Cell 8(6):633–638
74. Morigi M, Rota C et al (2010) Life-sparing effect of human cord blood-mesenchymal stem cells in experimental acute kidney injury. Stem Cells 28(3):513–522
75. Moschidou M, Mukherjee S, Blundell MP, Drews K, Jones GN, Abdulrazzak H, Nowakowska B, Phoolchund A, Lay K, Ramasamy TS, Cananzi M, ettersheim D, Sullivan M, Frost J, Moore G, Vermeesch JR, Fisk NM, Thrasher AJ, Atala A, Adjaye J, Schorle H, De Coppi P, Guillot PV (2012, in press) Valproic acid confers functional pluripotency to human amniotic fluid stem cells in a transgene-free approach. Mol Ther 20(10):1953–1967
76. Munoz JR, Stoutenger BR et al (2005) Human stem/progenitor cells from bone marrow promote neurogenesis of endogenous neural stem cells in the hippocampus of mice. Proc Natl Acad Sci U S A 102(50):18171–18176
77. Nambu M, Kishimoto S et al (2009) Accelerated wound healing in healing-impaired db/db mice by autologous adipose tissue-derived stromal cells combined with atelocollagen matrix. Ann Plast Surg 62(3):317–321
78. Nie C, Yang D et al (2011) Locally administered adipose-derived stem cells accelerate wound healing through differentiation and vasculogenesis. Cell Transplant 20(2):205–216
79. Odorico JS, Kaufman DS et al (2001) Multilineage differentiation from human embryonic stem cell lines. Stem Cells 19(3):193–204
80. Okita K, Nakagawa M et al (2008) Generation of mouse induced pluripotent stem cells without viral vectors. Science 322(5903):949–953
81. Olivier EN, Rybicki AC et al (2006) Differentiation of human embryonic stem cells into bipotent mesenchymal stem cells. Stem Cells 24(8):1914–1922
82. Pan HC, Yang DY et al (2006) Enhanced regeneration in injured sciatic nerve by human amniotic mesenchymal stem cell. J Clin Neurosci 13(5):570–575
83. Parolini O, Alviano F et al (2008) Concise review: isolation and characterization of cells from human term placenta: outcome of the first international workshop on placenta derived stem cells. Stem Cells 26(2):300–311
84. Pelagiadis I, Dimitriou H et al (2008) Biologic characteristics of mesenchymal stromal cells and their clinical applications in pediatric patients. J Pediatr Hematol Oncol 30(4):301–309
85. Piccoli M, Franzin C et al (2012) Amniotic fluid stem cells restore the muscle cell niche in a HSA-Cre, Smn(F7/F7) mouse model. Stem Cells 30(8):1675–1684
86. Pittenger MF, Mackay AM et al (1999) Multilineage potential of adult human mesenchymal stem cells. Science 284(5411):143–147
87. Prusa AR, Marton E et al (2003) Oct-4-expressing cells in human amniotic fluid: a new source for stem cell research? Hum Reprod 18(7):1489–1493
88. Prusa AR, Marton E et al (2004) Neurogenic cells in human amniotic fluid. Am J Obstet Gynecol 191(1):309–314
89. Puissant B, Barreau C et al (2005) Immunomodulatory effect of human adipose tissue-derived adult stem cells: comparison with bone marrow mesenchymal stem cells. Br J Haematol 129(1):118–129

90. Richards M, Fong CY et al (2002) Human feeders support prolonged undifferentiated growth of human inner cell masses and embryonic stem cells. Nat Biotechnol 20(9):933–936
91. Roelen BA, Dijke P (2003) Controlling mesenchymal stem cell differentiation by TGFBeta family members. J Orthop Sci 8(5):740–748
92. Rosen ED, Spiegelman BM (2000) Molecular regulation of adipogenesis. Annu Rev Cell Dev Biol 16:145–171
93. Roubelakis MG, Pappa KI et al (2007) Molecular and proteomic characterization of human mesenchymal stem cells derived from amniotic fluid: comparison to bone marrow mesenchymal stem cells. Stem Cells Dev 16(6):931–952
94. Sanchez-Ramos J, Song S et al (2000) Adult bone marrow stromal cells differentiate into neural cells in vitro. Exp Neurol 164(2):247–256
95. Sanchez L, Gutierrez-Aranda I et al (2011) Enrichment of human ESC-derived multipotent mesenchymal stem cells with immunosuppressive and anti-inflammatory properties capable to protect against experimental inflammatory bowel disease. Stem Cells 29(2):251–262
96. Sartore S, Lenzi M et al (2005) Amniotic mesenchymal cells autotransplanted in a porcine model of cardiac ischemia do not differentiate to cardiogenic phenotypes. Eur J Cardiothorac Surg 28(5):677–684
97. Schipper BM, Marra KG et al (2008) Regional anatomic and age effects on cell function of human adipose-derived stem cells. Ann Plast Surg 60(5):538–544
98. Schmidt D, Stock UA et al (2007) Tissue engineering of heart valves using decellularized xenogeneic or polymeric starter matrices. Philos Trans R Soc Lond B Biol Sci 362(1484):1505–1512
99. Sengenes C, Lolmede K et al (2005) Preadipocytes in the human subcutaneous adipose tissue display distinct features from the adult mesenchymal and hematopoietic stem cells. J Cell Physiol 205(1):114–122
100. Soldner F, Hockemeyer D et al (2009) Parkinson's disease patient-derived induced pluripotent stem cells free of viral reprogramming factors. Cell 136(5):964–977
101. Strakova Z, Livak M et al (2008) Multipotent properties of myofibroblast cells derived from human placenta. Cell Tissue Res 332(3):479–488
102. Sugaya K (2003) Potential use of stem cells in neuroreplacement therapies for neurodegenerative diseases. Int Rev Cytol 228:1–30
103. Sun B, Jeong YH et al (2007) Regulation of human umbilical cord blood-derived multipotent stem cells by autogenic osteoclast-based niche-like structure. Biochem Biophys Res Commun 357(1):92–98
104. Takahashi K, Ichisaka T et al (2006) Identification of genes involved in tumor-like properties of embryonic stem cells. Methods Mol Biol 329:449–458
105. Takahashi K, Yamanaka S (2006) Induction of pluripotent stem cells from mouse embryonic and adult fibroblast cultures by defined factors. Cell 126(4):663–676
106. Takeda Y, Mori T et al (2004) Can the life span of human marrow stromal cells be prolonged by bmi-1, E6, E7, and/or telomerase without affecting cardiomyogenic differentiation? J Gene Med 6(8):833–845
107. Takeuchi M, Takeuchi K et al (2007) Chromosomal instability in human mesenchymal stem cells immortalized with human papilloma virus E6, E7, and hTERT genes. In Vitro Cell Dev Biol Anim 43(3–4):129–138
108. Terai M, Uyama T et al (2005) Immortalization of human fetal cells: the life span of umbilical cord blood-derived cells can be prolonged without manipulating p16INK4a/RB braking pathway. Mol Biol Cell 16(3):1491–1499
109. Thomson JA, Itskovitz-Eldor J et al (1998) Embryonic stem cell lines derived from human blastocysts. Science 282(5391):1145–1147
110. Tisato V, Naresh K et al (2007) Mesenchymal stem cells of cord blood origin are effective at preventing but not treating graft-versus-host disease. Leukemia 21(9):1992–1999
111. Traktuev DO, Merfeld-Clauss S et al (2008) A population of multipotent CD34-positive adipose stromal cells share pericyte and mesenchymal surface markers, reside in a periendothelial location, and stabilize endothelial networks. Circ Res 102(1):77–85

112. Trivedi P, Hematti P (2007) Simultaneous generation of CD34+ primitive hematopoietic cells and CD73+ mesenchymal stem cells from human embryonic stem cells cocultured with murine OP9 stromal cells. Exp Hematol 35(1):146–154
113. Trivedi P, Hematti P (2008) Derivation and immunological characterization of mesenchymal stromal cells from human embryonic stem cells. Exp Hematol 36(3):350–359
114. Tsai MS, Hwang SM et al (2006) Clonal amniotic fluid-derived stem cells express characteristics of both mesenchymal and neural stem cells. Biol Reprod 74(3):545–551
115. Tsai MS, Lee JL et al (2004) Isolation of human multipotent mesenchymal stem cells from second-trimester amniotic fluid using a novel two-stage culture protocol. Hum Reprod 19(6):1450–1456
116. Urashima M, Hoshi Y et al (1995) Ex vivo expansion of umbilical cord blood hematopoietic progenitor cells by combinations of cytokines. Acta Paediatr Jpn 37(2):160–165
117. Waddington RJ, Youde SJ et al (2009) Isolation of distinct progenitor stem cell populations from dental pulp. Cells Tissues Organs 189(1–4):268–274
118. Wang JF, Wu YF et al (2004) Ex vivo expansions and transplantations of mouse bone marrow-derived hematopoietic stem/progenitor cells. J Zhejiang Univ Sci 5(2):157–163
119. Warren L, Manos PD et al (2010) Highly efficient reprogramming to pluripotency and directed differentiation of human cells with synthetic modified mRNA. Cell Stem Cell 7(5):618–630
120. Whyte MP, Kurtzberg J et al (2003) Marrow cell transplantation for infantile hypophosphatasia. J Bone Miner Res 18(4):624–636
121. Woltjen K, Michael IP et al (2009) piggyBac transposition reprograms fibroblasts to induced pluripotent stem cells. Nature 458(7239):766–770
122. Wright JT, Hausman GJ (1990) Adipose tissue development in the fetal pig examined using monoclonal antibodies. J Anim Sci 68(4):1170–1175
123. Wright JT, Hausman GJ (1990) Monoclonal antibodies against cell surface antigens expressed during porcine adipocyte differentiation. Int J Obes 14(5):395–409
124. Yan Y, Xu W et al (2009) Mesenchymal stem cells from human umbilical cords ameliorate mouse hepatic injury in vivo. Liver Int 29(3):356–365
125. Yen BL, Chien CC et al (2008) Placenta-derived multipotent cells differentiate into neuronal and glial cells in vitro. Tissue Eng Part A 14(1):9–17
126. Yu J, Hu K et al (2009) Human induced pluripotent stem cells free of vector and transgene sequences. Science 324(5928):797–801
127. Zhang HJ, Siu MK et al (2008) Oct4 is epigenetically regulated by methylation in normal placenta and gestational trophoblastic disease. Placenta 29(6):549–554
128. Zhao P, Ise H et al (2005) Human amniotic mesenchymal cells have some characteristics of cardiomyocytes. Transplantation 79(5):528–535
129. Zhao Y, Wang H et al (2006) Identification of stem cells from human umbilical cord blood with embryonic and hematopoietic characteristics. Exp Cell Res 312(13):2454–2464
130. Zhou T, Benda C et al (2011) Generation of induced pluripotent stem cells from urine. J Am Soc Nephrol 22(7):1221–1228
131. Zuk PA, Zhu M et al (2002) Human adipose tissue is a source of multipotent stem cells. Mol Biol Cell 13(12):4279–4295
132. Zuk PA, Zhu M et al (2001) Multilineage cells from human adipose tissue: implications for cell-based therapies. Tissue Eng 7(2):211–228

Adv Biochem Eng Biotechnol (2013) 130: 287–299
DOI: 10.1007/10_2012_142

Published Online: 28 July 2012

Role of the EU Framework in Regulation of Stem Cell-Based Products

Giovanni Migliaccio and Cristina Pintus

Abstract The use of stem cells for therapeutic purposes is regulated by two overlapping sets of rules. If used for transplantation, stem cells are covered by the collection, traceability and technical aspects of three European directives. When the stem cells are used as part of a medicinal product, they are covered by the legislation on pharmaceutical production and marketing authorization—in particular, by Regulation 1394/2007/EC.

Keywords Regulatory framework · Transplant · Advanced Therapy Medicinal Products (ATMP)

Contents

G. Migliaccio
Istituto Superiore di Sanità, Rome, Italy
e-mail: migliagi@iss.it

C. Pintus (✉)
Italian Medicines Agency (AIFA), Rome, Italy
e-mail: C.Pintus@aifa.gov.it

1 Introduction

The progress of technology to sustain cell proliferation in vitro has led to a panoply of possible new medical applications. Historically, blood transfusions might be considered as the beginning of the use of donor cells for therapeutic use. Further progress was achieved by the discovery of bone marrow-haematopoietic stem cells and their capacity to engraft and restore haematopoiesis in irradiated subjects after peripheral infusion. The understanding of HLA matching opened the way to transplantation first of bone marrow cells, then of organs; it also paved the way for scientific societies and National Competent Authorities (NCAs) to establish the guidelines in this specific field. The increasing use of cells and tissues in various therapeutic applications has created the need for the EU Commission to set regulatory standards regarding the minimal quality requirements for the collection of cells and tissues intended for human use [1–3].

The aim of this new legal framework was to harmonise the standards for donation, procurement, and testing; address the safety issues related to the clinical use of human material; and also facilitate the movement of these products across the Member States.

The important advantages for the public health of European citizens provoked more stringent regulations and increased costs of these products. However, these regulations covered only the most established uses of cells and tissues, such as bone powder, tendons, or heart valves and the transplantation of organs or suspensions containing stem cells (usually CD34- or CD133-positive cells from bone marrow or mobilized peripheral blood). A completely different type of therapy was already evolving in the 1990's based on a different interpretation of the mode of action of cells; it was referred to in Annex I of Directive 2001/83/EC [4], namely as medicinal products for gene therapy and somatic cell therapy. An update of the legislation defined advanced therapy medicinal products (ATMPs) in Regulation 1394/2007/EC [5], which also included tissue engineered products (TEPs).

Together with Directive 2001/20/EC on Good Clinical Practice [6] and Commission Directive 2005/28/EC [7] on the investigational medicinal products for clinical trials, the above-mentioned regulation had a strong impact on the research and development of stem cell-based products, placing them firmly inside the pharmaceutical framework. Herein, we describe the legal framework and its impact on stem cell based products.

2 Regulatory Definition of Stem Cell-Based Products in the EU

Until 2007, regulatory definitions of medical treatments and therapies with stem cells and differentiated adult cells were established in the European Community by several laws; however, the legal framework within the Member States was, and remains in part, not harmonized. Differences in the translation of the European directives in national legislation and related approved practices still exist between the different national authorities. It is therefore advisable to contact the NCAs in the early stages of a project to achieve clarifications about the national regulatory requirements related to production, preclinical research, and clinical investigation.

The first common legal framework for all European countries about manufacturing, clinical investigation and marketing authorization of products containing cells was Regulation (EC) 1394/2007 on ATMPs, which was published at the end of 2007 and came into force by December 2008 [5]. According to this European law, many innovative biotechnological cell-based products fall under the regulatory classification of medicinal products because their pharmacological, immunological or metabolic actions are considered to be the principal mode of action. Moreover, a new mode of action was identified for TEPs, which are supposed to act through regeneration, repair or replacement. Directive 2001/83/EC about medicinal products for human use [4] was amended. Its updated Annex I details the scientific and technical requirements for the quality, safety and efficacy data of all ATMPs. As indicated in Regulation (EC) 1394/2007 and in Directive 2001/83/EC ATMPs comprise four types of products:

1. **Gene therapy medicinal products**: Gene therapy medicinal products contain an active substance that contains or consists of a recombinant nucleic acid used in or administered to human beings for the purpose of regulating, repairing, replacing, adding or deleting a genetic sequence. Their therapeutic, prophylactic or diagnostic effects relate directly to the recombinant nucleic acid sequence they contain, or to the product of genetic expression of this sequence. It is important to note that Council Directive 98/81/EC [8] also applies to the containment of genetically modified micro-organisms and the appropriate measures regarding human health and environment regarding genetic therapy medicinal products. Gene therapy medicinal products do not include vaccines against infectious diseases.
2. **Somatic cell therapy medicinal products**: Somatic cell therapy medicinal products contain or consist (a) of cells or tissues that have been subject to substantial manipulation so that the biological characteristics, physiological functions or structural properties relevant for the intended clinical use have been altered or (b) of cells or tissues that are not intended to be used for the same essential function(s) in the recipient and the donor. They have properties for (or are used in or administered to human beings with an intention of) treating, preventing or diagnosing a disease through the pharmacological, immunological or metabolic actions of its cells and tissues.
3. **Tissue engineered products**: TEPs contain or consist of engineered cells that have properties for (or are administered to human beings with an intention of) regenerating, repairing or replacing a human tissue. A TEP may contain cells or tissues of human or animal origin, or both. The cells or tissues may be viable or non-viable. TEPs may also contain additional substances, such as cellular products, bio-molecules, bio-materials, chemical substances, scaffolds or matrices.
4. **Combined ATMPs**: Combined ATMPs must incorporate, as an integral part of the product, one or more medical devices within the meaning of article 1 of Directive 93/42/EEC [9] or one or more active implantable medical devices within the

meaning of article 1 of Directive 90/385/EEC [10], and its cellular or tissue part must contain viable cells or tissues, or its cellular or tissue part containing non-viable cells or tissues must be liable to act upon the human body with action that can be considered as primary to that of the devices referred to. Products that do not contain any viable cells and that do not act principally by pharmacological, immunological or metabolic actions are excluded from the definition of ATMPs.

The relevant change that the Regulation (EU) 1394/2007 [5] has introduced is the definition of "engineered cells" and the definition of their "non homologous use." Article 2 establishes that cells are considered "engineered" if they fulfill at least one of the following two conditions:

1. The cells or tissues have been subjected to "*substantial manipulation*" so that biological characteristics, physiological functions or structural properties relevant for the intended regeneration, repair or replacement are achieved. Substantial manipulation includes cell expansion, addition of lymphokines and growth factors to the culture medium, combination with medical devices such as biodegradable layers or tri-dimensional scaffolds on (or in) which cells are cultivated and genetically modified.

According to Annex I, "*non substantial manipulations*" include: cutting, grinding, shaping, centrifugation, soaking in antibiotic or antimicrobial solutions, sterilization, irradiation, cell separation, concentration or purification, filtering, lyophilisation, freezing, cryopreservation, and vitrification. If cells have been exposed to these techniques only, they do not qualify as ATMPs.

2. The cells or tissues *are not intended to be used for the same essential function* or functions in the recipient as in the donor. Examples of "*non homologous use*" are: injection of autologous bone marrow stem cells in damaged heart to cure infarcted tissue or use of autologous adipose stem cells to cure ulcers.

In some instances, a clear cut regulatory classification of a cell based product is not easy. For this reason, a formal statement for the classification of ATMPs is highly advisable and can be obtained by the Committee of Advanced Therapies (CAT) at the European Medicines Agency (EMA). The classification procedure [11] is free of charge and a list of the products that have been classified as ATMPs can be retrieved from the EMA website (http://www.ema.europa.eu).

Besides the newly defined ATMPs, other therapies and medical procedures making use of stem cells, progenitors or differentiated cells are nowadays considered to be common clinical practice, such as transplantation of bone marrow or blood transfusions, microimplants of cells for the treatment of facial wrinkles ("filling technique"), adipose tissue filling of breast after mastectomy, or transplant of skin layers in "deep burn" patients. All these products may claim to work through the presence of stem cells in the administered cell suspension, at least for their long-term efficacy. However, the stem cells are presented as working through a homologous activity, doing what is normally their function in presence of trauma or when tissue repair or maintenance is needed. Their collection and distribution is regulated by the following European laws.

(1) Directives 2004/23/EC [1], **2006/17/EC** [2] **and 2006/86/EC** [3]

These directives apply to the quality and safety standards of human cells, such as haematopoietic peripheral blood, umbilical-cord (blood) and bone marrow stem cells, reproductive cells, fetal tissues and cells, and adult and embryonic stem cells that undergo a "nonsubstantial manipulation." These directives cover, among other aspects, the testing of donors, processing, preservation, traceability, packaging and import of cells and tissues. The cells used to formulate an ATMP are considered to be raw materials and need to be collected by a cell and tissue establishment; the above-mentioned directives therefore apply in addition to Regulation (EU) 1394/2007.

Tissues and cells used as autologous grafts within the same surgical procedure are excluded from the above-mentioned directives about transplant of cells and tissues, as well as from Regulation (EC) 1394/2007.

Allogenic cells and tissues have to originate from voluntary and unpaid donations. Anonymity and privacy rules, as foreseen in Directive 95/46/EC [12], about the protection of personal and medical data of donors and recipients have to be respected. Careful documentation about the traceability of the product and of the donor and recipient has to be kept for 30 years by the tissue establishment and the investigator.

Tissues and cell establishments have to be accredited, designated and authorized by regional or central national competent health authorities.

(2) Directive 93/42/EEC [9] **and Directive 90/385/EEC** [10]

These directives deal with medical devices and active implantable medical devices, respectively. They apply in addition to Regulation (EC) 1394/2007 for ATMPs that are combined with matrices, scaffolds and other biodegradable materials (combined ATMPs).

The regulatory evaluation of the quality and safety of combined ATMPs is assessed as a whole product by the authorities competent for medicinal products. Former certifications released by a notified body for the medical device part are recognized; however, if deemed necessary, additional information about the medical device part can be requested by the NCAs.

If the cellular part does not exert the relevant action, such as haematopoietic stem cells used in femoral surgery to facilitate the implant, this product is considered to be a medical device.

(3) Directive 2002/98/EC [13]

This directive applies to human blood and blood components (plasma and blood cells), but not to blood stem cells. Quality, safety and efficacy requirements of haematopoietic stem cells are regulated by Directive 2004/23/EC [1].

3 Regulatory Bodies in EU

3.1 The European Medicines Agency

EMA is a decentralized body of the European Union and is responsible for the technical scientific evaluation of dossiers for European marketing authorization procedures ("centralized procedures") for medicinal products, including ATMPs. Five specific committees have been activated by EMA in the European Parliament for the evaluation of human, veterinarian, herbal, orphan, pediatric, and (since 2009) ATMPs (Committee for Advanced Therapies, CAT).

Once a final opinion on the submitted quality, safety and efficacy data of an ATMP is finalized by EMA, the European Commission (http://www.ec.europa.eu) adopts the decision for the marketing authorization, which is valid in all European Member States. The following pricing and reimbursement aspects are instead negotiated with the competent authorities of the single Member States.

With Regulation (EC) 1394/2007, a new type of evaluation procedure of ATMPs has been introduced by the European Commission: *the certification* [14, 15]. It consists of a submission of partial dossiers that may include quality and/or preclinical data only. The CAT reviews these data with respect to the regulatory requirements of a full marketing approval authorization; eventually, EMA will issue the certificate. The lower fees foreseen for the certification procedure and therefore the opportunities to enhance research-based initiatives are important incentives offered by the European Commission exclusively to small and medium entities (SME) and to public–private joint ventures.

EMA is also coordinating a network of NCAs. Adverse reactions/events reported from clinical trials and from post-marketing use of medicinal products are collected in the European database called "Eudravigilance" (http://eudravigilance.ema.europa.eu/highres.htm). Likewise, a European registry called "Eudract" (http://eudract.emea.europa.eu/index.html), which was also started and coordinated by EMA, is updated with information about all clinical trials approved in Europe. Regulators, ethical committees, clinical researchers, industry and citizens can retrieve, according to different access levels, much of the information about single clinical studies that are ongoing in Europe.

Clinical studies authorization and manufacturing of ATMPs, as well as the accreditation of GLP status to preclinical laboratories, are not under the remit of EMA but of the NCAs of the Member States.

3.2 The National Competent Authorities

Formal authorization for the different aspects related to manufacture and research of ATMPs have to be submitted to the NCAs. The NCAs are public entities that vary among the European countries and include Ministries of Health, National Medicines Agencies, National Scientific Institutes, Tissue Banks, local Ethical Committees, Notified Bodies. NCAs are responsible for the evaluation of quality and safety data of ATMPs and have to be contacted for: certification of cell factories that produce ATMPs, scientific and ethical approval of clinical trials, notification of centres that produce genetically modified organisms, registration of traceability information regarding donors and recipients of cells and tissues, reporting of adverse reactions, submission of financial and insurance information related to costs and risks of clinical studies, import or export of investigational products.

Furthermore, as foreseen by article 28 of Regulation (EC) 1394/2007, NCAs have been empowered to establish the regulatory requirements concerning production, clinical use, traceability and pharmacovigilance of ATMPs on a patient-named basis within the borders of the Member State. These ATMPs should be prepared on a non-routine basis and should be used under exclusive professional responsibility of a medical practitioner in a hospital within the same Member State and with an individual medical prescription for a custom-made product.

Because quality standards and the term "non-routine" preparation are not legally defined by the European Commission, a harmonized rule among the European countries for the ATMP treatment of single patients is still not available; so far, only three countries have introduced national provisions.

Finally, it is important to note that NCAs have the faculty to introduce more stringent protective measures concerning the use of products originating, for example, from embryonic stem cells or animal cells.

3.3 The European Directorate for the Quality of Medicines and HealthCare (http://www.edqm.eu/site/Homepage-628.htm)

The European Directorate for the Quality of Medicines and HealthCare (EDQM) establishes and provides standardized nomenclatures and quality standards for safe medicinal substances and products, which are published in the European Pharmacopoeia. The official standards concern the qualitative and quantitative composition and the tests to be carried out on medicines, on the raw materials used in production of medicines, and intermediates of synthesis. EDQM also develops guidance and standards in areas of blood transfusions, transplantation of organs, tissues and cells. The European Pharmacopoeia is legally binding in European member states. There are more than 2000 European monographs; since the 5th edition (2004–2007) on modern methods in microbiology, dedicated monographs on new biological therapies (cell therapy, gene therapy products) have been already included, with many more under discussion for implementation.

Some quality standards of the European Pharmacopoeia are applicable to ATMP manufacturing and testing. Table 1 depicts some of the monographs relevant for ATMPs.

4 European Legislation on Clinical Research and Manufacturing of Cell-Based Products

4.1 Clinical Research

The clinical use of ATMPs has to be in accordance with the overarching principles and ethical requirements established for medicinal products for human use in general. These are expressed in Directive 2001/20/EC [6], the European detailed

Table 1 Monographs of the European Pharmacopoeia relevant for ATMPs

5.1.7 "Viral safety" about how to measure and remove viral contaminants
5.2.3 "Cell substrates for the production of vaccines for human use"
5.2.8 "Minimising the risk of TSE"
2.6.27 "Microbial control of cellular products"
2.6.1 "Sterility of human haematopoietic stem cells"
2.7.29 "Viability of human haematopoietic stem cells"
5.1.6 "Alternative methods for control of microbiological quality"
2.6.27 "Microbiological control of cellular products"
5.14 "Gene transfer medicinal products for human use" about viral vectors, plasmids and bacterial cells for production

guidelines for good clinical practice in the conduct of clinical trials [16], and Commission Directive 2005/28/EC regarding the clinical use of investigational medicinal products [7].

The above-listed European laws and guidelines regulate many aspects related to clinical studies, such as the format of the application form (CTA) to request authorization by the NCAs, collection and notification of adverse reaction reports, quality requirements of the investigational medicinal product, and ethical considerations for paediatric patients and incapacitated adults participating in clinical trials and inspections.

At the end of 2009, a supplementary guideline for good clinical practice specific to ATMPs was issued by the European Commission [17]. It highlights the importance of keeping traceability documents concerning donation, procurement and testing of the cells used as starting material for the manufacturing of ATMPs and the need to link the donor/animal source to the patient. The guideline also requires that an alert card should be kept by the patient to report serious adverse events, even after a long follow-up period and even if arising in children of the recipient. Furthermore, the consent form should clearly state that certain ATMPs have an irreversible effect and that there are no antidotes to the treatment, when applicable.

A particularly important reference document should be consulted for the clinical use of ATMPs: EMA's first guideline on human cell-based medicinal products, which was issued in 2006 [18]. This document clarifies particular aspects of the clinical development of ATMPs, including the suitable pharmacodynamic and pharmacokinetic tests to be performed in relation to the intended use of the ATMPs or the assessment of the effect of the noncellular component in a combined ATMP. Moreover, it explains that dose-finding studies should be linked to the potency of the product and that clinical efficacy trials may be based on surrogate markers when the clinical endpoint can be observed only after a long follow-up period. Because of the variety of cell-based products and the different levels of risks for the patient, a comprehensive risk analysis that is based on risk factors identified through the whole product life cycle is highly recommended. Table 2 lists some of the general risk criteria highlighted in the EMA guideline; these should be taken into consideration when designing the clinical protocol and when choosing the most adequate tests to monitor the safety of ATMPs.

Table 2 General risk criteria (non exhaustive) listed in the *Guideline on human cell-based medicinal products*

Origin (autologous-allogeneic)
Ability to proliferate and/or differentiate
Ability to initiate an immune response (as target or effector)
Level of cell manipulation (in vitro/ex vivo expansion/activation/differentiation/genetic manipulation/cryo-conservation)
Mode of administration (e.g. ex vivo perfusion, local or systemic administration, surgery)
Duration of exposure or culture (short to permanent) or life span of cell
Combination product (cells and bioactive molecules or structural materials)
Availability of clinical data on or experience with similar products

4.1.1 Clinical Trials Facilitation Group

The approval of clinical trials can vary among European countries because different implementations of Directive 2001/20 [6]. This may sometimes cause delays in starting multi-center clinical trials across Europe. The EU Heads of Agencies have established the Clinical Trials Facilitation Group (CTFG) (http://www.hma.eu/77.html), which is an organization for the coordinated assessment of multinational clinical trial applications through the Voluntary Harmonisation Procedure (VHP). This procedure has been set up within the current legal framework for clinical trials and includes representatives of the NCAs, the European Commission, and EMA.

The aim of the CTFG is to harmonize the implementation of Directive 2001/20/EC and to share scientific assessment of protocols, quality data from investigational medicinal products, information to be submitted to Ethics Committees, and all of the documents needed for a clinical trial submission. The acceptability statement, released by the VHP within 60 days upon request, does not imply that the multinational clinical trial is authorized by the NCAs; however, usually unless there are further open points to be resolved, national approval to start the clinical trial should be released within 10 days.

4.1.2 Manufacture

The manufacture of ATMPs, as a general rule, should be in compliance with the principles of good manufacturing practice (GMP) of medicinal products [19] that the European Commission has published in addition to Directive 2003/94/EC [20], its Annex 1 on manufacture of sterile medicinal products, Annex 2 about manufacturing of biological medicinal products for human use, and Annex 13 related to manufacture of investigational medicinal products. Commission Directive 2005/28/EC [7] applies to the GMP of medicinal products and investigational medicines. Differences still exist regarding the production of investigational ATMPs because some Member States apply the cell and tissue directives technical requirements (minimal air quality [particulate and microbiological] defined as

Table 3 Some relevant EMA guidelines on GTMPs

Design modifications of gene therapy medicinal products during development
Quality, pre-clinical and clinical aspects of medicinal products containing genetically modified cells
Guidance on the quality, pre-clinical and clinical aspects of gene transfer medicinal products
Non-clinical studies required before first clinical use of gene therapy medicinal products
Follow-up of patients administered with gene therapy medicinal products
Scientific requirements for the environmental risk assessment of gene therapy medicinal products
Non-clinical testing for inadvertent germline transmission of gene transfer vectors
Development and manufacture of Lentiviral vectors
Quality, preclinical and clinical aspects of gene transfer medicinal products

static class A in class D particulate and microbiological levels) as an alternative to the more stringent GMP requirements (in operation level) [19] for the manufacture of medicinal products.

Specific requirements for the production of ATMPs were described for the first time in the *Guideline on Human Cell-Based Medicinal Products* issued by EMA in 2006 [18]. This document provides a regulatory definition of active substances, reagents, excipients, and other materials such as enzymes, cytokines, and sera used for culturing the cells; it identifies special requirements for the selection of raw materials and of matrixes and scaffolds used as components for a combined ATMP. Furthermore, it lists the in-process controls of critical steps for long-lasting manufacturing processes and complex manipulations in vitro. It defines the necessary tests to control contamination from adventitious microbial agents and emphasizes the need to assure the reproducibility of the process and the consistency of the final product. Release criteria and the acceptable degree of impurities in the final product should be established; the effect of storage and cryopreservation conditions should be evaluated against the expected potency of the product to be administered. The guideline underlines the importance of an adequate characterization of ATMPs; in addition, the cellular component should be described in terms of identity, purity, potency, viability, and suitability for the intended use.

More specific details about manufacturing and safety aspects of genetic ATMPs have been described in several guidelines issued by EMA, which are retrievable on EMA's website. Table 3 indicates some relevant EMA guidelines on genetic ATMPs (GTMPs).

5 Guidelines in EU

Within the European framework of pharmaceutical legislation, scientific guidelines do not have legal force; however, they should be considered as a harmonized Community position. Alternative approaches may be taken, provided that these are appropriately justified.

The introduction and general principles in Directive 2003/63/EC [21], the community code relating to medicinal products for human use, indicates that guidelines

Table 4 Some ICH guidelines applicable for ATMPs

ICH Q6B Note for guidance on specifications: test procedures and acceptance criteria for biotechnological/biological products (CPMP/ICH/365/96)
ICH Q5D Derivation and characterization of cell substrates used for production of biotechnological/biological products (CPMP/ICH/294/95)
ICH Q5A Guideline on quality of biotechnological products: viral safety evaluation of biotechnology product derived from cell lines of human or animal origin
ICH Q5E Comparability of biotechnological/biological products (CPMP/ICH/5721/03)
ICH S6 Preclinical safety evaluation of biotechnology derived products (CPMP/ICH/302/95)
ICH S7A Safety pharmacology studies for human pharmaceuticals (CHMP/ICH/529/00)

adopted by the EMA Committee for Medicinal Products for Human Use (CHMP) relating to quality, safety, and efficacy have to be taken into account. All the EMA guidelines specific for ATMPs can be retrieved at http://www.emea.europa.eu/htms/human/mes/advancedtherapies.htm.

European guidelines are also released by the European Commission and are published as *The Rules Governing Medicinal Products in the European Community*. (http://ec.europa.eu/health/documents/eudralex/index_en.htm).

As mentioned previously (see 3.3) with respect to the quality and the manufacturing process, several monographs of the European Pharmacopoeia should be considered as specific standards for cell-based products. In addition, several guidelines of the International Conference of Harmonisation (ICH) of Technical Requirements for Registration of Pharmaceuticals for Human Use should be considered for quality, safety and efficacy requirements of ATMPs (http://www.ich.org/products/guidelines.html). Table 4 lists a few of those requirements that are applicable to ATMPs.

6 The Overall Effect of the Regulatory Framework on the Clinical Use of Stem Cells

The Directives on cell and tissue donation do not seem to have particularly altered the use of these products because the industry was already using similar standards. However, the presence of a clear regulatory framework has helped the exchange of donations across Europe.

The application of the medicinal product requirements (European ATMP Regulation) to the development of products that contain cells in general or are based on stem cells has raised a number of opportunities as well as problems. The application of the "pharmaceutical industrial standard" to manufacturing has increased the time and expenses to develop a new product but has also forced the investigators to identify the mechanism of action and the conditions enabling efficacy in vivo. The resulting delay in the marketing application has been noted by industry and financial investors. On the other hand, the presence of a marketing approval (MA) allows the protection of the product itself and in some case it gives the exclusivity to market the product as any other registered drug.

However, the financial rewards of a MA have been limited by the presence of a number of products licensed at national level before the new ATMP Regulation became operative, which benefitted from a transitional period and stayed on the market until 2012 (so-called products legally on the market). Moreover, there is still a large grey area in the definition of the "identity" of the ATMP, which could lead to similar products being licensed at the same time.

A degree of fuzziness is inherent within a cell population with different cell populations proliferating and differentiating. This fact makes the definition of an "identity" and "consistency" across the production lots, as required by the new regulation, a complex task. A number of examples could be made for phenotypic markers or even ultrastructural components such as organelles like mitochondria, which vary in number between cells in the same population. The definition of the identity for cell populations produced in vitro will be the result of regulatory assessor experience more than simple regulatory terminology.

These uncertainties and the risk inherent in entering a novel field have slowed down the involvement and financial support of the pharmaceutical industry in the development of ATMPs. These products are currently under development mainly in academic and nonprofit organizations. For these organizations, the required GMP standards of production starting from the phase I clinical stage are quite expensive, which has delayed nonprofit-sponsored clinical research.

Overall, the initial impact of Regulation 1394/2007/EC has been overcome, but further development of cell-based products will require a continuous dialogue between researchers and regulatory agencies until new standards and mechanisms of financing are developed.

References

1. Directive 2004/23/EC of the European Parliament and of the Council of 31 March 2004 on setting standards of quality and safety for the donation, procurement, testing, processing, preservation, storage and distribution of human tissues and cells
2. Commission Directive 2006/17/EC of 8 February 2006 implementing Directive 2004/23/EC of the European Parliament and of the Council as regards certain technical requirements for the donation, procurement and testing of human tissues and cells
3. Commission Directive 2006/86/EC of 24 October 2006 implementing Directive 2004/23/EC of the European Parliament and of the Council as regards traceability requirements, notification of serious adverse reactions and events and certain technical requirements for the coding, processing preservation, storage and distribution of human tissues and cells
4. Directive 2001/83/EC of the European Parliament and of the Council of 6 November 2001 on the Community code relating to medicinal products for human use
5. Regulation (EC) No 1394/2007 of the European Parliament and of the Council on advanced therapy medicinal products and amending Directive 2001/83/EC and Regulation (EC) No 726/2004
6. Directive 2001/20/EC of the European Parliament and Council of 4 April 2001 on the approximation of the laws, regulations and administrative provisions of the Member States relating to the implementation of good clinical practice in the conduct of clinical trials on medicinal products for human use

7. COMMISSION DIRECTIVE 2005/28/EC of 8 April 2005 laying down principles and detailed guidelines for good clinical practice as regards investigational medicinal products for human use, as well as the requirements for authorisation of the manufacturing or importation of such products
8. Council Directive 98/81/EC amending Directive 20/219/EEC on the contained use of genetically modified micro-organisms
9. Council Directive 93/42/EEC of 14 June 1993 concerning Medical Devices
10. Council Directive 90/385/EEC of 20 June 1990 on the approximation of the laws of Member States relating to Active Implantable Medical Devices
11. Procedural advice on the of scientific recommendation on classification of advanced therapy medicinal products in accordance with article 17 of Regulation (EC) No 1394/2007 EMA/CAT/99623/2009
12. Directive 95/46/EC of the European Parliament and of the Council of 24 October 1995 on the protection of individuals with regard to the processing of personal data and on the free movement of such data
13. Directive 2002/98/EC of the European Parliament and of the Council setting standards of quality and safety for the collection, testing, processing, storage and distribution of human blood and blood components and amending Directive 2001/83/EC
14. Commission Regulation (EC) No 668/2009 implementing Regulation (EC) No 1394/2007 of the European Parliament and of the Council with regard to the evaluation and certification of quality and non-clinical data relating to advanced therapy medicinal products developed by micro, small and medium-sized enterprises
15. Procedural advice on the certification of quality and non-clinical data for small and medium sized enterprises developing advanced medicinal products. EMA/CAT/418458/2008
16. EudraLex—Volume 10 Clinical trials guidelines. "The rules governing medicinal products in the European Union"
17. Detailed guidelines on good clinical practice specific to advanced therapy medicinal products. European Commission ENTR/F/2/dn (2009) 35810
18. Guideline on Human Cell-based Medicinal Products, EMEA/CHMP/410869/2006
19. EudraLex—Volume 4. EU Guidelines on Good Manufacturing Practice
20. Directive 2003/94/EC laying down the principles and guidelines of good manufacturing practice in respect of medicinal products for human use and investigational medicinal products for human use
21. Directive 2003/63/EC amending Directive 2001/83/EC of the European Parliament and of the Council on the Community code relating to medicinal products for human use

Adv Biochem Eng Biotechnol (2013) 130: 301–302
DOI: 10.1007/10_2012_177

Published Online: 8 March 2013

Erratum to: Adult Mesenchymal Stem Cells Explored in the Dental Field

K. M. Fawzy El-Sayed, C. Dörfer, F. Fändrich, F. Gieseler, M. H. Moustafa and H. Ungefroren

The paragraph under DFSCs: "However, DFSCs exhibit telomerase activity, a characteristic feature of ESCs [77, 78, 85]. Telomerase is an enzyme that adds DNA sequence TTAGGG to the 5′ end in the telomere regions of the chromosomes. Normally the telomere region in each chromosome is shortened with every replication cycle (mitosis). Due to the action of telomerase in some cells expressing it, including ESCs and cancer cells, this region is not significantly shortened during mitosis and aging of the chromosomes is hindered, which principally confers immortality to the cells. Whether this expression is an advantage or

---

The online version of the original chapter can be found under 10.1007/10_2012_151.

K. M. Fawzy El-Sayed · C. Dörfer
Clinic for Conservative Dentistry and Periodontology, School of Dental Medicine, UKSH, Campus Kiel, Arnold-Heller Strasse 3 Hs. 26, 24105 Kiel, Germany
e-mail: christof.doerfer@uk-sh.de

F. Fändrich · H. Ungefroren (✉)
Clinic for Applied Cellular Medicine, UKSH, Campus Kiel, Arnold-Heller Strasse 3, Hs. 18, 24105 Kiel, Germany
e-mail: hendrik.ungefroren@uksh.de

F. Fändrich
e-mail: fred.faendrich@uksh-kiel.de

F. Gieseler · H. Ungefroren
First Department of Medicine, UKSH, Campus Lübeck, Ratzeburger Allee 160, 23538 Lübeck, Germany
e-mail: frank.gieseler@uksh.de

K. M. Fawzy El-Sayed · M. H. Moustafa
Department of Oral Medicine and Periodontology, Faculty of Oral and Dental Medicine, Cairo University, 1 Mathaf el Manial Street, Giza Egypt
e-mail: karim.fawzy@gmail.com

M. H. Moustafa
e-mail: mhosnymoustafa@gmail.com

may pose a potential risk for malignant tumor formation similar to the situation in ESCs in tissue engineering still needs to be extensively investigated." has not been published in its correct context.

It belongs to the SCAP section of the chapter which in its correct version should read as follows: "Stem cells from the apical papilla (SCAP) were first described in 2008 [78].Compared to DPSCs and BMMSCs, SCAP showed similar osteo/dentinogenic with lower adipogenic differentiation potential. SCAP further expressed a higher proliferation rate and mineralization potential compared to DPSCs [2]. Similar to other stem cell populations, SCAP expressed STRO-1 and CD146, were positive for CD34 and negative for CD45 as well as showed multiple dentinogenic markers including ALP, bone sialophosphoprotein, osteocalcin [2], and the growth factors TGFbetaRI and FGFR1 [78]. Compared to DPSCs, SCAP express lower levels of DSP, matrix extracellular phosphoglycoprotein (MEPE), transforming growth factor b receptor II (TGFbRII), FGFR3, Flt-1 (VEGF receptor 1), Flg (FGFR1), and melanoma-associated glycoprotein (MUC18) [30]. Upon stimulation with a neurogenic medium, SCAP expressed neurogenic markers as nestin and neurofilament M [78]. However, SCAP exhibit telomerase activity, a characteristic feature of ESCs [77, 78, 85]. Telomerase is an enzyme that adds DNA sequence TTAGGG to the 5′ end in the telomere regions of the chromosomes. Normally the telomere region in each chromosome is shortened with every replication cycle (mitosis). Due to the action of telomerase in some cells expressing it, including ESCs and cancer cells, this region is not significantly shortened during mitosis and aging of the chromosomes is hindered, which principally confers immortality to the cells. Whether this expression is an advantage or may pose a potential risk for malignant tumor formation similar to the situation in ESCs in tissue engineering still needs to be extensively investigated."

In addition the reference number 53 should be placed after "Stem cells from human exfoliated deciduous teeth (SHEDs) were identified in freshly exfoliated deciduous teeth containing living pulp remnants by Miura and colleagues" in the text, reference number 8 should be placed after "In contrast, Cordeiro and co-workers showed that when SHEDs were seeded in poly-L-lactide acid (PLLA)-scaffolds and transplanted into the subcutaneous tissue of immunodeficient mice, they differentiated into odontoblast like cells and into blood vessels that anastomosed with the host vasculature forming a continuous vascular supply to the newly implanted construct" and reference number 72 should be placed after "A study by Seo and colleagues initially identified and characterized human PDL-derived stem cells from extracted teeth as periodontal ligament stem cells (PDLSCs)" in the text.

Adv Biochem Eng Biotechnol (2013) 130: 303–303
DOI: 10.1007/10_2013_218

Published Online: 4 December 2013

Erratum to: Mesenchymal Stem Cells as Cellular Immunotherapeutics in Allogeneic Hematopoietic Stem Cell Transplantation

Claudia Papewalis, Daniela Topolar, Barbara Götz, Kevin Cieslak, Stefan Schönberger and Dagmar Dilloo

Erratum to: Advances in Biochemical Engineering/Biotechnology, DOI 10.1007/10_2012_158

The Author name 'Kevin Cieslak' was missed in the Authors list of the chapter "Mesenchymal Stem Cells as Cellular Immunotherapeutics in Allogeneic Hematopoietic Stem Cell Transplantation".

The online version of the original chapter can be found under DOI 10.1007/10_2012_158.

C. Papewalis · D. Topolar · B. Götz · K. Cieslak · S. Schönberger · D. Dilloo (✉)
Department of Pediatric Hematology and Oncology, Center for Pediatrics, University Hospital, Friedrich-Wilhelm University, Bonn Germany
e-mail: Dagmar.Dilloo@ukb.uni-bonn.de

Adv Biochem Eng Biotechnol (2013) 130: 305–310
DOI: 10.1007/978-3-642-37944-4

Index

MIX
Papier aus verantwortungsvollen Quellen
Paper from responsible sources
FSC® C105338

If you have any concerns about our products, you can contact us on
ProductSafety@springernature.com

In case Publisher is established outside the EU, the EU authorized representative is:
Springer Nature Customer Service Center GmbH
Europaplatz 3, 69115 Heidelberg, Germany

Printed by Libri Plureos GmbH
in Hamburg, Germany